Natural Fiber Composites

Composite Materials: Analysis and Design

Series Editor
Ever J. Barbero

PUBLISHED

Natural Fiber Composites

edited by **R.D.S.G. Campilho**

CRC Press
Taylor & Francis Group
Boca Raton London New York

CRC Press is an imprint of the
Taylor & Francis Group, an **informa** business

CRC Press
Taylor & Francis Group
6000 Broken Sound Parkway NW, Suite 300
Boca Raton, FL 33487-2742

First issued in paperback 2021

© 2016 by Taylor & Francis Group, LLC
CRC Press is an imprint of Taylor & Francis Group, an Informa business

No claim to original U.S. Government works

ISBN 13: 978-0-367-78331-0 (pbk)
ISBN 13: 978-1-4822-3900-3 (hbk)

Visit the Taylor & Francis Web site at
http://www.taylorandfrancis.com

and the CRC Press Web site at
http://www.crcpress.com

Contents

Preface

Composite materials have come to the fore a few decades ago because of their superior specific mechanical properties as a result of the increasing demand of both consumers and industries for highly performing materials and structures. However, the combination of the fibers with the aggregating material or matrix highly increases the complexity of the design process and usually leads to challenges in the composite engineering and, correspondingly, to more conservative solutions for a given application. Although the success of these materials is obvious, recently, a general consensus all around the world was reached regarding the negative influence of human beings on global warming and the environment. The best way in which the environment could be conserved is through the use of renewable and nontoxic natural materials, and all efforts should be undertaken to make them competitive. Actually, environmental awareness all around the world has led to the research and development of cheap and biodegradable materials that are concurrently available from nature. This triggered interest in more sustainable materials that could be processed with lower energy consumption, such as natural fiber composites. Recycling of natural fiber composites and natural fiber reinforcement of waste materials are other steps used for saving resources and the environment. Although the use of these materials dates back to civilization itself, it is clear that renewed incentives for their use are emerging. Thus, scientists and engineers have become more interested in the study of natural fibers and their composites. The replacement of conventional materials and artificial composites with natural fiber composites can thus become a reality, contributing towards the creation of a sustainable economy. On the other hand, concerns on the availability of petrochemicals in the future can also trigger the use of natural fiber composites. On account of large research efforts in fiber extraction and chemical treatments, fiber–matrix adhesion, or processing conditions, natural fiber composites are currently a viable replacement for glass composites in many applications in terms of both mechanical strength and a lower price. Actually, by treating the fibers with coupling agents, engineering the fiber orientation of the natural fiber components, devising extraction techniques to increase the fiber length, and combining with the best possible matrix, very interesting characteristics have been found. These achievements and the superior environmental performance are important drivers for the growing use of natural fiber composites in the near future. Despite all these advantages, some features still prevent a more widespread use of these materials, such as the strength prediction during structural loading and uncertainties about long-term performance. However, it is expected that a lot of useful information previously gathered for artificial composites can be applied to these materials.

This book is comprised of 12 independently written chapters covering the most relevant topics related to introductory knowledge on natural fiber composites, material properties, treatment and processing, modeling, design, and applications. The first chapter is introductory, giving an overview of natural fiber composites, and each of the next few chapters deals with a specific issue of paramount importance that is required to understand and to be able to analyze and design structural components in such materials. The initial chapters discuss issues such as the characterization of natural fibers, matrices, and respective composites. At this stage, relevant information is provided on how to choose the best possible set of materials for a specific application given the design requirements. Methods that enhance the performance and processing techniques follow these initial discussions, enabling us to understand how to improve the strength of the fabricated composites and also which is the most suitable processing technique, respectively. Testing should always be considered during design as a safeguard against design mistakes and to study how the structures behave under service conditions. Environmental issues are not forgotten, and many related aspects are discussed. The last chapters focus on modeling, design issues, and applications. Modeling aims at providing the necessary tools to design natural fiber composites back at the office, as well as at reducing prototype testing to a minimum. Design is related to the overall design process and tools that are used to bring the product to life. Joint design is also included as structures in general usually require some means of joining either because of their dimensions or due to their complex shape, which prevents construction in a single piece. The chapter on applications overviews past, present, and potential applications of these materials based on their characteristics showing cases of success that substantiate the future bet on natural fiber composites. Together, this set of subjects aims at enabling the reader to analyze and design natural fiber composite structures in a scientifically supported manner with the assurance of using state-of-the-art information and methods.

This book has an internationally recognized team of contributors with each one writing about their specific field of knowledge and, thus, providing the best overview of each particular subject. As the editor of this book, it was a great pleasure for me to work with the expert contributors in this book.

R.D.S.G. Campilho
ISEP
Portugal

Series Preface

Fifty years after their commercial introduction, composite materials are of widespread use in many industries. Applications such as aerospace, windmill blades, highway bridge retrofit, and many more require designs that assure safe and reliable operation for 25 years or more. Using composite materials, virtually any property, such as stiffness, strength, thermal conductivity, and fire resistance, can be tailored to the user's needs by selecting the constituent material, their proportion and geometrical arrangement, and so on. In other words, the engineer is able to design the material concurrently with the structure. Also, modes of failure are much more complex in composites than in classical materials. Such demands for performance, safety, and reliability require that engineers consider a variety of phenomena during the design. Therefore, the aim of the **Composite Materials: Analysis and Design** book series is to bring to the design engineer a collection of works written by experts on every aspect of composite materials that is relevant to their design.

Variety and sophistication of material systems and processing techniques has grown exponentially in response to an ever-increasing number and type of applications. Given the variety of composite materials available as well as their continuous change and improvement, understanding of composite materials is by no means complete. Therefore, this book series serves not only the practicing engineer but also the researcher and student who are looking to advance the state-of-the-art in understanding material and structural response and developing new engineering tools for modeling and predicting such responses.

Thus, the series is focused on bringing to the public existing and developing knowledge about the material–property relationships, processing–property relationships, and structural response of composite materials and structures. The series scope includes analytical, experimental, and numerical methods that have a clear impact on the design of composite structures.

Ever Barbero
West Virginia University, Morgantown

Editor

R.D.S.G. Campilho was born in 1979. In 2003, he graduated in mechanical engineering at the Instituto Superior de Engenharia do Porto (ISEP), Porto, Portugal. He completed his MS degree in 2006 and his doctoral degree in 2009, both at Faculdade de Engenharia da Universidade do Porto, Porto, Portugal. He currently serves as an assistant professor at ISEP, where he teaches mechanical engineering. He is an active researcher in numerical modeling, finite element methods, cohesive zone models for fracture behavior, natural and artificial composite materials, and adhesive joint design.

Contributors

María Virginia Alonso
Department of Chemical Engineering
Complutense University of Madrid
Madrid, Spain

Alireza Ashori
Department of Chemical Technologies
Iranian Research Organization for
    Science and Technology (IROST)
Tehran, Iran

Nadir Ayrilmis
Department of Wood Mechanics and
    Technology
Faculty of Forestry
Istanbul University
Istanbul, Turkey

Z.N. Azwa
Faculty of Health, Engineering and
    Sciences
University of Southern Queensland
Toowoomba, Australia

R.D.S.G. Campilho
Departamento de Engenharia Mecânica
Instituto Superior de Engenharia do
    Porto
Instituto Politécnico do Porto
Porto, Portugal

Lucas F.M. da Silva
Departamento de Engenharia Mecânica
Faculdade de Engenharia
Universidade do Porto
Porto, Portugal

H.N. Dhakal
Advanced Polymer and Composites
    (APC) Research Group
School of Engineering
University of Portsmouth
Portsmouth, United Kingdom

Juan Carlos Domínguez
Department of Chemical Engineering
Complutense University of Madrid
Madrid, Spain

Carlos A. Fuentes
Department of Materials Engineering
    (MTM)
KU Leuven
Leuven, Belgium

Roberts Joffe
Division of Material Sciences
Lulea University of Technology
Lulea, Sweden

W. Karunasena
Faculty of Health, Engineering and
    Sciences
University of Southern Queensland
Toowoomba, Australia

J. MacMullen
Advanced Polymer and Composites
    (APC) Research Group
School of Engineering
University of Portsmouth
Portsmouth, United Kingdom

Bo Madsen
Department of Wind Energy
Section of Composites and Materials
  Mechanics
Technical University of Denmark
Roskilde, Denmark

A.C. Manalo
Faculty of Health, Engineering and
  Sciences
University of Southern Queensland
Toowoomba, Australia

M.R. Mansor
Faculty of Mechanical Engineering
Universiti Teknikal Malaysia
  Melaka
Melaka, Malaysia

Samrat Mukhopadhyay
Department of Textile Technology
IIT Delhi
New Delhi, India

Mercedes Oliet
Department of Chemical
  Engineering
Complutense University of Madrid
Madrid, Spain

Liva Pupure
Division of Material Sciences
Lulea University of Technology
Lulea, Sweden

Dipa Ray
Irish Centre for Composites Research
  (ICOMP)
Mechanical, Aeronautical and
  Biomedical Engineering
  Department
Materials and Surface Science Institute
University of Limerick
Limerick, Ireland

S.M. Sapuan
Department of Mechanical and
  Manufacturing Engineering
Universiti Putra Malaysia
Selangor, Malaysia

Shinichi Shibata
University of the Ryukyus
Nishihara, Japan

Le Quan Ngoc Tran
Singapore Institute of Manufacturing
  Technology
Agency for Science, Technology and
  Research (A*STAR)
Singapore

Aart Willem Van Vuure
Department of Materials Engineering
  (MTM)
KU Leuven
Leuven, Belgium

Janis Varna
Division of Material Sciences
Lulea University of Technology
Lulea, Sweden

Ignace Verpoest
Department of Materials Engineering
  (MTM)
KU Leuven
Leuven, Belgium

B.F. Yousif
Faculty of Health, Engineering and
  Sciences
University of Southern Queensland
Toowoomba, Australia

Z.Y. Zhang
Advance Polymer and Composites
  (APC) Research Group
School of Engineering
University of Portsmouth
Portsmouth, United Kingdom

1 Introduction to Natural Fiber Composites

R.D.S.G. Campilho

CONTENTS

1.1   INTRODUCTION

Composite materials have come to the fore a few decades ago because of their superior specific mechanical properties, as a result of the increasing demand of both consumers and industries for highly performing materials and structures. However, the combination of the fibers with the aggregating material or matrix highly increases the complexity of the design process and usually leads to challenges in the composite engineering and, correspondingly, to more conservative solutions for a given application. Although the success of these materials is obvious, recently, a general consensus all around the world was reached regarding the negative influence of the human being on global warming and the environment. The best way in which the environment could be conserved is by using renewable and nontoxic natural materials, and all efforts should be undertaken to make them competitive. Actually, the environmental consciousness all around the world has led to the research and development of the next generation of materials, products, and processes [1]. Within this scope, it is necessary to develop cheap and biodegradable materials that are concurrently available from nature. This awareness triggered interest in more sustainable materials that are able to be processed with lower energy consumption, such as

1

natural fiber composites. As a result, natural fiber composites are under intense investigation because of their potential as alternatives for synthetic fibers. This century, in particular, has witnessed major improvements in sustainable technology and biocomposites, and the interest in these issues is still increasing. Recycling of natural fiber composites and natural fiber reinforcement of waste materials are other steps for conserving resources and the environment. Because of these issues, biocomposites are gaining industrial interest in a world focused on environmental outcomes.

The use of these materials dates back to civilization itself and, for many centuries, natural fibers have been used as raw material. Natural fibers were initially used around 3000 years ago along with clay in Egypt, and they have been used ever since. Recently, it is clear that renewed incentives for their use are emerging. Thus, scientists and engineers have become more interested in the study of natural fibers and their composites. The replacement of conventional materials and synthetic composites with natural fiber composites can thus become a reality, contributing to the creation of a sustainable economy. On the other hand, concerns on the availability of petrochemicals in the future could also trigger the use of natural fiber composites due to the induced pressure from the global market. Because of this, natural polymers are also gaining ground as matrix materials and are taking their market share. It should, however, be noticed that biodegradability is not the sole attribute of natural materials: some synthetic materials can be biodegradable, whereas some natural materials may not be. Obviously, an ideal natural fiber composite is fully biodegradable under controlled conditions and is composed only of short-cycle renewable plants. On account of large research efforts in fiber extraction and chemical treatments, fiber–matrix adhesion, or processing conditions, natural fiber composites are currently considered a viable replacement for glass composites in many applications in terms of both mechanical strength and a lower price. Actually, by treating the fibers with coupling agents, engineering the fiber orientation of the natural fiber components, devising extraction techniques to increase the fiber length, and combining with the best possible matrix, very interesting characteristics are achievable. Other advantages include the large availability, renewability of raw materials, flexibility during processing, low cost, low density, and, because of this, high specific strength and stiffness. Compared with synthetic fibers, energy requirements for processing are much lower, and energy recovery is also possible. Kim et al. [2] showed that natural fiber composites have a higher energy absorption rate under impact loadings than glass-reinforced composites. These achievements and the superior environmental performance are important drivers for the growing use of natural fiber composites in the near future, and they enable these materials to be attractive to industrial companies. Despite all these advantages, some features still prevent a more widespread use of these materials, such as the strength prediction during structural loading, uncertainties about the long-term performance, moisture absorption, lower fire resistance, lower mechanical properties and durability, limited processing temperatures, larger scatter in the cost and properties than synthetic composites, and some difficulties in the use of well-known fabrication processes [3]. However, it is expected that a lot of useful information previously gathered for synthetic composites can be applied to these materials. In fact, many efforts are being made to address the mentioned limitations, with attention to surface treatments for the fibers and interfacial improvement with the matrix. Natural fiber composites with a thermoplastic matrix (e.g., polyethylene [PE], polypropylene [PP], or polyvinyl chloride [PVC]) are also recent solutions. There is equal potential for biodegradable polymers to replace synthetic ones in the near future, at least in applications that do not require a long lifespan, and these matrices have recently seen an important increase in industrial applications. Regarding the production volumes, the main products are starch-based plastics, poly(lactic acid) (PLA), and microbial synthesis polymers or polyhydroxyalkanoates (PHA) [4]. As a result of intensive research and development, these materials became competitors with conventional engineering materials in some fields of application, with new compositions and manufacturing processes emerging. The research interest in natural fiber composites has been consistent over the past two decades, but this has not yet translated into a large range of industrial applications.

In the industry, several companies are becoming increasingly interested in using materials that weigh less, are durable and ecologically efficient, and present interesting mechanical properties. Within this scope, natural fibers are highly valued since they come at a low cost, are recyclable and biodegradable, can be easily processed, and have a very low density. Because of this, the use of natural fiber reinforcement will likely highly increase over the next few years. According to the technical report by Lucintel [5], the global market on natural fiber composites had reached US$289.3 million in 2010, with a compound annual growth rate (CAGR) of 15% from 2005 onward. By 2016, the natural fiber market should reach nearly US$550 million, with CAGR being reduced to around 11%. In terms of applications, the global market for natural fibers is mainly divided into two: wood and nonwood fibers. Wood fibers are typically used in the construction industry, and this application is more widespread in North America. On the other hand, nonwood natural fiber applications thrive in Europe, with tremendous growth mainly in the automotive industry, by using thermoplastic and thermoset-based natural fiber composites, because of issues such as raw material renewability, environmentally sound materials, good sound insulation properties, and fuel saving, on account of the smaller component weight. This usage was made possible by large investments and development in using compression molding as the adopted process in the European automotive industry. Automotive applications include door interior panels, package trays, trunk liners, and seat backs. More specific examples are interior vehicle parts such as door trim panels made of natural composites with polypropylene matrix, or exterior parts, for example, engine or transmission covers, with polyester reinforced with natural fibers [6]. This change was triggered by the European Union End-of-Life Vehicle Directive (2000), stipulating that 80 wt.% of a waste vehicle should be reused or recycled. On account of this directive, the use of these materials has been increasing in the past years. For vehicle applications, using thermoplastic matrices rather than thermoset gives some advantages, such as increased design possibilities, since fabrication by injection molding and extrusion become feasible, in addition to the possibility of recycling. In civil engineering, natural fiber composites can also play an important role because of the lower weight and cost of natural fiber reinforcement plates, compared with carbon- or glass-based composites. Natural fiber fabrics are easier to handle, with advantages in column wrapping for posterior cure with temperature, and are acoustic insulators.

However, according to Dittenber and GangaRao [3], there is a major ecological benefit of using natural fiber composites in construction, since these materials enable the fabrication of large and biodegradable structures only with natural resources and with a reduced amount of embodied energy. Extruded natural fiber composites for decking applications are used in the United States because of the generous thickness of the plates, which allows overcoming limitations with regard to the mechanical properties. Regarding the global usage of natural fibers, Europe is the largest consumer, and Asia is becoming a big market for natural fibers because of the increasing demand in both China and India. In the near future, a fragmentation of the natural fibers market is expected because of emerging economies [5]. Bio-based plastics also follow this increasing tendency of natural fiber composites, with past growth rates of 38% between 2003 and 2007 (worldwide), reaching 48% in Europe alone. The fabrication capacity of bio-based plastics increased from 0.36 million tons in 2003 to 2.33 million tons in 2013, and it is expected to increase further to 3.45 million tons in 2020 [7]. Global markets both now and in the future should be very competitive, striving to get the best possible materials, and those companies that show innovation in this area will perform the best. On account of their potential, natural materials can play a very important role in the near future for the success of industries.

1.2   NATURAL FIBERS

Nonrenewable resources are becoming scarcer on the planet, and a generalized awareness exists regarding renewable resources and products. Because of this, different natural fibers, or species of natural plants that can result in natural reinforcement fibers, are always appearing. There are three ways in which natural fibers can be used: in textiles, paper, and fabrics; for biofuel; and as reinforcement material for composites. As for reinforcement, natural fibers can eventually be used to replace glass fibers in some applications, providing composite parts to be used in the automotive industry, construction, and packaging. Natural fibers can be categorized according to their origin (Figure 1.1): lignocellulosic materials, animals, or minerals [6,8–9]. Lignocellulosic fibers, also known as cellulose-based fibers, can be divided into wood and nonwood or plant fibers. Wood fibers are undoubtedly the most abundant. Plant fibers also have an important market share, and these consist of cellulose, hemicellulose, lignin, and pectin [6]. Many of the fiber properties can be approximated by the relative content of these constituents. Nonwood lignocellulosic fibers are divided into seed fibers, leaf fibers, bast or stem fibers, fruit fibers, and stalk fibers. Most industrial fibers are from bast (e.g., hemp, flax, kenaf, and jute). These fibers are collected from the phloem that surrounds the stem and exist in plants of a certain required height; this enables fibers with high stiffness to maintain stability. Fibers from leafs (e.g., sisal) are also common as raw materials but generally suffer from lower stiffness. Figure 1.2 shows some examples of natural fibers and natural fiber fabrics [9].

The plants that originate fibers can be viewed as primary or secondary, as a function or role of the fiber in the plant. Primary plants, such as jute, hemp, kenaf, or sisal, are grown with the sole objective of providing fibers for industrial usage. Secondary plants (e.g., pineapple, oil palm, or coir) have a different main purpose, such as that

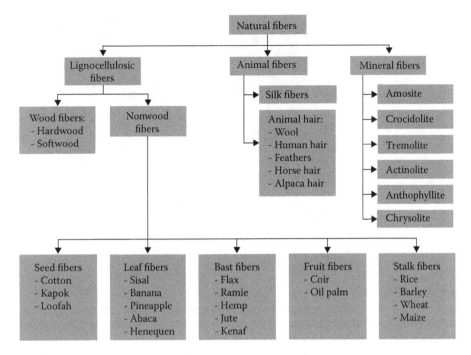

FIGURE 1.1 Classification of natural fibers. (From Saxena, M. et al., *Advances in composite materials – Analysis of natural and man-made materials*, Rijeka: InTech, 121–162, 2011; Technologies and products of natural fiber composites. CIP-EIP-Eco-Innovation-2008: Pilot and market replication projects – ID: ECO/10/277331; Majeed, K. et al., *Materials and Design*, 46, 391–410, 2013 [6,8–9].)

of a human food source. In general, lignocellulosic natural fibers such as flax, hemp, henequen, sisal, coconut, jute straw, palm, bamboo, rice husk, wheat, barley, oats, rye, cane (sugar and bamboo), reeds, kenaf, ramie, oil palm, coir, banana fiber, pineapple leaf, papyrus, wood, or paper have been used as reinforcement in thermosetting and thermoplastic resin composites [10]. Fabricated products in natural composites include door and trunk liners, parcel shelves, seat backs, interior sunroof shields, and headrests [11]. Table 1.1 details the most commercially used natural fibers, in terms of annual worldwide production. Natural fibers usually have a diameter on the order of 10 μm and are, by themselves, a composite material, since they are composed by a primary cell wall and three secondary cell walls. The cell walls include microfibrils that are randomly oriented. The angle of the microfibrils with respect to the fiber axis has a major role in the fiber properties, given that smaller angles give high strength and stiffness, whereas larger angles provide ductility [12]. Since fibers are bundled together by lignin and fixed to the stem by pectin (both of which are weaker than cellulose), these constituents must be removed for the fibers to attain the maximum reinforcement effect. Fibers are still used in bundles connected by lignin, since this is less time consuming, but the overall strength is smaller than using the isolated fibers. The length of the fibers also plays an important role in the composite strength, especially when the interfacial adhesion is weak. Compared with glass or carbon

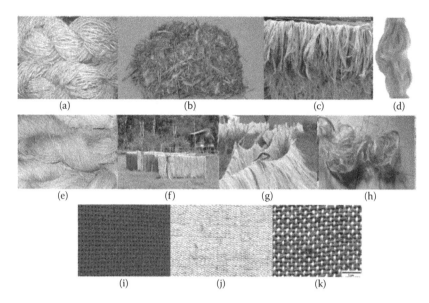

FIGURE 1.2   Natural fibers: banana (a), sugarcane bagasse (b), curaua (c), flax (d), hemp (e), jute (f), sisal (g), and kenaf (h). Natural fiber fabrics: jute (i), ramie–cotton (j), and jute–cotton (k). (From Majeed, K. et al., *Materials and Design,* 46, 391–410, 2013 [9].)

TABLE 1.1
Worldwide Production of Most Used Commercial Natural Fibers

Fiber Type	World Production (10^3 ton)
Sugarcane bagasse	75,000
Bamboo	30,000
Jute	2,300
Kenaf	970
Flax	830
Grass	700
Sisal	378
Hemp	214
Coir	100
Ramie	100
Abaca	70

fibers, natural fibers benefit from lower density, less tool wear during machining, no health hazards, biodegradability, availability of natural and renewable sources, and lower cost per unit volume basis [13–14]. Natural fibers also provide a higher degree of design flexibility, because they will bend rather than break during processing. However, their specific stiffness and strength do not match those of synthetic fibers, and they suffer from high moisture absorption and poor wettability to some resins. Natural fibers generally work well as reinforcements of inorganic polymers, synthetic polymers, and natural polymers because of their high strength and stiffness as well as low density [15]. Typical strength and stiffness values for flax fibers are actually close to those of *E*-glass fibers [16], which, in turn, gives higher specific properties on account of the smaller density. However, being materials of a natural origin, the scatter in mechanical properties is higher than for synthetic fibers, because of variations in the fiber structure emerging from changing climate conditions during growth (where the fibers are sourced), area of growth, age of the plant, processing methods, and fiber modifications [7,17]. The lack of standardized procedures for testing natural fibers also helps in the scattering of properties. Table 1.2 shows the main factors related to the stage of production that affect the fiber properties [3].

Other drawbacks include the difficulty to create a strong bond between the fibers and matrix, and the moisture absorption, with consequences on the composite strength. Many other factors influence the behavior of fibers, such as their length, physical properties (e.g., dimensions, defects, structure, and cell wall thickness), cellulose content, and spiral angle of the cell layers. Some variations in the chemical composition also exist between plants of the same species; among the plant constituents (stalk and root); and between world region, age, environmental conditions, and soil characteristics. Table 1.3 compares the most relevant mechanical properties of some nonwood lignocellulosic fibers and synthetic fibers (for comparison) [3,18]. Because of the high degree of variability of natural fibers and testing methods, the mechanical properties have a large scatter. Another feature is their hollow nature, which not only offers the potential for reduced weight but is also a challenge for waterproofing [19]. For comparison purposes, the most typical values for each quantity

---

TABLE 1.2

Factors Related to the Production of Natural Fibers That Affect Fiber Properties

Stage	Factors Affecting Fiber Properties
Plant growth	Plant species
	Crop cultivation
	Crop location
	Fiber location in plants
	Climate
Harvesting	Fiber ripeness, which affects:
	Cell wall thickness
	Fiber coarseness
	Fiber–structure adhesion
Fiber extraction	Decortication process
	Type of retting method
Supply	Transportation conditions
	Storage conditions
	Age of fibers

Source: Dittenber, D.B., and GangaRao, H.V.S., *Composites: Part A,* 43, 1419–1429, 2012 [3].

TABLE 1.3

Typical Physical and Mechanical Properties of Natural and Synthetic Fibers

Fiber	Density (g/cm³)	Length (mm)	Diameter (μm)	Tensile Strength (MPa)	Tensile Modulus (GPa)	Specific Modulus (approx.)	Elongation (%)	Cellulose (wt.%)	Hemicellulose (wt.%)	Lignin (wt.%)	Pectin (wt.%)	Waxes (wt.%)	Micro-fibrillar Angle (degrees)	Moisture Content (wt.%)
E-glass	2.5–2.59	—	<17	2000–3500	70–76	29	1.8–4.8	—	—	—	—	—	—	—
Abaca	1.5	—	—	400–980	6.2–20	9	1.0–10	56–63	20–25	7–13	1	3	—	5–10
Alfa	0.89	—	—	35	22	25	5.8	45.4	38.5	14.9	—	2	—	—
Bagasse	1.25	10–300	10–34	222–290	17–27.1	18	1.1	32–55.2	16.8	19–25.3	—	—	—	8.7–12
Bamboo	0.6–1.1	1.5–4	25–40	140–800	11–32	25	2.5–3.7	26–65	30	5–31	—	—	—	8.0
Banana	1.35	300–900	12–30	500	12	9	1.5–9	63–67.6	10–19	5	—	—	—	—
Coir	1.15–1.46	20–150	10–460	95–230	2.8–6	4	15–51.4	32–43.8	0.15–20	40–45	3–4	—	30–49	7.85–8.5
Cotton	1.5–1.6	10–60	10–45	287–800	5.5–12.6	6	3–10	82.7–90	5.7	<2	0–1	0.6	—	—
Curaua	1.4	35	7–10	87–1150	11.8–96	39	1.3–4.9	70.7–73.6	9.9	7.5–11.1	—	—	—	—
Flax	1.4–1.5	5–900	12–600	343–2000	27.6–103	45	1.2–3.3	62–72	18.6–20.6	2–5	2.3	1.5–1.7	5–10	8–12
Hemp	1.4–1.5	5–55	25–500	270–900	23.5–90	40	1–3.5	68–74.4	15–22.4	3.7–10	0.9	0.8	2–6.2	6.2–12
Henequen	1.2	—	—	430–570	10.1–16.3	11	3.7–5.9	60–77.6	4–28	8–13.1	—	0.5	—	—
Isora	1.2–1.3	—	—	500–600	—	—	5–6	74	—	23	—	1.09	—	—
Jute	1.3–1.49	1.5–120	20–200	320–800	8–78	30	1–1.8	59–71.5	13.6–20.4	11.8–13	0.2–0.4	0.5	8.0	12.5–13.7
Kenaf	1.4	—	—	223–930	14.5–53	24	1.5–2.7	31–72	20.3–21.5	8–19	3–5	—	—	—
Nettle	—	—	—	650	38	—	1.7	86	10	—	—	4	—	11–17
Oil Palm	0.7–1.55	—	150–500	80–248	0.5–3.2	2	17–25	60–65	—	11–29	—	—	42–46	—
Piassava	1.4	—	—	134–143	1.07–4.59	2	7.8–21.9	28.6	25.8	45	—	—	—	—
PALF	0.8–1.6	900–1500	20–80	180–1627	1.44–82.5	35	1.6–14.5	70–83	—	5–12.7	—	—	14.0	11.8
Ramie	1.0–1.55	900–1200	20–80	400–1000	24.5–128	60	1.2–4.0	68.6–85	13–16.7	0.5–0.7	1.9	0.3	7.5	7.5–17
Sisal	1.33–1.5	900	8–200	363–700	9.0–38	17	2.0–7.0	60–78	10.0–14.2	8.0–14	10.0	2.0	10–22	10–22

Source: Dittenber, D.B., and GangaRao, H.V.S., *Composites: Part A,* 43, 1419–1429, 2012; Barbero, E.J., *Introduction to composite materials design.* Boca Raton: Taylor & Francis, 2011 [3,18].

can be approximated to the average of the presented range. The specific modulus values were obtained by the average stiffness and density, and the most attractive fibers from this point of view are curaua, flax, hemp, jute, pineapple leaf fiber (PALF), and ramie. Values in the same order of magnitude are found between wood and nonwood fibers. The most commonly used synthetic matrix materials used with natural fibers are PP, polyester, polyurethane, and epoxy. Most of the components made of natural fiber composites are fabricated by press-molding, even though a large range of processes are currently feasible [20]. Figure 1.3a compares the specific modulus of some natural fibers, and also E-glass fibers, showing in some of the cases a possibly higher performance of natural fibers, more specifically for ramie, PALF, kenaf, jute, hemp, flax, curaua, and bamboo. On the other hand, a much larger scatter can also be found for natural fibers, because of the bigger variations in stiffness and density. Figure 1.3b shows the evaluation of the cost per weight of some natural fibers and E-glass. In this scenario, all natural fibers behave better or at least identically to E-glass [3].

Mineral-based natural composites (i.e., asbestos) are naturally occurring mineral fibers (silicate-based minerals) or modified fibers that are processed from minerals. Asbestos is minerals that originate from nature in the form of fiber bundles. Mineral fibers are basically divided into six fibrous materials: amosite, crocidolite, tremolite, actinolite, anthophyllite, and chrysolite asbestos. The amosite asbestos, also known as brown or gray asbestos because of the presence of magnesium and iron, can be used as building materials, fire retardants, or thermal insulation products. Crocidolite or blue asbestos is not typically used commercially. Tremolite is formed by metamorphism of dolomite and quartz sediments. When heated, it is converted to diopsite and becomes toxic. Actinolite can be found in metamorphic rocks, and it is formed by the metamorphism of rocks with magnesium and dolomite shales. Chryso LITE or white asbestos are extremely soft silicate minerals of phyllosilicates. The fibers are extremely strong and long hollow cylinders. In 2006, asbestos mining reached 2.3 million tons [6]. At that time, Russia held the biggest extraction

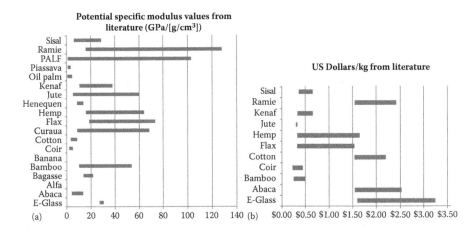

FIGURE 1.3 Comparison of the specific modulus of the most common natural fibers and E-glass fibers (a) and of the cost per weight of some natural fibers and E-glass (b). (From Dittenber, D.B., and GangaRao, H.V.S., *Composites: Part A*, 43, 1419–1429, 2012 [3].)

share of 40.2%, followed by China with 19.9%, Kazakhstan with 13.0%, Canada with 10.3%, and Brazil with 9.9%. Typical applications of asbestos fibers require properties such as inflammability, thermal, electrical, and sound insulation, adsorption capacity, wear and friction properties, and chemical inertness. Mineral fibers are usually combined with cement or woven to produce fabrics or mats.

Animal fibers are composed of proteins, for example, silk, wool, human hair, and feathers. Wool distinguishes itself from the other fibers by being crimped, elastic, and growing in staples [21]. In general, fibers taken from animals include sheep wool and goat, alpaca, or horse hair. Silk fibers come from natural proteins, and they can be woven into textile fabrics. The most widely known form is extracted from larvae cocoons of the mulberry silkworm. Silk fibers have a triangular prism-like structure, allowing silk fabrics to refract light in different angles and thus to produce different colors. Human hair is a filamentous biomaterial growing from follicles in the human dermis. It is primarily composed of a protein called keratin (approximately 95%). Feathers are highly complex integumentary structures that are produced in vertebrates, originate from follicles in the epidermis or outer skin layer that produce keratin proteins [22], and constitute the characteristic plumage of birds.

1.3   BIOPOLYMERS

The recent advances in biopolymers are triggered by the international interest to develop materials that are eco-friendly and do not depend on petroleum, because these resources are depleting and new solutions must be found. For instance, government institutions in countries such as the United States are establishing goals for production to account for a minimum amount of biomaterials. This is a challenge, because the property improvement of biopolymers is costly, and these can currently cost approximately 10 times more than common resins (PLA and starch-based resins are the cheapest ones). Until this cost problem is addressed, a possibility is the combination of natural fibers with a petroleum-based resin to make a composite that is not fully eco-friendly, but is partially disposable through incineration, and can eventually give good life cycle assessment (LCA) indicators. A partial solution to this problem would be milling the semi-biocomposites into small particles and their respective use as powder reinforcements in polymer mortars [23]. Currently, fabrication of natural fiber composites with natural polymers is feasible with biopolymers such as rubber, starch, soy protein, and PLA. Test results showed that soy protein generally behaves the best and rubber behaves the worst, due to issues of interfacial strength between the matrix and natural fibers. Starch polymers are easy to handle during the fabrication process, but they are sensitive to moisture. However, Mohanty et al. [24] showed that proper additives can partially eliminate this limitation and give the composite a good resistance to humidity and they also act as compatibilizers with jute fibers. Ochi [25] studied a biocomposite of hemp-reinforced starch, reporting an improvement of the tensile and flexure strength of the composites with increasing fiber content (until wt.70%). Values of 365 and 223 MPa were obtained for the tensile and flexure strengths, respectively. A comparison between hemp/starch and flax/starch composites was carried out by Nättinen et al. [26], showing that, for a fiber content of 10%, the mechanical behavior was similar (strength of 7.9 MPa, modulus of 0.68 GPa, and impact strength of 6.8 KJ/m^2 for the hemp composite, compared with 7.6, 0.60, and 12.8 for the flax composite). Some experiences with rubber seed oil–based polyurethane are also available in the literature, along with other types of oil (e.g., tung oil, peanut oil, walnut oil, or linseed oil [3]). Despite these options, there are two types of bio resins, soy-based and PLA, which offer the most cost and performance potential to replace petroleum-based polymers.

Biopolymers based on soy resins are one of the most researched nowadays, in the form of either soy protein concentrates or soy protein isolates, obtained by purification of defatted soy flour [3]. These polymers are characterized by reduced strength and sensitivity to degradation by humidity and, on account of this, they can be mixed with other polymers and thus produce soy-based matrices [27–28]. Actually, this is a very good option, although the individual characteristics and cost of soy matrices (and biopolymers in general) do not allow them to be a replacement for other solutions. Within this scope, some successful attempts obtained very interesting improvements in resistance to moisture and mechanical properties in soy matrices and their respective composites [29]. In the work by Mohanty et al. [30], natural composites made of soy bioplastic and short hemp fiber as reinforcement were tested, and the tensile modulus and strength of a 30% fiber reinforcement improved up to nine times the matrix strength. PLA is the other kind of bio resin that is already used in several applications, and it has a large industrial market. Processing of this material includes a few steps: raw material originating from dextrose or other renewable land materials, fermentation to convert into lactic acid, and polymerization. This material is completely biodegradable by a process of hydrolysis, forming lactic acid and, eventually, carbon monoxide. Thus, the use of PLA can reduce pollution, if replacing components made of harmful materials. A few years ago, PLA was very expensive, although some advances made it more affordable, in such a way that currently the bottleneck in its use is the supply capacity. Despite this fact, availability is expected to increase due to worldwide awareness to this material and creation of processing facilities. Porras and Maranon [31] experimentally characterized a full biocomposite made of bamboo fabrics as reinforcement and PLA as resin. An examination of the composite by scanning electron microscopy showed a strong bond between the fibers and resin, and mechanical testing revealed excellent energy absorption, which made these composites viable for use in some structural applications. Baghaei et al. [32] produced PLA reinforced with hemp composites, with hemp content between 10 and 45 wt.%. The natural fiber coupons were fabricated by compression molding and characterized regarding the mechanical performance, porosities, and thermal characteristics. The mechanical tests revealed tensile and flexural strengths that were approximately 2 and 3.3 times those of the neat PLA (considering fiber content of 45 wt.%). The impact characteristics improved approximately two times those of the PLA, but the tests showed that very small fiber contents actually reduced the impact properties.

1.4 BIOCOMPOSITES

Biocomposites can be made of natural fibers with synthetic resins, natural resins with synthetic fibers, or both natural components. These materials have been used for decades, with application in aircrafts since the 1940s [9]. Nowadays, the use of these materials extends to the construction industry, vehicle parts, household applications, and others. Natural fiber composites have a number of interesting characteristics, such as lower environmental impact, CO_2 neutrality, and lower CO_2 emissions than synthetic composites when composted or incinerated. In addition, they weigh less and are cheaper. Studies regarding their use as load-bearing components are also encouraging [33]. One of the differences among synthetic composites is the large property variation, because of the following reasons: dissimilar testing protocols, moisture conditions, physical properties, cell dimensions, chemical composition, microfibrillar angle, structure, defects, scatter in the mechanical properties of the fibers and matrix, and fiber–matrix interaction. The tendency for moisture absorption of natural fibers is also a major issue, as it highly influences the mechanical properties of the composites. There is a clear relationship between the moisture content of the natural fiber and the noncrystalline regions and voids. This issue was studied in detail in the work by Rowell [34]. The equilibrium moisture content of the fibers for a specific air humidity (i.e., the real moisture content of the fibers after exposure to a given amount of humidity) also has a major effect on the composite properties. For example, at the same air relative humidity of 65%, abaca fibers have a moisture content of around 15%, compared with 7% for flax. The transcrystallinity at the interface of natural fibers also affects their composite strength. Some surface treatments (stearation) can induce this effect. These issues were addressed by Zafeiropoulos et al. [35] for flax/isotactic PP with as is, dew-retted, duralin-treated, and stearic acid-treated fibers, showing a more than 100% improvement of the interfacial shear strength for the treated fibers.

In general, modification of the fiber surface can improve adhesion to the matrix. On the other hand, a weak interface reduces the efficiency of the stress transfer between the fibers and matrix, leading to premature damage in the composites and lower strength. The treatment methods are basically divided into physical and chemical methods. The former changes the structural and surface properties of the fibers and promotes the mechanical bonding of the matrix, although it does not change the chemical composition. Stretching, calendaring, and thermotreatment are examples of physical methods that are applicable to natural fibers. The corona treatment is an example of a physical process for surface activation that changes the surface energy of the fibers [36]. Another possibility is the plasma treatment, which induces different surface modifications depending on the gas, by modification of the surface energy and creation of surface cross-links [37]. Chemical treatments act by improving the adhesion with a third material between the fibers and matrix. This material promotes the compatibility between the fibers with hydrophilic behavior and the hydrophobic matrix. One chemical method is the silane treatment, which gives hydrophilic properties to the interface by using silanes that are used as primers to promote adhesion [38]. The largely used alkaline treatment or mercerization disrupts the hydrogen bonding in the fiber structure, thus increasing the fiber anchorage [39]. The acetylation treatment makes the surface of natural fibers more hydrophobic by coating the OH groups of fibers [40]. Another possibility by which the strength of natural fiber composites could be improved is the maleated coupling. The maleic anhydride not only acts on the surface but also improves the interfacial bonding [41]. Finally, the enzyme treatment is environmentally friendly and cost-effective, and it acts by promoting reactions on the fiber surface that improve adhesion [42]. Table 1.4 gives, as

TABLE 1.4

Mechanical Properties of Hemp-Reinforced Natural Fiber Composites and Different Resins

Matrix	Tensile strength (MPa)		Young's modulus (GPa)		Source
	Resin	Composite	Resin	Composite	
PLA	47.5–51	75–85 (30% hemp fibers)	3.5–5	8–11 (30% hemp fibers)	[43]
PP	22.8–35.46	28.1–45.33 (40% hemp fibers)	1.07–1.1	3.5–3.72 (40% hemp fibers)	[44]
Polystyrene	34.1±0.68	40.4±0.65 (22.5% hemp fibers)	—	—	[45]
Epoxy	25	60±5 (30% hemp fibers)	0.7	3.6±0.4 (30% hemp fibers)	[46]
Polyester	12.5±2.5	60±5 (35% hemp fibers)	1.1±0.2	1.75±0.5 (35% hemp fibers)	[47]
Unsaturated polyester	25±5	65±2.5 (30% hemp fibers)	1.5±1	8.75±1.25 (30% hemp fibers)	[48]

an example, the mechanical properties of natural fiber composites with hemp fibers and different matrices [43–48]. The main conclusion to be drawn here is that, notwithstanding the matrix material, the addition of the natural fibers highly improves the strength and stiffness of the resulting material. It is also visible that, due to the chemical reactions between the hydroxyl groups on the fiber surface and the thermoplastic resin, composites with thermoplastic resins excel those with thermoset resins. Table 1.5 compares the mechanical properties of different natural fiber composites as a function of the fiber loading [49–52]. Overall, the introduction of the reinforcement in the polymer significantly improves the Young's modulus as the wt.% content of the fibers is increased. The tensile strength of the composites increases as well, except for the results of the palm leaf fibers/PP composite. In general, the strength improvements are more modest. The impact strength increased for the ramie fibers/PP composite, whereas there is no available data for the other composite systems.

A merit comparison between glass and natural fiber composites (on average) is shown in Figure 1.4. Price can be similar between both composites, but glass composites excel in mechanical performance while having a significant recyclability penalty. Thus, if natural fibers are to replace glass fibers for a given application, this will have to occur in a way in which the mechanical properties are safeguarded. Natural fiber composites do not match glass composites in terms of mechanical properties, as opposed to what occurs with specific properties, especially the stiffness. Therefore, natural fibers are more suitable for providing stiffness in applications that are neither under moisture nor under any adverse environmental conditions. Addressing the issues of moisture and performance of natural fiber composites is possible, as previously mentioned, but this will require new approaches. Actually, in the same manner that metal parts or structures cannot be replaced by synthetic fibers without any design modifications, because of

TABLE 1.5

Mechanical Properties of Natural Fiber Composites with Different Resin and Matrix Combinations

	Fiber Content (wt.%)	Young's Modulus (MPa)	Tensile Strength (MPa)	Impact Strength (kJ/m²)	Source
Ramie fibers/PP	0	1300	35	2.8	[49]
	10	1400	42	3.0	
	20	1600	51	4.2	
	30	2250	66	4.7	
Palm truck fibers/high-density PE	0	475	17.5	—	[50]
	20	750	17	—	
	30	975	18	—	
	40	1500	20	—	
Palm leaf fibers/PP	0	800	27.5	—	[51]
	7	700	23.5	—	
	15	650	21	—	
	28	675	17	—	
Pineapple leaf fibers/polycarbonate	0	1100	67.5	—	[52]
	5	1150	67	—	
	10	1450	66	—	
	20	2000	71	—	

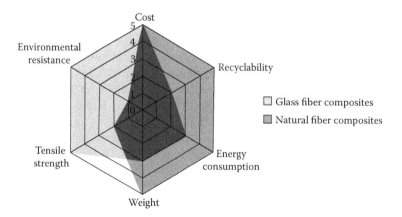

FIGURE 1.4 Merit comparison of glass and natural fiber composites (on average).

the intrinsic differences in properties and fabrication processes, the replacement of glass by natural fiber composites also requires new designs and solutions to obtain the best performance. The immediate applications of natural fiber composites are restricted to limited performance parts, where these materials can really excel because of their bio characteristics and, eventually, cost advantage (or at least nondisadvantage). Examples are selected components for the automotive industry with low strength requirements, such as panels and trims, which also improve the bio credentials of vehicles. Apart from this, any component whose performance demands are within the reach of natural fiber composites can potentially be fabricated in these materials. Examples are wood parts, since the consumer demands for water and moisture absorption for wood components are usually low. Replacement of unreinforced plastics is also a chance for development, and the use of low-cost plant fillers is ongoing in the electronics industry. The replacement of the fillers by fibers can give significant performance improvements. At the moment, there are still many challenges to be overcome for natural fiber composites to be able to be applied in all current glass applications (to be discussed in Section 1.8). Nonetheless, the eventual success of such replacement surely relies on the ongoing and future research and the development of new designs that favor the mechanical properties of these materials.

1.5 BENEFITS AND APPLICATIONS OF NATURAL FIBER COMPOSITES

In past decades, natural fibers and natural fiber composites received attention from researchers in several industries, such as in civil construction, automotives, and biomedicine [53], mostly based on three factors: reduction of costs, weight reduction, and sustainability. The mechanical behaviour of lignocellulosic fibers (non-wood or plant) and their composites, either with biological or synthetic materials as matrix, have been studied extensively by the scientific community in parallel to industrial use in vehicles and construction. Actually, natural fibers of flax, hemp, sisal, or jute can replace glass or other kinds of synthetic fibers in epoxy, polyester, PVC, PE, or PP matrices, with the following benefits:

- Lower costs because of reduced cost of raw materials, smaller cycle times, lower weights, and reduction in the fuel consumption (vehicle parts)
- Identical mechanical properties of glass-reinforced parts, with fabrication advantages such as smaller tool wear, good sound insulation, and geometrical stability
- Eco-friendliness, renewability of the raw materials, recyclability, no toxicity, and CO_2 neutrality.

Many literature examples exist on the use of natural fiber composites in automotive applications, mainly for interior vehicle parts [1,54–55], with either thermoplastic or thermoset matrices. The selected materials for these applications should meet requirements of minimum strength and strain to failure, impact and flexural properties, sound insulation, fire resistance, processing characteristics (dwell time and temperature), odor, dimensional stability, and energy absorption under crash conditions. Bledzki and Gassan [16] reported an application of jute, coffee bag wastes and PP bags in trim parts of Brazilian trucks after recycling. Saxena et al. [54] concluded that using natural fiber composites in vehicle applications as trim parts, panels, shelves, and brake shoes can give an advantage of 10% in weight, fabrication process energy savings of 80%, and an overall reduction of 10% in the cost of parts. Moreover, around 6000 natural fiber composite parts could be introduced in vehicles with this potential advantage. In locomotives, components such as the gear casing, doors and side panels, interior furnishing and seating, luggage racks, berths, chair backings, modular toilets, and roof panels in natural fiber composites can also bring benefits to weight, cost, corrosion resistance, and weight-reduction-driven fuel consumption savings. For civil engineering applications, natural fiber composites from bast fibers are, in general, the best, whereas flax gives the best balance between strength and stiffness to cost and weight. Jute-reinforced composites are very common, but their strength and stiffness does not match flax. Because of the specific stiffness advantages compared with glass composites, natural fiber composites are an excellent solution for reinforcement of existing infrastructures. In general, in the development of natural fiber composites, which are biodegradable, the replacement of synthetic materials such as glass-reinforced composites without compromising their distinctive characteristics is currently a big challenge and will continue to be so.

Natural animal fiber composites are scarcely used in industrial or other applications. Animal fibers, such as wool or spider silk, are made of proteins and find useful applications in bioengineering and medicine. Wool is the most used natural fiber, although it suffers from low fracture resistance, which is its biggest limitation. A major application of wool fibers is the fabrication of rock wool fabrics or panels that are used in the construction industry on account of their good fire resistance and sound absorption. Silk fibers are characterized by their stability even when exposed to varying environmental conditions; they have a low weight, and their composites are very tough and impact resistant. Some applications of these fibers were reported in automotive, aerospace, and sport equipment industries [6]. Feathers meet application in cement-bonded feather boards, which are resistant to decay and termite attacks due to the keratin. These feather boards can be employed in paneling, ceilings, and insulation, although not as structural components. Animal feather composites can compete with conventional materials with regard to a few specific applications.

There is also a history of application of mineral fibers or asbestos in corrugated panels (e.g., roofing compounds), gaskets, pipeline wrapping, sheets, rods, shaped moldings, and thermal and/or electrical insulation. Fabrics of mineral fibers also find application in parts that involve friction, such as brake or clutch pads, because they are durable to friction and are heat and oil resistant. Mineral fibers can be fabricated with biosynthetic matrices to produce a large variety of products. Chrysotile with rubber matrix finds application in packings, gaskets, and heavy-duty insulation parts as compressed boards. More specific applications include reinforcement agents in coatings and adhesives. Mineral fibers have a significant limitation, related to health hazards, including lung, eye, and skin diseases, which causes numerous deaths under working conditions. Because of this and environmental concerns, these fibers are being less used.

1.6   LIFE CYCLE ASSESSMENT OF NATURAL FIBER COMPOSITES

Natural fiber composites have become feasible alternatives to glass composites since the 1990s, and some of them are highly attractive for their use in vehicle and leisure parts. Traditionally, thermoset matrices are generally used, but thermoplastics such as PP have recently attracted attention because of processing and recyclability issues [56]. As previously referred to, natural fiber composites are considered to have a number of environmental advantages. Since the environmental benefit of these materials is a driver for the increased use of these materials, a comprehensive evaluation of the environmental impact of natural fiber composites for all product stages, between raw material extraction and end-of-life disposal, should be carried out. According to the definition by Duflou et al. [57], the LCA analysis balances the environmental costs and benefits of different materials for a given application, while considering the different phases of the product. More specifically, the LCA methodology gives an assessment of the sustainability of materials that quantifies the effect on the environment of the raw materials and their extraction or production, energy consumption for fabrication, impact during life, waste generation and recycling, and incineration or disposal after their lifespan (Figure 1.5). This comparison should be made on an equivalent functional basis, since natural fiber composites are usually lighter, even compared with synthetic composites.

A potential difference between synthetic and natural fiber composites that stands out immediately is the energy requirements to fabricate the fibers. Glass fibers are produced at around 1550°C, because of the high melting temperature of glass, which makes this a major issue. Regarding the matrix, energy consumption can be in the form of mineral oil extraction, separation, refinement, and polymerization. When considering bio materials as a matrix, for example, PLA, PHA, or modified starch, different results can be expected. Several LCA studies are available in the literature, namely comparisons with synthetic composites, for which natural fiber composites typically aim to substitute. In the study by Mohanty et al. [24], the authors stated that the required energy to fabricate natural fibers, by weight, is between 20% and 25% that of synthetic fibers. Different investigators used LCA analyses to conclude that, in the whole fabrication process, natural fiber composites only spend approximately 60% of the energy used by synthetic composites [58]. The presented value considers the effect of material extraction or harvest, further processing, transportation, and composite fabrication. Other studies showed that the energy needed to fabricate a natural fiber fabric is only 30%–40% of that required to fabricate a glass mat [59]. Wötzel et al. [60] presented a comparative LCA study of a panel for an Audi A3, considering the original acrylonitrile butadiene styrene (ABS) copolymer part and a hemp fiber (66 wt.%)/epoxy composite. The study is somehow incomplete, since it does not account for some important aspects, such as the component use and end-of-life disposal, but it models the inputs, energy use, and pollution until the part fabrication. The authors concluded that the natural composite uses 45% less energy

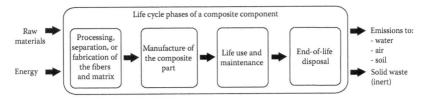

FIGURE 1.5 Typical phases of a composite part, with impact on energy and emissions.

and emissions are lower. Nonetheless, emissions of some polluting substances such as nitrates and phosphates are higher because of their fertilizer application in hemp crops, although this is not very significant. Schmidt and Beyer [61] focused on an insulation component of a Ford vehicle, originally made of ethylene propylene diene copolymer (EPDM), PP, and glass fibers. This component was weighted against a tentative new design by replacing hemp fibers (30 wt.%) with glass. This was a more in-depth study, as it covered all previously mentioned stages of the product. The natural fiber component showed significant advantages regarding the cumulative energy demand (CED) (savings of 88.9 MJ), CO_2 pollution (8.18 kg), and generic emissions. Corbiere-Nicollier et al. [62] evaluated transport pallets made of either the original glass-reinforced PP or china reed fiber-reinforced PP. For an equivalent performance, the natural fiber component required 53 wt.% of fibers, as compared with 42 wt.% for the original counterpart. The entire life cycle was assessed, ending with incineration in both cases. Overall, the natural composite showed significant advantages regarding the environmental impact, except nitrate emissions.

Duflou et al. [57] suggested three indicators to study LCA: (1) the CED, which is a global environmental factor and a major driver; (2) greenhouse-gas (GHG) emissions, because of the climate change and global warming implications, measured in CO_2 equivalents or CO_2e; and (3) aggregate environmental impact score, usually expressed in milli-ecopoints (mPT). By dividing the LCA analysis in the production, use, and end-of-life phases, Duflou et al. [57] proposed a comprehensive comparison between different natural fiber and glass-reinforced composites. Table 1.6 evaluates the indicators CED, GHG, and mPT for production of different matrices, fibers, and composite fabrication [63–66]. The matrix advantage is mainly with regard to the GHG and ecopoints, whereas CED reductions are much smaller, except for linseed oil monomer (ELO). Major reductions in CED and GHG can be found for natural fibers with respect to glass fibers, which are their main competitor, whereas mPT data are inconclusive. The values for the different fabrication processes are mainly indicative, as these quite vary between material choices. The LCA analysis for the use phase has some specificities that cannot be neglected for the sake of a realistic analysis (e.g., different lifespan of materials and weight). In vehicles, weight reduction has a double impact, because it reduces both fuel consumption and emissions, and because of this the major share of fiber-reinforced materials (approx. 44%) goes to transportation systems [67]. Shifting vehicle parts from glass to natural fiber composites brings weight benefits between 22% and 27% [68]. Table 1.7 provides a comparison of CED and GHG during the use life for vehicles and other parts in traditional materials and their equal functionality equivalents in synthetic and natural fiber composites [62,69–72]. Carbon fiber composites provide a massive saving related to steel, aluminum, and even glass composites, because of the weight savings for the same function. Replacement of glass with natural fibers is also recommended. A limitation of this analysis is that it considers an equal lifespan between synthetic and natural fiber composites, since analyses of the use life of these materials are not available, and lifespan is highly dependent on the moisture level in the composite. Table 1.8 is related to the CED and GHG impact of different end-of-life disposal strategies (SMC refers to Sheet Molding Compound glass composites, and GMT

TABLE 1.6

Comparison of CED, GHG, and mPT for Different Materials and Fabrication Processes

Material	CED (MJ/kg)	GHG (kg of CO_2e/kg)	Ecopoints (mPT/kg)
Matrix			
Epoxy	76–137	4.7–8.1	734
Unsaturated polyester (PES)	62.8–78	2.3	644
PP	73.4	2.0	276
Modified starch (Mater-BI®)	54.8	1.3	275
PLA (Ingeo 2009™)	67.8	1.3	312
PHA (generic)	59–107	0.7–4.4	—
Linseed oil monomer (ELO)	19	1.2	—
Reinforcement			
Carbon fiber (generic)	286–704	22.4–31	833
Carbon nanofiber (CNF)	654–1807	70–92	—
Glass fiber (generic)	45	2.6	264
Flax fiber	9.6–12.4	0.4	350
Hemp fiber	6.8–13.2	1.6	—
Jute fiber	3.8–8.0	1.3–1.9	—
Sugarcane bagasse	11.7	—	—
Composite Fabrication Process			
Sheet molding compound	3.5–3.8	—	13
Resin transfer molding	12.8	—	46
Pultrusion	3.1	—	11
Autoclave	21.9	—	—
Injection molding	21.1–29.9	0.5–1.2	126

Source: Suzuki, T., and Takahashi, J., Prediction of energy intensity of carbon fiber reinforced plastics for mass-produced passenger cars. Proceedings of JISSE-9, Tokyo, Japan, 29 November–11 December, 2005; Patel, M., *Energy*, 28, 721–740, 2007; Boustead, I., *Eco-profiles of the European Plastics Industry. Association of Polymer Manufacturers in Europe*, Brussels, Belgium, 2005; Ecolizer 2.0. 2003–2011. *Openbare Vlaamse Afvalstoffenmaatschappij*, Mechelen, Belgium [63–66].

refers to Glass-Mat Reinforced thermoplastics) [70,72–75]. The recycling technique depends on the material, and it can include mechanical or thermal methods for composites and remelting and recasting for metals. Accumulation in landfills is also an option for composites, actually the most common a few years ago, but it is not eco-efficient because it does not allow recovering the embodied energy of the materials. Moreover, composites still need treatments to reduce the environmental impact of the wastes. In general, fiber composites are incinerated, which allows the embodied energy to be recovered. Glass composites can equally be incinerated, but glass fibers

TABLE 1.7

Comparison of CED and GHG during Their Use Life for Vehicles and Other Parts

Vehicle Part	Original Material	Composite Replacement	CED Change (GJ/Part)	GHG Change (kg of CO_2e/Part)
Propeller shaft	Steel	Carbon and glass fiber/epoxy	−3.7	−227
	Aluminum		−2.5	−158
Car closure panel	Steel	Carbon fiber/epoxy	−26.9	−2096
	Aluminum		−6.8	−531
	Glass fiber/ poly(ethylene terephthalate - PET)		−13.1	−1023
Car door	Steel	Glass fiber/PP	−2.0	−150
	Aluminum		+0.8	+67
Car interior	Talc/PP	Bagasse/PP	−19.3	−206
Transport pallet	Glass fiber/PP	China reed/PP	−0.6 to −2.3	—

Source: Corbiere-Nicollier, T. et al., *Resources, Conservation and Recycling,* 33, 267–287, 2001; Song, Y.S. et al. *Composites: Part A,* 40, 1257–1265, 2009; Puri, P. et al., *International Journal of Life Cycle Assessment,* 14, 420–428, 2009; Luz, S.M. et al., *Resources, Conservation and Recycling,* 54, 1135–1144, 2010; Schexnayder, S.M. et al., Environmental Evaluation of New Generation Vehicles and Vehicle Components. Report ORNL/TM-2001–266, Oak Ridge National Laboratory, Oak Ridge, 2001 [62,69–72].

are incombustible, which leads to an energy consumption of approximately 1.7 MJ per glass fiber weight [76]. Natural fiber composites have the logical advantage of being combustible and, thus, helping the process. Recycling can essentially be carried out in four ways: (1) mechanical recycling, (2) chemical treatment, (3) pyrolysis, and (4) fluidized-bed processing [57]. Natural fiber composites, in particular, can be recycled by using many techniques, without a significant loss of mechanical properties. In the work by Bourmand and Baley [77], a sisal/PP composite showed reductions of only 10.1% and 17.2% in tensile modulus and tensile strength, respectively, after seven cycles, in opposition to a glass/PP composite that showed 40.1% and 52.5% property losses, respectively. Biodegradation is another scenario that is used to dispose biocomposites [78]. Table 1.9 compares natural fiber composites against the original counterparts in vehicle applications using the CED change during the production, use, and end-of-life phases [60,71]. Detailed information about the components and use scenario is given in the source references. For the three parts, the end-of-life strategy is incineration (with energy recovery). Natural fibers consistently provide less energy in incineration on account of the lower required mass for the same application. On the other hand, natural fiber composites behave a lot better in the production and use phases, resulting in a lower accumulation of CED during the entire life.

In conclusion, there is a so large number of variables that are necessary to include in the study and so many different materials and applications that a conclusive generic study that these materials are actually more eco performing in all applications actually does not exist. Nevertheless, in comparison to general materials, natural composites spend approximately 20 MJ less of energy for 1 kg of material and

TABLE 1.8
CED and GHG Impact of Different End-of-Life Disposal Strategies

	Landfill		Recycling		Incineration with Energy Recovery	
	CED (MJ/kg)	GHG (kg of CO_2e/kg)	CED (MJ/kg)	GHG (kg of CO_2e/kg)	CED (MJ/kg)	GHG (kg of CO_2e/kg)
SMC	—	—	7	0.4	−7.5	0.9
GMT	0.09	0–0.02	11	0.9	−25.2	1.9
Carbon fiber composite	0.11	0.02	10–15	—	−31.7 to −34	3.2–3.4
Natural fiber composite	—	—	—	—	−12 to −34	2.3–2.9
Steel	—	—	11.7– 19.2	0.5–1.2	—	—
Aluminum	—	—	2.4–5.0	0.3–0.6	—	—

Source: Puri, P. et al., *International Journal of Life Cycle Assessment*, 14, 420–428, 2009; Schexnayder, S.M., Environmental Evaluation of New Generation Vehicles and Vehicle Components. Report ORNL/TM-2001–266, Oak Ridge National Laboratory, Oak Ridge, 2001; Leterrier, Y., *Comprehensive Composite Materials*, 1073–1102, Oxford: Pergamon, 2000; Hedlund-Aström, A., Model for end of life treatment of polymer composite materials. Royal Institute of Technology, Stockholm, Sweden, 2005; Duflou, J.R. et al., *CIRP Annals: Manufacturing Technology*, 58, 9–12, 2005 [70,72–75].

TABLE 1.9
CED Comparison during the Production, Use, and End-of-Life Phases between Original and Natural Fiber Replacement Components

Part	Original	Natural Fiber Composite Replacement	CED Change (MJ/Part)			Source
			Production	Use	End-of-Life	
Car interior	Talc/PP	Bagasse/PP	−222	−19313	+62.3	[71]
Side panel of small vehicles	ABS	Hemp/epoxy	−59	−71	+27	[60]
Side panel of large vehicles	ABS	Hemp/epoxy	−59	−118	+27	[60]

prevent the release of 1 kg or more of CO_2 for 1 kg of material into the atmosphere [79]. Thus, despite not being fully conclusive, the discussed and other available studies highly reinforce the need to develop these materials.

1.7 POTENTIAL OF NATURAL FIBER COMPOSITES AND DRIVERS FOR CHANGE

Recent advances in natural fiber composite technology enabled the development of materials that exhibited attractive performance and sustainability. To date, these materials have been applied mostly in vehicle products and some construction applications, with the previously mentioned advantages. If these new materials are to be generalized to other sectors of industry, as, for example, household products or goods, there are basic inherent properties that they must accomplish: performance for the desired function, usability, reliability, and durability. The discussion is divided into three main areas of actuation: mechanical properties, environment, and cost-effectiveness.

1.7.1 MECHANICAL PROPERTIES

Currently, natural fiber–related technology is being improved to provide better mechanical characteristics of the bio-based components. With this large effort, it will be possible for biocomposites to exploit other fields of application that are not currently in use. But for this to happen, the knowledge of the materials, fabrication processes, and design methods must reach a much higher degree of confidence. These issues, together with proper standardization for these materials, can give them a distinctive edge over conventional materials. At the moment, large efforts are being made to make biocomposites a solution for load-bearing parts in construction. In fact, some authors tested the use of cellular plates and beams as structural parts made from hemp, jute, and flax fibers in polyester resin [80] in the construction industry (house building). The components were experimentally tested, and the results showed that the cellular arrangement of the natural fibers can improve the composite mechanical properties just enough to compete with other engineering materials (e.g., glass fiber composites or common construction materials) and make them viable to load-bearing applications. This line of research is to be followed in the future to make civil construction a strong application of natural fiber composites. Applications in other sectors of industry rely on additional improvements. However, based on the current state of the art, some limitations of these materials still need to be addressed for them to be considered competitive against synthetic composites. It was previously mentioned that natural composites are a cost-effective solution compared with other materials, but it is also true that if a 100% biological and recyclable solution is needed, costs increase and this also needs further research efforts. Moreover, ecological superiority over synthetic composites is not yet fully true because of the fabrication techniques that consume large amounts of energy. Another feature to be improved is the resistance to moisture and temperature, and here a long path exists, knowing that there are limits to the materials themselves. For example, currently it is possible to make a part fully biodegradable with the proper choice of a bio matrix, although biodegradation would be high. Significant improvements in some key aspects of bio materials such as large nonlinearity/relaxation, long-term performance, and small impact resistance can occur by improving the processing of the fibers and composite fabrication. It can be concluded that new frontiers will emerge for these materials when the following characteristics are met with a significant degree of comparison with the other materials: durability, dimensional stability, environment resistance, and fire resistance [7].

One of the material-related fields that has endured major enhancements recently is nanotechnology. Common natural fibers contain a small amount of nanocrystalline cellulose. The artificial fabrication of natural fibers with this structure could produce fibers with 10% of the strength of carbon nantubes, but with a cost approximately 1000× less [7]. Research on this field mainly uses wood pulp to produce nanocellulose, but other non-lignocellulosic products can be used with this purpose: hemp [81], wheat [82], or flax [83]. Some authors [33] obtained cellulose nanofibers with a mixed chemical/mechanical technique and combined them with a starch polymer. Preparation of the nanofibers enabled cleaning the fiber surface of hemicelluloses, lignin, and pectin, and also the defibrillation of nanofibers from the initial fiber bundles. It is also possible to fabricate microfibrillated cellulose from wheat and soy by cryocrushing, disintegration, and fibrillation, producing fibers with a diameter between 30 and 40 nm [84]. Many other works used similar techniques to produce these nanomaterials from soy, root crops, wood, seaweed, cotton, hemp, cereals, and sea squirts, among others. Composites made with these nanofibers experience a major improvement in their tensile strength and stiffness. Nanotechnology can also be used differently to improve natural fiber composites by application of coatings, diminish the effects of biodegradation, or increase the fire resistance of the materials. With the recent efforts under way, it is a matter of time until nanoconcepts give natural fiber composites the performance, durability, value, service life, and utility that makes them more competitive, while maintaining their ecologic features.

1.7.2 ENVIRONMENT

Natural fiber composites fit in the concepts of sustainable economy, since synthetic materials are replaced by bio-based and renewable ones. These materials also have the potential to be more cost-effective for identical structural characteristics, and there is the opportunity to produce or grow the fiber plants in controlled facilities or farms. Compared with synthetic resins and fibers (or even conventional materials) that these materials can potentially replace, the carbon footprint will be tremendously reduced. Synthetic fibers and resins have posed difficulties with regard to their disposal for decades, accounting for approximately 20% of the total landfill space, depending on the country. This is a strong motivation for the replacement of synthetic composites, since landfill capacity is scarce and overcrowded. In terms of saving the environment, it is more urgent to replace the matrix than the fibers by natural equivalents, since petroleum-based resins take hundreds of years to degrade [85]. Recycling is an opportunity, although recycled petroleum-based resins lose some characteristics by incorporation of external substances, which affects the adhesion between fibers and matrix. On the other hand, PLA can be reconverted practically without affecting its performance. Natural fiber composites benefit the environment in three ways, compared with synthetic fibers: (1) less pollution during fabrication, (2) lower fuel consumption and CO_2 emissions during transport to the construction sites (if applicable), and (3) absence or significant reduction of the disposal and energy-consuming disposal efforts. However, the CO_2 advantage arising from the fiber processing is highly variable, such that accounting for the environmental advantage of natural fiber composites is not the easiest of tasks [86].

1.7.3 COST-EFFECTIVENESS

The price of materials is a major issue for the assessment of natural fiber composites position in the market and overall potential. Specifically for the case of civil construction, the cost of raw materials has a share between 60% and 75% of the total costs and, thus, the proper choice of materials is the major issue to act on [3]. On the other hand, the choice for these materials also relies on environmental issues, such as disposableness of the materials, lack of raw materials for synthetic fibers and matrices, and public opinion. Some countries even have specific laws on the use of recycled or natural materials [85]. Mohanty et al. [24] predict a sustained yearly increase of natural fibers and bio resins of approximately 60% for construction and of approximately 30% for vehicles. Satyanarayana et al. [87] presented slightly smaller numbers, between 10% and 22% overall. Because of this, it is likely that worldwide production of natural fibers and bio resins will have difficulties in meeting industry expectations very soon. This opens a window of opportunity for other regions than the United States and Europe to bet on these materials and India, for instance, is nowadays a large producer. However, the costs of the bio matrix materials are still too high to enable them to compete with plastics. Fabrication processes should be kept as simple as possible, to reduce costs and pollution. Surface treatments are equally required for the composites to be able to compete with established materials in the industry, and these should be improved to allow natural fiber composites to match or at least to approach the mechanical properties of synthetic fiber composites [24]. It was previously shown that a large price variation exists with regard to natural fibers, because of different reasons, and a good route involves choosing natural fibers that not only satisfy the desired properties but will also always be cost-effective, disregarding these variations. Since natural fibers and bio resins are under intense research, unlike synthetic composites that are established in the market, room for improvement exists. Once the cost is lowered to acceptable levels, the use of these materials is prone to become widespread in many fields of the industry. Currently, a consensus does not exist regarding the future of natural fiber composites, with the remaining doubt being whether these materials would be able to replace synthetic fibers in many applications.

1.8 CHALLENGES IN THE USE OF NATURAL FIBER COMPOSITES

It is a general consensus among natural fiber composite researchers that significant challenges exist regarding their use as load-bearing components. In the beginning, since the mechanical strength was not so valued, these materials were limited in their application to nonstructural parts. Other limitations were the environmental degradation (especially to moisture) and low resistance to impact. Research in fabrication processes, fiber–matrix compatibility, and other areas has shown that these limitations could be overcome. Moreover, recent indicators show that there is room for significant improvements by application of coatings, fiber surface treatment technology, new resins, and incorporation of additives. Many other challenges exist, which are listed as follows [3].

The biggest challenge is perhaps the hydrophilic nature of natural fibers, which makes them prone to water absorption and, as a consequence, leads to poor adhesion to hydrophobic polymer matrices. Many reported cases exist of failure of natural fiber composites under water or moisture exposure, namely by delamination and fiber swelling. Natural matrices are also more sensitive to water absorption than are petroleum-based ones. When immersed in water, their behavior is even worse, and this was checked by Singh and Gupta [88], finding a strength reduction in sisal-reinforced composites that was approximately 31% larger by immersion in water than by exposure at 95% relative humidity. Based on the current state of the art, overcoming this limitation should involve one of the following two procedures: fiber or matrix modification. Fiber modification usually involves the process of alkalization, which reduces the cellulose capacity to bond hydrogen and, in turn, cancels the bonding of open hydroxyl groups with water molecules. This process concurrently eliminates hemicellulose, which helps in making the fibers less hydrophilic. Issues to account for in a proper treatment are the alkali concentration, the exposure time, need for water washing after the treatment under penalty of fiber degradation over time, and production of waste products. Another alternative seems to be the Duralin steam process [86]. This treatment promotes the creation of aldehyde and phenolic functionalities and promotes a significant reduction of moisture uptake in water environments, as well as diffusion in the material. This process has some advantages: increase of the fiber strength, ductility and resistance to fungi, and dimensional stability. Matrix modification is not often used as treatment, because resins are typically hydrophobic and are not very prone to absorb moisture. Despite this fact, some soy-based resins can be affected by moisture and, thus, solutions exist to overcome this limitation: additives and coatings. One solution for additives was proposed by Kumar and Zhang [29], who treated the soy proteins by immersion in benzilic acid, allowing the improvement of strength and moisture resistance, while keeping the matrix fully biodegradable. On the contrary, the use of coatings can also be effective if the coating material is chosen correctly and well applied. A coating example is a natural lignin-based coating [89].

Durability and behavior prediction over long periods of time are two major concerns with natural fiber composites, especially under adverse environments/moisture. From an extensive literature review, a major gap of knowledge is undoubtedly the fatigue behavior of these materials, and this definitely needs addressing before these materials go under usage for load-bearing components. Opposed to this, investigations exist on durability under wet conditions. One of the main considerations is the growth of bacteria. Singh et al. [90] tested natural fiber composites of jute-reinforced phenolic resin after water exposure and detected hyphae fungi. Singh and Gupta [91] showed that prolonged ultraviolet exposure leads to matrix cracking and fibrillation, accompanied by strength reductions of more than 50%. To reduce these effects, polyurethane coating was applied with success. The use of coatings and fiber treatments as bleaching or using silanes seems to be a field to be explored, with some good preliminary results [92].

The strength and stiffness of natural fiber composites usually falls between 100 and 200 MPa and between 1 and 4 GPa, respectively, which is markedly low for structural parts. A possibility to overcome this limitation, although without improvements to the materials, is taking advantage of design (e.g., sandwich solutions). Other modifications to improve mechanical properties and also the durability involve adhesion or wetting, and the use of hybrid solutions, even though this last one cancels or reduces the bio advantage. Regarding the fibers/matrix adhesion, many treatments exist that change the fiber surface and improve wetting. The most straightforward way in which adhesion could be increased is by drying the fiber before mixing with the matrix to cancel bonding of H_2O molecules to the fiber surface. Adhesion can be improved by additives, coupling agents, or alkalization. Additives can be applied to fibers (e.g., calcium chloride or sodium carbonate nanoparticles) or matrices (e.g., maleated polyolefins). Malkapuram et al. [19] used maleated polyolefins to improve adhesion. Others [93] mixed soy protein concentrate powder and microbamboo fibrils, thus obtaining good results. Coupling agents are compatibilizers between fibers and matrix that improve the bond by the removal of weak boundary layers and creation of a cross-linked region. Covalent bonds are then created between the fibers and matrix [19]. The use of silane coupling agents is very common between synthetic fibers and bio resins; however, it is not so common between both bio components, although trialkoxysilanes have been used [92]. Maleated propylene is highly effective in terms of mechanical properties and is easy to apply, giving stiffness and strength improvements of more than 100% [94]. Careful examinations of the fracture surfaces showed a drastic reduction of fiber pullout after treatment, with the fiber fracture coinciding with the macro cracks of the material. Other coupling agents are stearic acid, isocyanates, and triazine. Finally, alkalization or the alkali treatment reduces the absorption and modifies the fiber surface, which results in the improvement of mechanical properties. This treatment promotes the removal of cementitious materials and increases the surface roughness. The main improvements are the tensile and flexural mechanical properties, although with a reduction of the impact strength. This treatment removes the nonstructural components of the fibers, which increases their specific properties. Some authors [95] found an improvement of approximately 50% in mechanical properties. Chang et al. [96] tested the combination between ultrasonic and alkali treatments, resulting in extraction of low molecular constituents and depolymerization of macromolecules. Improvements in the use of the alkali treatment alone were found.

The variability in the fiber properties, already discussed in terms of origin, is a major setback of natural fiber composites if they are going to be used as structural parts, because of the associated design uncertainties. Thus, quality assurance protocols are needed so that the differences in properties between different batches are within tolerable limits [86]. Variability of cost and availability also exist. Some costs of natural fibers actually vary with the climate conditions. Another scenario is an eventually poor season for a particular plant, which make the respective fibers less available and lead to a higher cost. To reduce these problems to a minimum, the most relevant plants should be cropped in several regions around the globe.

Hybridization, or a combination of natural fibers with glass fibers, also constitutes a possible way in which the mechanical properties could be improved. Some studies are available on this solution that reported improvements of strength and reduction of moisture absorption [97]. This is, however, a partial solution, since most of the bio advantages vanish. A combination of natural and synthetic matrices is also possible to allow an increase of mechanical properties, but with the same inconvenience.

The improvement of fire resistance is another point requiring review. Strength and stiffness of natural fiber composites can be significantly depreciated at temperatures of 120°C. Currently, very few of these materials pass civil construction fire resistance tests. Usually, the flammability is higher for fibers constituted mainly by cellulose, compared with those with a higher content of hemicellulose [98]. Fire resistance is also increased with the following factors: silica or ash in the fiber composition, and fiber structure with higher crystallinity and small polymerization. Among the synthetic resins to be used along with natural fibers, phenolic resins are probably the best choice with regard to fire resistance. Otherwise, coatings and/or additives can also be used. Coatings of ceramics, intumescents, ablatives or glass, or chemical additives are possible solutions. The most effective coatings for natural fiber composites are intumescents, which act by forming a cellular surface; when heated, these protect the base material. Fillers such as tack or nanoparticles also improve the resistance of these materials to fire by being heat barriers [98]. A lignin coating that promoted the appearance of char was also successfully tested [89]. An integrated solution would be the combination of a cellulosic material that promotes formation of char with an intumescent coating.

Processing difficulties and limitations are also present in the fabrication of these materials. The maximum operating temperature is around 180°C [25], which brings some restrictions to the chosen matrix and fabrication methods. Regarding the bio resins, they are not usually processed at very high temperatures, which is a good indicator, as it makes the process feasible with respect to degradation of the fibers. Hand layup is the simplest process, and it does not require heat. However, it is limited in the wt.% of natural fibers. Compression molding, on the other hand, is the most widespread technique for natural fiber composites, achieving approximately 80 wt.% of fibers [87]. Other processes were tested for natural fiber composites, such as resin transfer molding (RTM), vacuum molding, and vacuum-assisted RTM. Pultrusion is widespread for glass and carbon composites, but it is new for natural fiber composites. Some authors [99] showed the feasibility of this technique, although hybrid solutions with glass fibers gave better results. It was shown that the obtained composites had good mechanical properties, although adhesion could still be improved. Others [58] tested production methods such as compounding or injection molding, but mechanical properties were limited by the fiber length and the room for improvement is big.

1.9   CONCLUSIONS

Natural fiber composites have been receiving a lot of attention in recent decades from the industry and researchers because of their potential in civil engineering, vehicles, packaging, and other fields. This is because of some desirable characteristics such as low cost, low density (and, thus, high specific properties), eco friendliness, processing advantages, and reduction of CO_2 emissions in the whole life cycle. Despite these facts, some limitations exist, such as low mechanical properties and toughness; property variations; sensitivity to temperature, moisture, and UV radiation; and fire resistance. Some of these limitations are related to the typical short life in outdoor applications. Plant fibers are the most widely used, and a huge variety is available, with distinct physical and mechanical properties. Most of the research to date has focused on the feasibility of using these materials for industry applications, with promising results. Mechanical properties and durability are the main focus of research. Properties are also prone to be enhanced by proper treatments, and many solutions are already available for improvement. Flax, hemp, and ramie fibers show particularly impressive specific mechanical properties. The fiber–matrix adhesion is the main factor that dictates the mechanical properties, and a lot of research can be found in this area. Moisture absorption can also be minimized by coating the fibers or modifying their surface. Improved fire resistance can be implemented by intumescent coatings. The variability in the fiber properties is more difficult to prevent, but the creation of standards and quality assurance protocols can overcome this limitation before the raw materials reach the fabrication process. Natural composites can be fabricated by nearly all processes available for synthetic composites, with few or no modifications. However, to produce high-quality natural fibers, a several-step fabrication process may be required, which negatively affects the costs. Animal fibers are still an emerging area, and asbestos or mineral fibers exhibit some attractive properties, such as thermal or acoustic insulation, but their carcinogenic nature prevents many of the possible applications. Bio resins are also important, although some limitations exist regarding their compatibility with natural fibers. If this issue is properly taken care of, a full ecological solution can be achieved. Natural composites are a big challenge to materials science and, although nowadays they may not be as cost efficient as synthetic composites, their cost is decreasing and their performance is improving such that, in the near future, beyond being the most ecological solution, they will also be the cheapest one. However, it should not be assumed that natural composites are more eco-friendly than all other materials (e.g., wood). Despite this fact, life cycle analyses showed that there is a huge benefit when compared to other materials, energy requirements between 20% and 40%. Natural fiber composites will definitely get more and more attention and applications, especially in Europe, because of public opinion and legislation on the subject. The use of these materials is essential to safeguard nonrenewable resources and the planet, and research and improvements in this field constitute an ongoing process.

REFERENCES

1. Mohanty, A.K., Misra, M., Drzal, L.T., Selke, S.E., Harte, B.R., and Hinrichsen, G. 2005. Natural Fibers, Biopolymers, and Biocomposites: An Introduction. In *Natural Fibers, Biopolymers, and Biocomposites*, ed. Mohanty, A.K., Misra, M., and Drzal, L.T., 1–38. Boca Raton: CRC Press.
2. Kim, W., Argento, A., Lee, E., Flanigan, C., Houston, D., Harris, A., and Mielewski, D.F. 2011. High strain-rate behavior of natural fiber-reinforced polymer composites. *Journal of Composite Materials* 46:1056–1065.

3. Dittenber, D.B., and GangaRao, H.V.S. 2012. Critical review of recent publications on use of natural composites in infrastructure. *Composites: Part A* 43:1419–1429.

4. Shen, L., Haufe, J., and Patel, M.K. 2009. Product overview and market projection of emerging bio-based plastics. PRO-BIP; Final Report, Report No: NWS-E-2009-32. Utrecht, The Netherlands: Utrecht University.

5. Natural fiber composites market trend and forecast 2011–2016: Trend, Forecast and Opportunity Analysis. Technical report, Lucintel, 2011.

6. Saxena, M., Pappu, A., Sharma, A., Haque, R., and Wankhede, S. 2011. Composite materials from natural resources: Recent trends and future potentials. In *Advances in Composite Materials–Analysis of Natural and Man-Made Materials*, ed. Tesinova, P., 121–162. Rijeka, Croatia: InTech.

7. Faruk, O., Bledzki, A.K., Fink, H.P., and Sain, M. 2012. Biocomposites reinforced with natural fibers: 2000–2010. *Progress in Polymer Science* 37:1552–1596.

8. Technologies and products of natural fiber composites. CIP-EIP-Eco-Innovation-2008: Pilot and market replication projects – ID: ECO/10/277331.

9. Majeed, K., Jawaid, M., Hassan, A., Abu Bakar, A., Abdul Khalil, H.P.S., Salema, A.A., and Inuwa, I. 2013. Potential materials for food packaging from nanoclay/natural fibers filled hybrid composites. *Materials and Design* 46:391–410.

10. Varghese, S., Kuriakose, B., and Thomas, S. 1994. Stress relaxation in short sisal fiber-reinforced natural rubber composites. *Journal of Applied Polymer Science* 53:1051–1060.

11. Suddell, B.C., and Evans, W.J. 2003. The increasing use and application of natural fiber composites materials within the automotive industry. Proceedings of the Seventh International Conference on Woodfiber-Plastic Composites, Forest Products Society, Madison, USA.

12. John, M.J., and Anandjiwala, R.D. 2008. Recent developments in chemical modification and characterization of natural fiber-reinforced composites. *Polymer Composites* 29:187–207.

13. Wambua, P., Ivens, J., and Verpoest, I. 2003. Natural fibers: Can they replace glass in fiber reinforced plastics? *Composites Science and Technology* 63:1259–1264.

14. Herrera-Franco, P.J., and Valadez-González, A. 2005. A study of the mechanical properties of short natural-fiber reinforced composites. *Composites: Part B* 36:597–608.

15. Bledzki, A.K., and Gassan, J. 1996. Effect of coupling agents on the moisture absorption of natural fiber-reinforced plastics. *Angewandte Makromolekulare Chemie* 236:129–138.

16. Bledzki, A.K., and Gassan, J. 1999. Composites reinforced with cellulose based fibers. *Progress in Polymer Science* 24:221–274.

17. Mediavilla, V., Leupin, M., and Keller, A. 2001. Influence of the growth stage of industrial hemp on the yield formation in relation to certain fiber quality traits. *Industrial Crops and Products* 13:49–56.

18. Barbero, E.J. 2011. *Introduction to Composite Materials Design*. Boca Raton, FL: Taylor & Francis.

19. Malkapuram, R., Kumar, V., and Negi, Y.S. 2009. Recent development in natural fiber reinforced polypropylene composites. *Journal of Reinforced Plastics and Composites* 28:1169–1189.

20. Plackett. D. 2002. The natural fiber-polymer composite industry in Europe technology and markets. Proceedings of the Progress on Woodfiber-Plastic Composites Conference 2002, University of Toronto, Toronto, Canada.

21. D'Arcy, J.B. 1986. *Sheep Management & Wool Technology*. Kensington: NSW University Press.

22. Hornik, C., Krishan, K., Yusuf, F., Scaal, M., and Brand-Saberi, B. 2005. Dermo-1 misexpression induces dense dermis, feathers, and scales. *Developmental Biology* 277:42–50.

23. Grozdanov, A., Avella, M., Buzarovska, A., Gentile, G., and Errico, M.E. 2010. Reuse of natural fiber reinforced eco-composites in polymer mortars. *Polymer Engineering & Science* 50:762–766.

24. Mohanty, A.K., Misra, M., and Drzal, L.T. 2001. Surface modifications of natural fibers and performance of the resulting biocomposites: An overview. *Composite Interfaces* 8:313–343.

25. Ochi, S. 2006. Development of high strength biodegradable composites using Manila hemp fiber and starch-based biodegradable resin. *Composites: Part A* 37:1879–1883.

26. Nättinen, K., Hyvärinen, S., Joffe, R., Wallström, L., and Madsen, B. 2010. Naturally compatible: Starch acetate/cellulosic fiber composites. I. Processing and properties. *Polymer Composites* 31:524–535.

27. Ochi, S. 2008. Mechanical properties of kenaf fibers and kenaf/PLA composites. *Mechanics of Materials* 40:446–452.

28. Haq, M., Burgueno, R., Mohanty, A.K., and Misra, M. 2009. Processing techniques for biobased unsaturated-polyester/clay nanocomposites: Tensile properties, efficiency, and limits. *Composites: Part A* 40:394–403.

29. Kumar, R., and Zhang, L. 2009. Soy protein films with the hydrophobic surface created through non-covalent interactions. *Industrial Crops and Products* 29:485–494.

30. Mohanty, A.K., Tummala, P., Liu, W., Misra, M., Mulukutla, P.V., and Drzal, L.T. 2005. Injection molded biocomposites from soy protein based bioplastic and short industrial hemp fiber. *Journal of Polymers and the Environment* 13:279–285.

31. Porras, A., and Maranon, A. 2012. Development and characterization of a laminate composite material from polylactic acid (PLA) and woven bamboo fabric. *Composites: Part B* 43:2782–2788.

32. Baghaei, B., Skrifvars, M., and Berglin, L. 2013. Manufacture and characterization of thermoplastic composites made from PLA/hemp co-wrapped hybrid yarn prepregs. *Composites: Part A* 50:93–101.

33. Alemdar, A., and Sain, M. 2008. Biocomposites from wheat straw nanofibers: Morphology, thermal and mechanical properties. *Composites Science and Technology* 68:557–565.

34. Rowell, R.M. 2008. Natural fibers: Types and properties. In *Properties and Performance of Natural-Fiber Composites*, ed. Pickering, K., 3–66. Cambridge, UK: Woodhead Publishing.

35. Zafeiropoulos, N.E., Baillie, C.A., and Matthews, F.L. 2001. The effect of transcrystallinity on the interface of green flax/polypropylene composite materials. *Advanced Composite Materials* 10:229–236.

36. Gassan, J., Gutowski, V.S. 2000. Effects of corona discharge and UV treatment on the properties of jute-fiber epoxy composites. *Composites Science and Technology* 60:2857–2863.

37. Marais, S., Gouanvé, F., Bonnesoeur, A., Grenet, J., Poncin-Epaillard, F., Morvan, C., and Metayer, M. 2005. Unsaturated polyester composites reinforced with flax fibers: Effect of cold plasma and autoclave treatments on mechanical and permeation properties. *Composites: Part A* 36:975–986.

38. Xu, Y., Kawata, S., Hosoi, K., Kawai, T., and Kuroda, S. 2009. Thermomechanical properties of the silanized-kenaf/polystyrene composites. *Express Polymer Letters* 3:657–664.

39. Bisanda, E.T.N. 2000. The effect of alkali treatment on the adhesion characteristics of sisal fibers. *Applied Composite Materials* 7:331–339.

40. Bledzki, A.K., Mamun, A.A., Lucka-Gabor, M., and Gutowski, V.S. 2008. The effects of acetylation on properties of flax fiber and its polypropylene composites. *Express Polymer Letters* 2:413–422.

41. Mohanty, S., Nayak, S.K., Verma, S.K., and Tripathy, S.S. 2004. Effect of MAPP as a coupling agent on the performance of jute–PP composites. *Journal of Reinforced Plastics and Composites* 23:625–637.
42. Saleem, Z., Rennebaum, H., Pudel, F., and Grimm, E. 2008. Treating bast fibers with pectinase improves mechanical characteristics of reinforced thermoplastic composites. *Composites Science and Technology* 68:471–476.
43. Sawpan, M.A., Pickering, K.L., and Fernyhough, A. 2011. Improvement of mechanical performance of industrial hemp fiber reinforced polylactide biocomposites. *Composites: Part A* 42:310–319.
44. Li, Y., Pickering, K.L., and Farrell, R.L. 2009. Analysis of green hemp fiber reinforced composites using bag retting and white rot fungal treatments. *Industrial Crops and Products* 29:420–426.
45. Vilaseca, F., López, A., Llauró, X., PèLach, M.A., and Mutjé, P. 2004. Hemp strands as reinforcement of polystyrene composites. *Chemical Engineering Research and Design* 82:1425–1431.
46. Hautala, M., Pasila, A., and Pirila, J. 2004. Use of hemp and flax in composite manufacture: A search for new production methods. *Composites: Part A* 35:11–16.
47. Rouison, D., Sain, M., and Couturier, M. 2006. Resin transfer molding of hemp fiber composites: Optimization of the process and mechanical properties of the materials. *Composites Science and Technology* 66:895–906.
48. Mehta, G., Drzal, L.T., Mohanty, A.K., and Misra, M. 2006. Effect of fiber surface treatment on the properties of biocomposites from nonwoven industrial hemp fiber mats and unsaturated polyester resin. *J Appl Polym Sci* 99:1055–1068.
49. Feng, Y., Hu, Y., Zhao, G., Yin, J., and Jiang, W. 2011. Preparation and mechanical properties of high-performance short ramie fiber-reinforced polypropylene composites. *Journal of Applied Polymer Science* 122:1564–1571.
50. Mahdavi, S., Kermanian, H., and Varshoei A. 2010. Comparison of mechanical properties of date palm fiber-polyethylene composite. *Bioresources* 5:2391–2403.
51. Bendahou, A., Kaddami, H., Sautereau, H., Raihane, M., Erchiqui, F., and Dufresne, A. 2008. Short palm tree fibers polyolefin composites: Effect of filler content and coupling agent on physical properties. *Macromolecular Materials and Engineering* 293:140–148.
52. Threepopnatkul, P., Kaerkitcha, N., and Athipongarporn, N. 2009. Effect of surface treatment on performance of pineapple leaf fiber–polycarbonate composites. *Composites: Part B* 40:628–632.
53. Saheb, D.N., and Jog, J.P. 1999. Natural fiber polymer composites: A review. *Advances in Polymer Technology* 18:351–363.
54. Saxena, M., Pappu, A., Haque, R., and Sharma, A. 2011. *Sisal Fiber Based Polymer Composites and Their Applications Cellulose Fibers: Bio- and Nano-Polymer Composites*. Berlin, Germany: Springer.
55. Xin, X., Xu, C.G., and Qing, L.F. 2007. Friction properties of sisal fiber reinforced resin brake composites. *Wear* 262:736–741.
56. Mohanty, A.K., Drazl, L.T., and Misra, M. 2002. Engineered natural fiber reinforced polypropylene composites: Influence of surface modifications and novel powder impregnation processing. *Journal of Adhesion Science and Technology* 16:999–1015.
57. Duflou, J.R., Deng, Y., Acker, K.V., and Dewulf, W. 2012. Do fiber-reinforced polymer composites provide environmentally benign alternatives? A life-cycle-assessment-based study. *MRS Bulletin* 37: April.
58. Mutnuri, B., Aktas, C.J., Marriott, J., Bilec, M., and GangaRao, H. 2010. Natural fiber reinforced pultruded composites. Proceedings of COMPOSITES 2010, Las Vegas, Nevada; 9–11 February.

59. Mueller, D.H., and Krobjilowski, A. 2003. New discovery in the properties of composites reinforced with natural fibers. *Journal of Industrial Textiles* 33:111–130.

60. Wötzel, K., Wirth, R., and Flake, R. 1999. Life cycle studies on hemp fiber reinforced components and ABS for automotive parts. *Angewandte Makromolekulare Chemie* 272:121–127.

61. Schmidt, W.P., and Beyer, H.M. 1998. Life cycle study on a natural fiber reinforced component. SAE Technical paper 982195. SAE Total Life-cycle Conference. Graz, Austria; December 1–3.

62. Corbiere-Nicollier, T., Laban, B.G., Lundquist, L., Leterrier, Y., Manson, J.A.E., and Jolliet, O. 2001. Lifecycle assessment of biofibers replacing glass fibers as reinforcement in plastics. *Resources, Conservation and Recycling* 33:267–287.

63. Suzuki, T., and Takahashi, J. 2005. Prediction of energy intensity of carbon fiber reinforced plastics for mass-produced passenger cars. Proceedings of JISSE-9, Tokyo, Japan; 29 November–11 December.

64. Patel, M. 2007. Cumulative energy demand (CED) and cumulative CO2 emissions for products of the organic chemical industry. *Energy* 28:721–740.

65. Boustead, I. 2005. *Eco-profiles of the European Plastics Industry.* Association of Polymer Manufacturers in Europe, Brussels, Belgium.

66. Ecolizer 2.0. 2003–2011. Openbare Vlaamse Afvalstoffenmaatschappij, Mechelen, Belgium.

67. Anandjiwala, R.D., and Blouw, S. 2007. Composites from bast fibers: Prospects and potential in the changing market environment. *Journal of Natural Fibers* 4:91–109.

68. Joshi, S.V., Drzal L., Mohanty, A., and Arora, S. 2004. Are natural fiber composites environmentally superior to glass fiber reinforced composites? *Composites: Part A* 35:371–376.

69. Song, Y.S., Youn, J.R., and Gutowski, T.G. 2009. Life cycle energy analysis of fiber-reinforced composites. *Composites: Part A* 40:1257–1265.

70. Puri, P., Compston, P., and Pantano, V. 2009. Life cycle assessment of Australian automotive door skins. *International Journal of Life Cycle Assessment* 14:420–428.

71. Luz, S.M., Caldeira-Pires, A., and Ferrão P.M.C. 2010. Environmental benefits of substituting talc by sugarcane bagasse fibers as reinforcement in polypropylene composites: Ecodesign and LCA as strategy for automotive components. *Resources, Conservation and Recycling* 54:1135–1144.

72. Schexnayder, S.M., Das, S., Dhingra, R., Overly, J.G., Tonn, B.E., Peretz, J.H., Waidley, G., and Davis, G.A. 2001. Environmental Evaluation of New Generation Vehicles and Vehicle Components. Report ORNL/TM-2001–266, Oak Ridge National Laboratory, Oak Ridge.

73. Leterrier, Y. 2000. Life cycle engineering of composites. In *Comprehensive Composite Materials*, ed. Anthony, K., and Carl, Z., 1073–1102. Oxford, UK: Pergamon.

74. Hedlund-Aström, A. 2005. Model for end of life treatment of polymer composite materials. Royal Institute of Technology, Stockholm, Sweden.

75. Duflou, J.R., De Moor, J., Verpoest, I., and Dewulf, W. 2005. Environmental impact analysis of composite use in car manufacturing. *CIRP Annals - Manufacturing Technology* 58:9–12.

76. Shen, L., and Patel, M. 2008. Life cycle assessment of polysaccharide materials: A review. *Journal of Polymers and the Environment* 16:154–167.

77. Bourmaud, A., and Baley, C. 2007. Investigations on the recycling of hemp and sisal fiber reinforced polypropylene composites. *Polymer Degradation and Stability* 92:1034–1045.

78. Kumar, R., Yakubu, M., and Anandjiwala, R. 2010. Biodegradation of flax fiber reinforced poly lactic acid. *Express Polymer Letters* 4:423–430.

79. Patel, M., and Narayan, R. 2005. How sustainable are biopolymers and biobased products? The hope, the doubt, and the reality. In *Natural fibers Fibers, Biopolymers, and Biocomposites*, ed. Mohanty, A.K., Misra, M., and Drzal, L.T., 833–854. Boca Raton, FL: Taylor & Francis.
80. Burgueno, R., Quagliata, M.J., Mehta, G., Mohanty, A.K., Misra, M., and Drzal, L.T. 2005. Sustainable cellular biocomposites from natural fibers and unsaturated polyester resin for housing panel applications. *Journal of Polymers and the Environment* 13:139–149.
81. Wang, B., Sain, M., and Oksman, K. 2007. Study of structural morphology of hemp fiber from the micro to the nanoscale. *Applied Composite Materials* 14:89–103.
82. Kaushik, A., Singh, M., and Verma, G. 2010. Green nanocomposites based on thermoplastic starch and steam exploded cellulose nanofibrils from wheat straw. *Carbohydrate Polymers* 82:337–345.
83. Liu, D.Y., Yuan, X.W., Bhattacharyya, D., and Easteal, A.J. 2010. Characterisation of solution cast cellulose nanofiber-reinforced poly(lactic acid). *Express Polymer Letters* 4:26–31.
84. Alemdar, A., and Sain, M. 2008. Isolation and characterization of nanofibers from agricultural residues – wheat straw and soy hulls. *Bioresource Technology* 99:1664–1671.
85. Fowler, P.A., Hughes, J.M., and Elias, R.M. 2006. Biocomposites: Technology, environmental credentials and market forces. *Journal of the Science of Food and Agriculture* 86:1781–1789.
86. Bismarck, A., Mishra, S., and Lampke, T. 2005. Plant fibers as reinforcement for green composites. In *Natural Fibers, Biopolymers, and Biocomposites*, ed. Mohanty, A.K., Misra, M., and Drzal, L.T., 37–108. Boca Raton, FL: Taylor & Francis.
87. Satyanarayana, K.G., Arizaga, G.G.C., and Wypych, F. 2009. Biodegradable composites based on lignocellulosic fibers – an overview. *Progress in Polymer Science* 34:982–1021.
88. Singh, B., and Gupta, M. 2005. Natural fiber composites for building applications. In *Natural Fibers, Biopolymers, and Biocomposites*, ed. Mohanty, A.K., Misra, M., and Drzal, L.T., 261–290. Boca Raton, FL: Taylor & Francis.
89. Doherty, W., Halley, P., Edye, L., Rogers, D., Cardona, F., Park, Y., and Woo, T. 2007. Studies on polymers and composites from lignin and fiber derived from sugar cane. *Polymers for Advanced Technologies* 18:673–678.
90. Singh, B., Gupta, M., and Verma, A. 2000. The durability of jute fiber-reinforced phenolic composites. *Composites Science and Technology* 60:581–589.
91. Singh, B., and Gupta, M. 2005. Performance of pultruded jute fiber reinforced phenolic composites as building materials for door frame. *Journal of Environmental Polymer Degradation* 13:127–137.
92. Xie, Y., Hill, C.A.S., Xiao, Z., Militz, H., and Mai, C. 2010. Silane coupling agents used for natural fiber/polymer composites: A review. *Composites: Part A* 41:806–819.
93. Huang. X, and Netravali, A. 2009. Biodegradable green composites made using bamboo micro/nano-fibrils and chemically modified soy protein resin. *Composites Science and Technology* 69:1009–1015.
94. Pickering, K. 2008. Introduction. In *Properties and Performance of Natural-Fiber Composites*, ed. Pickering, K., XV–XVII. Boca Raton, FL: CRC Press.
95. De, D., Adhikari, B., and De, D. 2007. Grass fiber reinforced phenol formaldehyde resin Composite: Preparation, characterization and evaluation of properties of composite. *Polymers for Advanced Technologies* 18:72–81.
96. Chang, W.P., Kim, K.J., and Gupta, R.K. 2009. Ultrasound-assisted surface-modification of wood particulates for improved wood/plastic composites. *Composite Interfaces* 16:687–709.

97. Ray, D., and Rout, J. 2005. Thermoset biocomposites. In *Natural Fibers, Biopolymers, and Biocomposites*, ed. Mohanty, A.K., Misra, M., and Drzal, L.T., 291–346. Boca Raton, FL: Taylor & Francis.

98. Chapple, S.C., and Anandjiwala, R. 2010. Flammability of natural fiber-reinforced composites and strategies for fire retardancy: A review. *Journal of Thermoplastic Composite Materials* 23:871–893.

99. Peng, X., Fan, M., Hartley, J., and Al-Zubaidy, M. 2011. Properties of natural fiber composites made by pultrusion process. *Journal of Composite Materials* 46:237–246.

2 Natural Fibers and Their Characterization

Z.N. Azwa, B.F. Yousif, A.C. Manalo,
and W. Karunasena

CONTENTS

2.1 INTRODUCTION

To understand the contribution of bamboo fibers in a composite system, the characteristics of the fiber should be comprehensively evaluated. One of the major issues regarding the use of natural fibers in the polymer composite industry is the incompatibility between the hydrophilic fibers and the hydrophobic polymer resin, which weakens interfacial adhesion, leading to ineffective load transfer from the matrix resin to the reinforcing fibers. One of the most recent and effective techniques for improving interfacial adhesion between natural fibers and polymer matrices is alkali treatment, which involves the immersion of fibers in sodium hydroxide (NaOH) solution.

Although several studies have shown the potential of using bamboo fibers as reinforcement in natural composites, limited work is conducted to comprehensively study the influence of the NaOH on the structural, physical, and tensile characteristics of bamboo fiber. Such studies are essential to understand these important characteristics of bamboo fibers, leading to their mechanical improvement, producing composites with enhanced properties and, eventually, expanding the use of bamboo fibers in the polymer composite industry. This chapter aims at evaluating the effect of alkali treatment on the structural, physical, and tensile characteristics of bamboo fibers, as well as the interfacial and tensile properties of the corresponding bamboo fiber/polyester composites. The morphological changes of the fibers and the impact they have on the interfacial adhesion of the fiber–matrix interface were studied using scanning electron microscopy (SEM) to understand the mechanical behavior of the composites. The single fiber tensile test (SFTT) and single fiber fragmentation test (SFFT) were used to study the tensile and interfacial behaviors of the fibers, respectively.

2.2 INFLUENCE OF FIBER TREATMENT ON THE STRUCTURAL PROPERTIES OF FIBERS

The changes in the structure of bamboo fibers due to the effect of alkali treatment are discussed in this section. Figure 2.1 shows the lateral views from optical micrographs and the SEM of bamboo fibers treated with NaOH at 0, 4, 6, and 8 wt.%. For untreated bamboo fibers, the optical micrograph shows an almost smooth fiber surface. The SEM indicates that the fiber surface is mostly covered with a cell wall and only some fibrils are visible. Through alkalization, it can be seen that the surfaces of all treated bamboo fibers are rougher than those of the raw fibers. It was further observed that different concentrations of NaOH provide different levels of treatment on the fibers. At 4% NaOH treatment, some cell walls can still be observed but most of them have been removed from the fiber surface and more microfibrils have been exposed. The fiber surface is further roughened at 6% NaOH treatment, and some voids were observed from the optical micrograph. This was confirmed by the SEM wherein voids in the matrix of the fiber are present. At 8% NaOH treatment, more matrixes were removed, exposing the fibrils to the alkali attack. It can be observed that the surface of the fibrils experienced some surface disintegration.

To explain these mechanisms further, it is important to understand the arrangement of the structural components of bamboos. The structure of bamboos from macro to nanolevel is represented in Figure 2.2 [1]. Bamboo fiber itself is a composite, consisting of two main components: parenchyma tissues made up of a lignin–hemicellulose matrix and sclerenchyma tissues that consist of vascular bundles made up of many phloem fibers [2]. The chemical composition of bamboo fibers is 45.70% cellulose, 25.90% hemicelluloses, 24.95% lignin, and 1.65% ash [3]. However, through the steam explosion process, it was suggested by Shao et al. [4] that xylan, which is the major hemicellulose of bamboo cell walls, was significantly removed during the treatment. The strength of natural fibers is believed to be influenced by its cellulosic component [5–6]; thus, this may suggest that nearly 50% of the composition of a bamboo fiber contributes to its strength.

Raw bamboo fibers are protected by the outer layer, which is the cell wall. This can be observed in Figure 2.1 where a thick layer covers most of the fiber's perimeter, providing a smooth surface profile. However, some exposed fibril was also visible, which may imply that some areas on the fiber surface were affected by the steam explosion process. As NaOH is introduced in the fiber treatment, these cell walls were removed, exposing more fibrils, which in Figure 2.2 refer to the cellulose grains. At 6% NaOH, the

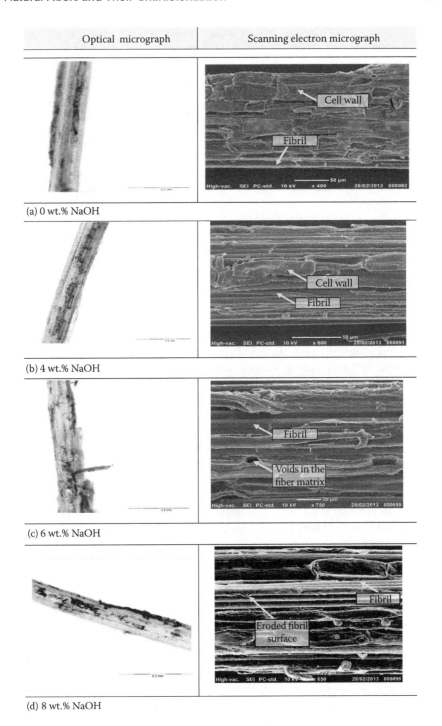

| Optical micrograph | Scanning electron micrograph |

(a) 0 wt.% NaOH

(b) 4 wt.% NaOH

(c) 6 wt.% NaOH

(d) 8 wt.% NaOH

FIGURE 2.1 Morphology of treated and untreated bamboo fibers. (a) 0 wt.% NaOH, (b) 4 wt.% NaOH, (C) 6 wt.% NaOH, and (d) 8 wt.% NaOH.

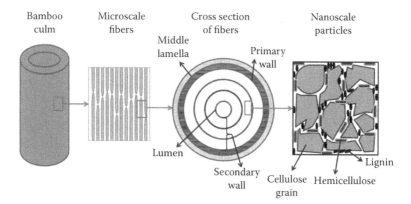

FIGURE 2.2 Structure of bamboo from macro to nanolevel. (Adapted from Zou, L., Jin, H., Lu, W.Y., and Li, X., Nanoscale structural and mechanical characterization of the cell wall of bamboo fibers. *Materials Science and Engineering: C*, 2009. 29(4): p. 1375–79. With permission [1].)

alkali may have been strong enough to erode the hemicellulose–lignin matrix, which explains the presence of voids in Figure 2.1c. This was also reported by Wong et al. [7], where hemicelluloses, lignin, waxes, oil, and surface impurities of the bamboo fibers were soluble in NaOH aqueous solution, causing the cell wall to open up as these substances were dissolved. As the NaOH concentration was further increased to 8%, more hemicellulose–lignin matrix was removed from the fiber, exposing more fibrils. These fibrils are also more vulnerable to alkali attack, which may explain the rougher fibril surface observed in Figure 2.1d. Once the cellulosic component of the fiber is affected by the alkali, the strength of the fiber might be compromised. Many studies vary the concentration of NaOH from 1 to 10 wt.% for fiber alkalization, as higher concentrations worsen the mechanical properties of the fiber [7].

Studying the morphology of natural fibers is important for predicting their interaction with polymer resins in a composite system [8]. The smooth surface texture of untreated bamboo fibers provides poor friction and weak interlocking with polymers, thus probably resulting in a composite with low strength. In comparison with other natural fibers, similar findings were also observed with regard to the changes in the fiber morphology due to a higher concentration of NaOH treatment. Hemp and kenaf fibers treated at 6% NaOH were observed through SEM to be roughened with obvious removal of wax, oil, and other surface impurities in comparison to the raw fibers [9]. Sydenstricker et al. [10] confirm that at 5 and 10% NaOH, sisal fibers are seriously physically affected with a rougher surface as compared with fibers treated at 1 and 2% NaOH, whereby the integrity of the fiber is still preserved. Surface impurities on Borassus fruit fiber were observed to be completely removed at 15% NaOH treatment, but with an increase in the presence of pores. The authors highlighted that if these pores exceed the desired amount, the mechanical properties of the fibers might be reduced [11]. However, some natural fibers exhibit a different reaction toward NaOH treatment. Raffia fibers become smoother and cleaner as the concentration of the solution is increased from 0 to 10% NaOH [12]. Untreated coir fiber contains globular protrusions and cuticles on its surface, which are removed at 7% NaOH, making the fiber smoother [13]. Thus, it is noteworthy to highlight that different natural fibers undergo different morphological changes due to alkalization. It is important to understand how NaOH treatment and its various concentrations alter the fiber surface to predict its consequent interlocking with the polymer resin.

2.3 INFLUENCE OF FIBER TREATMENT ON THE PHYSICAL PROPERTIES OF FIBERS

From the morphology study in Section 2.1, it is expected that the alkali treatment may have an impact on the physical and mechanical properties of the bamboo fibers. The physical observations discussed in this section include the visual appearance as well as the changes in diameter and density of fibers treated with NaOH at 0, 4, 6, and 8 wt.% through immersion for 24 hours.

2.3.1 VISUAL APPEARANCE OBSERVATION

Figure 2.3 shows the color changes experienced by the bamboo fibers. The treated fibers are darker as compared with the untreated ones. Thus, it is possible to visually differentiate between raw fibers and treated fibers. However, there was no obvious difference in appearance between the 4, 6, and 8% NaOH-treated fibers. Other researchers had also observed some changes in color due to exposure of fibers to treatments. The lignin component of a natural fiber is said to be responsible for its changes in color [14]. Elenga et al. [12] had measured the changes in the color of Raffia textilis fiber using a spectrocolorimeter. It was found that alkalization increased the yellowness and redness, and it reduced the lightness of the fibers. This can also be visually observed in the bamboo fibers studied. It was also agreed by

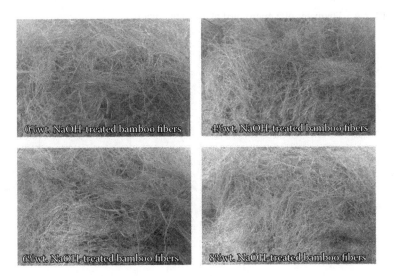

FIGURE 2.3   Changes in appearance of bamboo fibers due to alkalization.

the authors that the lightness did not vary significantly with the NaOH concentration. Khan and Ahmad [15] also reported a reduction in lightness of jute fibers by 10 to 20% due to alkalization, which was a result of lignin removal. However, some natural fibers experienced an increase in lightness after treatment. Rout et al. [16] reported that by bleaching coir fibers, the fiber changes color from dark brown to silvery white, which they attributed to the removal of lignin. Sisal fibers were also reported by Sydenstricker et al. [10] to become lighter in color and rougher at treatments higher than 5% NaOH.

2.3.2 DIAMETER DISTRIBUTION

The diameter of the cellulose fibril of the bamboo fiber was measured using SEM and is shown in Figure 2.4. The diameter is approximately 11.3 μm. In a nanoscale study conducted by Zou et al. [1], the authors observed that the diameter of the microfibrils or the building blocks of bamboo fibers is about 5 to 20 μm. This is in agreement with the measurement performed in this study.

The average diameter measured from 40 fibers for each treatment concentration is shown in Figure 2.5. The average diameter decreases as the fibers undergo alkalization till 6% NaOH. This corresponds to the SEM in Figure 2.1, which exhibits

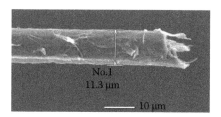

FIGURE 2.4   Diameter of a microfibril of a bamboo fiber.

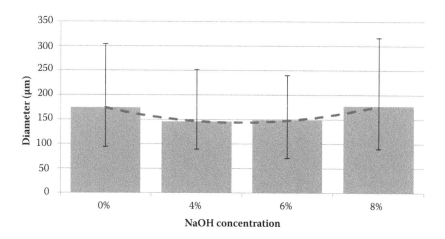

FIGURE 2.5   Average of fiber diameter for treated and untreated fibers.

the removal of the cell wall of the fibers as well as other surface impurities, thus reducing the diameter of alkalized fibers. However, at 8% NaOH, the average diameter was observed to increase, which can be attributed to the partial separation of fibrils with the removal of hemicellulose and lignin that acts as a matrix to bind these fibrils together. Overall, the average diameter of the bamboo fibers used in this study is 161.52 μm, with a range from 89 to 317 μm, measured from 160 fibers. This measurement is within the range of 88 to 330 μm presented by Prasad and Rao [17] for bamboo fibers.

Figure 2.6 shows the distributions of fiber diameters at different levels of treatment. The frequency distribution for all fiber conditions is Gaussian. Fibers treated with 4 and 6% NaOH (Figure 2.6b and c) present a narrower range of distribution due to the removal of the cell walls and surface impurities, whereas alkalization at 8% NaOH (Figure 2.6d) presents a wider distribution, similar to the untreated fibers, which is probably due to the weakening of the interaction between the fibrils in the fiber bundle that leads to partial fibrillation [8]. This phenomenon can be demonstrated in Figure 2.7 where the erosion of the cell wall and partial removal of the fiber matrix lead to the separation of the individual fibrils, dispersing them further from one another.

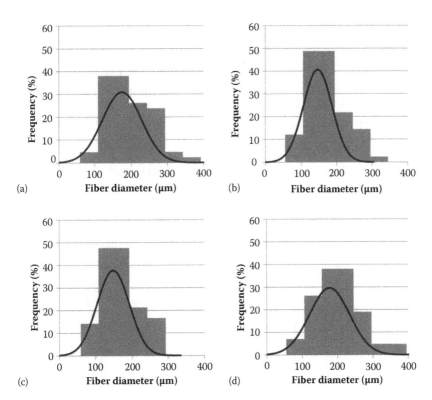

FIGURE 2.6 Distributions of fiber diameters: (a) 0%, (b) 4%, (c) 6%, and (d) 8% sodium hydroxide (NaOH).

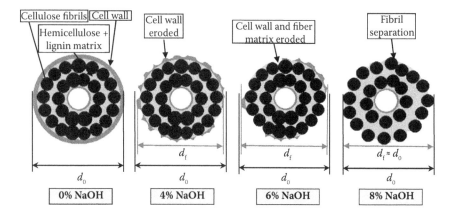

FIGURE 2.7 Schematic diagram of diameter change of a bamboo fiber due to alkalization. Note: d_0 = initial diameter, d_f = final diameter.

2.3.3 CHANGES IN DENSITY

The density of the fibers was measured using a helium pycnometer. As the concentration of the treatment is increased, the density of the fibers is slightly decreased as shown in Figure 2.8. Without treatment, raw bamboo fibers have a density of 1.47 g/cm^3, which reduced to 1.39 g/cm^3 at 8% NaOH treatment. This can be attributed to the extraction of soluble components of the fiber such as hemicelluloses, lignin, and other impurities as well as to the corrosion of the cell wall. From the morphology observations, the amount of voids on treated fibers is higher than on untreated fibers. Thus, this can contribute to the lower density of treated fibers.

A helium pycnometer measures the absolute density of a fiber, which does not account for the lumen and pores; therefore, it measures only the solid matter of the fibers [18]. Different natural fibers react differently to alkalization in terms of their densities as presented by Mwaikambo [18] whereby some fibers experienced

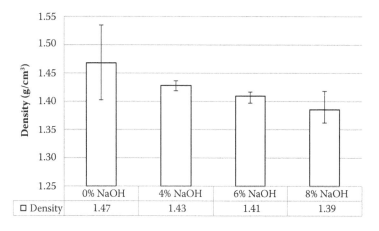

FIGURE 2.8 Density of bamboo fibers at different concentrations of alkali treatment.

a reduction in their density, some experienced an increase, and others exhibited no changes. The decrease in density due to alkalization was also observed in sisal fibers [10,19] and bamboo fibers [7]. However, some fibers undergo an increase in density such as reported in Borassus fruit fibers [11], hemp, and kenaf [9], which may be caused by cell wall densification of these fibers.

2.4 TENSILE PROPERTIES OF FIBERS

Untreated and treated bamboo fibers with a 10 mm gauge length were subjected to tensile loading at a rate of 1 N/min until complete breakage was attained according to ASTM D3379 [20]. The SFTT performed on the fibers yields typical stress–strain curves as presented in Figure 2.9. Initially, the curves show increased strain at very low stress, which can be neglected, as it may be caused by the clamps adjusting to the full length of the fibers. Once this has been established, it can be observed that all fibers behave in a brittle manner whereby a sudden halt in linear stress increment is achieved at maximum load. Generally, the highest maximum stress was obtained by 6% NaOH-treated fibers, followed by 4, 0, and 8% NaOH-treated fibers. A minimum of ten fibers for each fiber sample were tested. The tensile properties of the fibers were extracted from the stress–strain curves, and the average values were obtained.

Figure 2.10 shows the average tensile strength, whereas Figure 2.11 shows the average modulus of elasticity of untreated and treated bamboo fibers. The tensile strength and elastic modulus of bamboo fibers are higher for treated fibers for approximately 6% NaOH and decrease at 8% NaOH. The increase in tensile strength and elastic modulus is the highest at 6% NaOH treatment with 181 and 47% improvement, respectively. The increase in tensile strength due to fiber treatments was also observed in Agave fibers [21], kenaf fibers [22], and Borassus fruit fibers

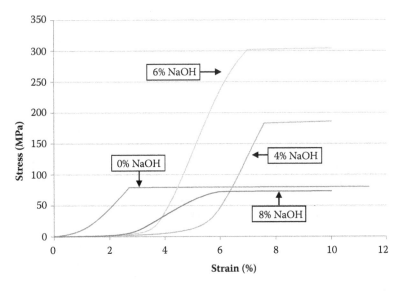

FIGURE 2.9 Stress–strain curves of bamboo fibers at different concentrations of NaOH treatment.

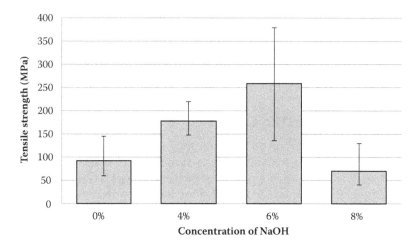

FIGURE 2.10   Tensile strength of bamboo fibers at different NaOH concentrations.

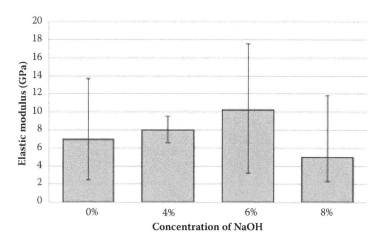

FIGURE 2.11   Elastic modulus of bamboo fibers at different NaOH concentrations.

[11]. Furthermore, it was indicated by Boopathi et al. [11] that alkalization causes the fiber matrix to partially dissolve and allows better alignment of fibrils, which may contribute to the increase in tensile strength. Kalia et al. [23] highlighted that alkalization affects the chemical composition of natural fibers, the degree of polymerization, and the molecular orientation of the cellulose crystallites due to the removal of cementing substances such as lignin and hemicellulose, leading to a long-lasting effect on the strength and stiffness of the fibers. The improvement in strength among NaOH-treated bamboo fibers (1, 3, and 5% by weight) was also reported by Wong et al. [7], and it was predicted that a higher concentration of caustic soda causes a decrease in micro-fibril angle, whereby natural fibers with low micro-fibril angles exhibit high strength. However, at a higher concentration of NaOH, which in this case is 8%, fibers are weakened by higher matrix removal and damage in cellulosic fibrils [11]. This was observed through SEM, as shown in Figure 2.1, where some

damage can be observed of the 8% NaOH-treated fiber on the fibril surface, which reduced the tensile strength even further, till 24% compared with the untreated fibers. The stiffness was also observed to decrease by 29%.

The increase in the stiffness of the treated fibers as highlighted by the values of the elastic modulus was also observed by some researchers. The authors attributed this to the densification of the fiber cell walls due to the high removal of hemicellulose through alkalization and also to the formation of new hydrogen bonds in between the chain of cellulose fibrils [24–25]. However, based on the density measurement on the bamboo fibers, alkalization had caused the density of treated fibers to decrease with the presence of voids. The improved stiffness of natural fibers was mentioned by Kalia et al. [23] to be attributed to the crystalline region (cellulosic) of the fiber. Therefore, in this case, the possible improvement in fibril alignment that causes an increase in the tensile strength of fibers may also contribute to the improvement in stiffness. With better alignment, less deformation can be expected as the fibers are being stretched in the tensile direction.

2.5   MORPHOLOGY OF FRACTURED FIBERS

Figures 2.12 through 2.15 show the comparison between the fractured surfaces of untreated and treated fibers obtained through SEM. In Figure 2.12, it can be seen that the untreated fiber experienced a brittle-like failure, where the fracture occurs all the way through the cross-section of the fiber. This may be contributed by the presence of the fiber matrix, which ensures cohesion between fibrils. It can be noted that brittle materials have low values of elastic modulus. The failure of 4 and 6% NaOH-treated fibers can be seen as a result of individual fibril breakages

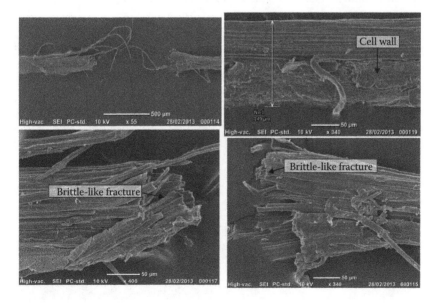

FIGURE 2.12 Scanning electron microscopy (SEM) of fractured untreated fibers due to single fiber tensile test (SFTT).

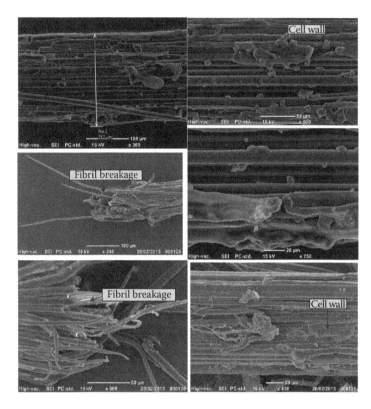

FIGURE 2.13   SEM of fractured 4% NaOH-treated fibers due to SFTT.

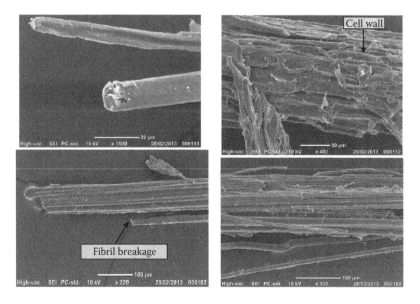

FIGURE 2.14   SEM of fractured 6% NaOH-treated fibers due to SFTT.

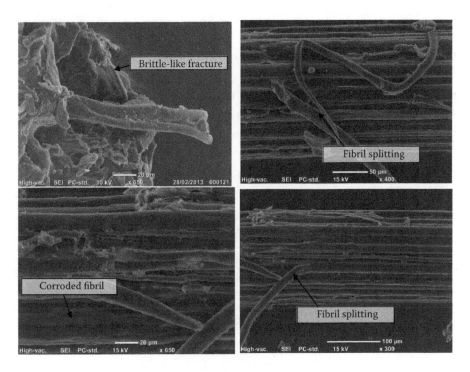

FIGURE 2.15 SEM of fractured 8% NaOH-treated fibers due to SFTT.

as presented in Figures 2.13 and 2.14. With the partial removal of the fiber matrix, the tensile load is mainly transferred along the fibrils, with less influence from neighboring fibrils, thus making it less brittle than the untreated fiber. The stiffness decreases for 8% NaOH-treated fibers due to the damage on the fibrils. It was observed that the fractured surfaces for 8% NaOH-treated fibers are similar to those for the untreated fibers. The absence of the cell wall with some mid-splitting of fibrils can be clearly seen in Figure 2.15, which may contribute to the decrease in the modulus of elasticity.

2.6 COMPARISON OF BAMBOO FIBERS WITH OTHER NATURAL FIBERS

The characteristics of the untreated bamboo fibers studied in this chapter are presented in Table 2.1 [17,26], along with other natural fibers. The bamboo fibers are physically comparable to hemp fibers but with lower tensile properties. This corresponds to the amount of cellulosic component present in the fibers, which plays a major role in the tensile strength of the fibers [6]. The percentage of cellulose in hemp, jute, bamboo, and coir fibers is 74.40, 66.25, 45.70, and 43.44, respectively

TABLE 2.1
Comparison of Several Characteristics of Untreated Bamboo Fibers Along with Other Natural Fibers

Properties	Hemp	Coir	Jute	Bamboo
Diameter (μm)	80–300 [17]	100–460 [17]	54 [26]	94–304
Density (g/cm³)	1.45 [17]	1.29 [26]	1.39 [26]	1.40–1.53
Tensile strength (MPa)	227–700 [17]	120–304 [26]	307–1000 [26]	60–145
Elastic modulus (GPa)	9–20 [17]	4–6 [26]	13–54 [26]	3–14

Sources: Ratna Prasad, A.V. and Mohana Rao, K., Mechanical properties of natural fiber reinforced polyester composites: Jowar, sisal and bamboo. *Materials and Design*, 2011. 32(8–9): p. 4658–63; Defoirdt, N., Biswas, S., Vriese, L.D., Tran, L.Q.N., Acker, J.V., Ahsan, Q., et al., Assessment of the tensile properties of coir, bamboo and jute fiber. *Composites Part A: Applied Science and Manufacturing*, 2010. 41(5): p. 588–95. [17,26]

[3,27,28]. Generally, the tensile properties of fiber increase with an increasing amount of cellulose.

2.7   INTERFACIAL PROPERTIES OF BAMBOO FIBER/POLYESTER COMPOSITE

The SFFT was performed to determine the interfacial shear strength between reinforcing fibers and a polymer resin in a composite system. This study is based on the Kelly and Tyson [29] technique, basically subjecting the specimen to an increasing tensile load, which is transferred to the fiber through the fiber–matrix interface. Each test specimen for SFFT consists of a single fiber embedded in unsaturated polyester resin with a gauge length of 12 mm as shown in Figure 2.16. Four sets of SFFT samples were fabricated according to the following percentages of NaOH treatment: 0, 4, 6, and 8. The interfacial shear strength (τ_{IFSS}) was computed using the following equation [7,30]:

$$\tau_{IFSS} = \frac{F_{deb}}{d_f \pi L_e} \qquad (2.1)$$

where F_{deb} is the debonding force, d_f is the fiber diameter, and L_e is the fiber embedment length. The interfacial shear strength for each set was taken from the average values of six specimens.

The typical load–displacement curves obtained from the SFFT for bamboo fibers treated at different concentrations of NaOH-reinforced polyester composites are presented in Figure 2.17. It can be observed that on loading, the force increased at a similar rate for all NaOH concentrations, with 0 and 4% NaOH-treated fibers

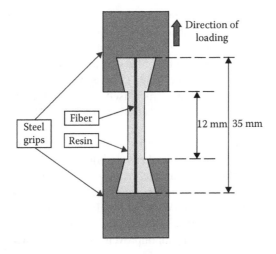

FIGURE 2.16   Test setup for single fiber fragmentation test (SFFT).

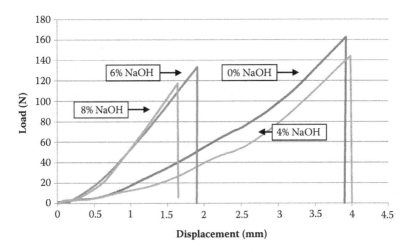

FIGURE 2.17   Load–displacement curves of SFFT at different concentrations of NaOH.

experiencing higher displacement at the beginning of the test. This may indicate that the fiber–matrix interface for 0 and 4% NaOH-treated fibers has weaker bonding with the polyester, which may lead to higher sliding of the fiber throughout the hole as compared with the 6 and 8% NaOH-treated fibers. The fiber–matrix interface provides the platform for load transfer from the polyester to the bamboo fibers and across this interface, in an equilibrium state, a frictional shear force acts equally in the opposite direction of the applied force [7]. On reaching the peak load, debonding occurs, causing the load to drop drastically, thus indicating the release of stored energy.

The mean interfacial shear strengths (IFSS) calculated from Equation 2.1 for treated and untreated bamboo fiber/polyester composites, with F_{deb} values taken from maximum loads in Figure 2.17, are presented in Figure 2.18. The IFSS for 0, 4, 6, and 8% NaOH-treated bamboo fibers, at an embedment length of 12 mm, are 23, 25, 28, and 22 MPa, respectively. The trend shows an increment in IFSS as the concentration of treatment is increased to 6% NaOH, with an optimum improvement of 22%. At a higher concentration of 8% NaOH, a reduction of IFSS was observed. The raw, untreated fibers have a low IFSS value, which reflects their hydrophilic nature, which may be incompatible with the hydrophobic polyester [31–32]. The incompatibility weakens its bonding strength with the matrix; together with the smoother fiber surface (as observed in Figure 2.12b) as well as with the presence of surface impurities, it renders poor interlocking between the two components of the composite. With the rougher fiber surface and the removal of surface impurities achieved through alkalization, the IFSS increased for 4 and 6% NaOH-treated fibers, indicating an improvement in the mechanical interlocking between the fibers and the matrix. The rougher surface and the presence of voids on the treated fibers also provide a higher surface area for contact and adhesion, which increases the frictional shear force. The IFSS obtained for 6% NaOH-treated fibers is higher compared with those for 4% NaOH-treated fibers, and they experienced less initial displacement, as observed in Figure 2.17. This may suggest that the fiber achieves its optimum surface condition for fiber–matrix adhesion, and together with its improved tensile strength, it shows the best mechanical performance at 6% NaOH treatment. This improvement in interfacial strength values was also observed in untreated and NaOH-treated coir fibers with poly(butylene succinate) matrix, where the authors observed an increase in roughness of the surface of the fibers after alkalization [13].

However, there is a limit to the concentration of NaOH in bamboo fiber treatment. The IFSS for 8% NaoH-treated fibers was reduced to a slightly higher value than that

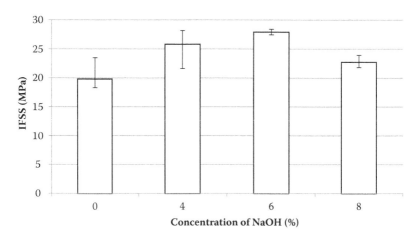

FIGURE 2.18   Effect of alkalization on the interfacial shear strength of bamboo–polyester composites.

of the untreated fibers. As observed in Figure 2.15d, these fibers have a considerably rougher surface due to the exposure of the individual fibrils as a result of complete removal of cell walls, as compared with the untreated fibers. In Figure 2.17, the displacement due to tensile loading for 8% NaOH-treated fibers was similar to that for 6% NaOH-treated fibers. This may suggest that the interfacial adhesion between the matrix and the fiber is better than that for the untreated fibers. However, as observed in Figures 2.1 and 2.15c and d, the corroded fibril and fibril splitting suggest that at 8% NaOH, the concentration of the caustic soda may be too corrosive for the bamboo fibers, which results in a damaging effect on the fibers. This was also proved in the tensile test performed whereby the tensile strength of the fiber was reduced. It can be assumed that although there was better interfacial adhesion between the fiber and the matrix for 8% NaOH-treated fibers, the damaged fibers may have contributed to the failure of the composite, thus affecting its IFSS value. The behavior of the fiber–matrix adhesion can be simplified as demonstrated in Figure 2.19, whereby better fiber–matrix interlocking is expected as the fiber surface is roughened by alkali treatment.

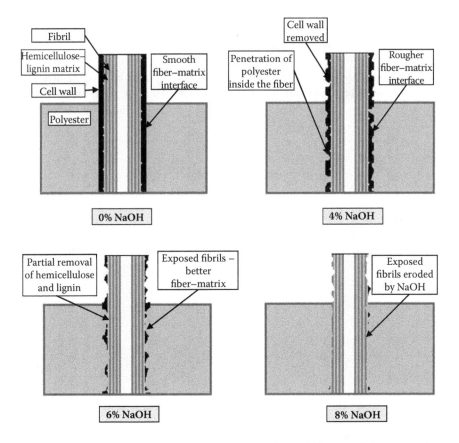

FIGURE 2.19 Schematic diagram of fiber–matrix interface at different NaOH concentrations.

2.8 TENSILE PROPERTIES OF BAMBOO FIBER/ POLYESTER COMPOSITES

The experimental investigation is conducted on a 20% fiber volume fraction of randomly oriented bamboo fiber/polyester composite samples and neat polyester (NP) resin, with dimensions shown in Figures 2.20 and 2.21. The samples include untreated (BPC-0) and 4, 6, and 8 wt.% NaOH-treated composites (BPC-4, BPC-6, and BPC-8, respectively), as well as NP. The tensile tests were conducted as per ASTM D638 [33], whereby the samples were subjected to tensile loading until failure and the stress–strain diagrams were plotted. All tests were carried out with five replicates for each sample type at a cross-head speed of 1.3 mm/min. The elastic moduli were experimentally determined from the slopes of the linear portion of the stress–strain curves. The theoretical values of the elastic moduli were also calculated for comparison.

2.8.1 STRESS–STRAIN DIAGRAM

Figures 2.22 through 2.25 represent the stress–strain curves of the bamboo fiber/ polyester composites with different concentrations of NaOH treatment, whereas the stress–strain curves for NP are presented in Figure 2.26. Each set of tests was conducted on five specimens, and only the best curves were used in determining the tensile strength and elastic modulus.

The results for the tensile strength are summarized in Figure 2.27. It is observed that NP has the highest tensile strength at 33.79 MPa. This is in agreement with the value provided by the manufacturer, which is at 30.9 MPa. With the addition of

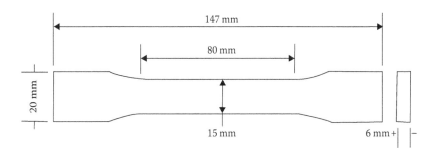

FIGURE 2.20   Neat polyester (NP) specimen dimension.

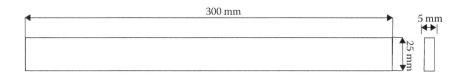

FIGURE 2.21   Bamboo fiber/polyester composite specimen dimension.

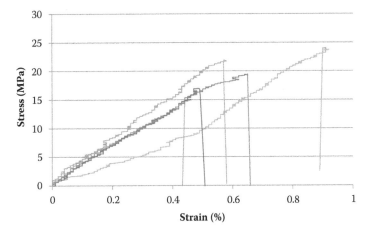

FIGURE 2.22   Stress–strain curves for composites with 0 wt.% NaOH treatment.

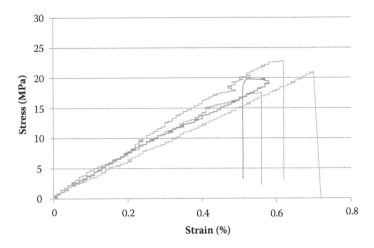

FIGURE 2.23   Stress–strain curves for composites with 4 wt.% NaOH treatment.

bamboo fibers, a decrease of approximately 10 MPa was observed in the bamboo fiber/polyester composites. It was highlighted by Manalo et al. [34] that only the fibers that are oriented perpendicular to the load provide reinforcement in a composite laminate fabricated with randomly oriented fibers. The maximum tensile strength achieved by samples BPC-0, BPC-4, and BPC-6 is very much comparable, with only a slight increase in the average values for the treated samples. A mere 6% increase in strength was observed for BPC-6 compared with BPC-0. However, there was an obvious drop in tensile strength at a higher NaOH concentration as demonstrated by the BPC-8 sample, with an 18% reduction observed.

The improvement in tensile strength of the fibers as well as the interfacial properties between the bamboo fibers and polyester matrix for fibers treated at 4 and 6% NaOH

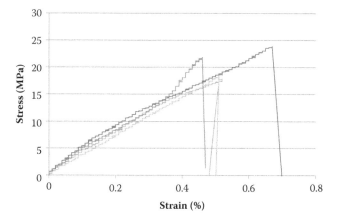

FIGURE 2.24   Stress–strain curves for composites with 6 wt.% NaOH treatment.

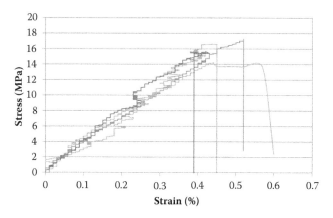

FIGURE 2.25   Stress–strain curves for composites with 8 wt.% NaOH treatment.

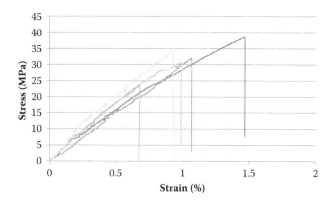

FIGURE 2.26   Stress–strain curves for NP.

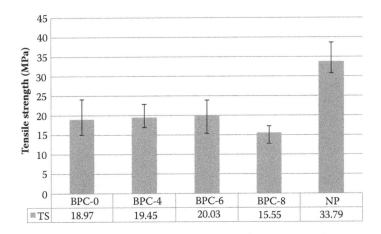

FIGURE 2.27   Tensile strength of specimens at different concentrations of NaOH treatment.

may have contributed to the slight increase in average tensile strength of their corresponding composites, with BPC-6 having the highest tensile strength value. Generally, the mechanical performance, such as the tensile properties of a fiber/polymer composite, depends on the strength and modulus of the reinforcement, strength and toughness of the matrix, and the effectiveness of the interfacial stress transfer between the fibers and the matrix [19]. Other researchers have also reported on the effect of alkali treatment on the improvement of the mechanical properties of natural fiber/polymer composites. Nam et al. [13] observed an increase in tensile strength of coir/poly(butylene succinate) composite treated at 5% NaOH. The increase in strength was observed for all fiber mass content and is also higher than the NP polymer. It is interesting to note that these fibers were uniformly arranged. Rong et al. [19] attributed the increased strength of sisal fibers and better wetting of the fibers by the epoxy resin as the reason for better strength of sisal/epoxy composite treated with 2% NaOH. The decrease in strength for BPC-8 is expected to be due to the degradation experienced by the fiber caused by the high level of corrosiveness of the NaOH. The weakened fiber fails to sustain the load transferred from the matrix, thus causing an early failure of the composite.

The elastic modulus for all samples is summarized in Figure 2.28. The elastic modulus for the NP resin as provided by the supplier is 3.1 GPa. This is comparable to the value obtained from this experiment, which is approximately 3.3 GPa. With the addition of untreated bamboo fibers, the stiffness of the composite was observed to decrease slightly to 3.18 GPa. However, this behavior is improved with alkali treatment as exhibited by all treated composites. BPC-6 achieved the highest modulus of elasticity, which is 14% higher than that of the NP, whereas BPC-8 experienced a slightly lower average stiffness but with a higher standard deviation. The influence of the damaged fibers but with better fiber–matrix adhesion may vary the impact on the stiffness depending on the distribution of the fibers. Many authors agree that the efficiency of the interfacial adhesion may contribute to a higher elastic modulus. The Young's modulus of bamboo fiber–poly (lactic acid) composite was reported by Lee and Wang [35] as increasing rapidly from 2666 to 2964 MPa with the addition of a

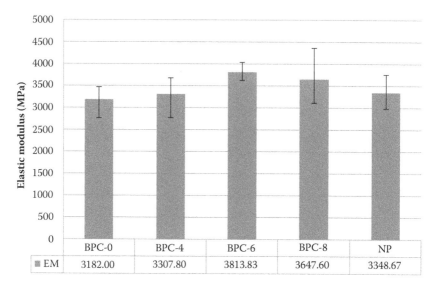

FIGURE 2.28 Elastic modulus of specimens at different concentrations of NaOH treatment.

bio-based coupling agent. Aziz and Ansell [9] indicated that the flexural modulus of a composite depends on the type of chemical bonds between the fiber surface and resin matrix. Their study on hemp/polyester and kenaf/polyester composites shows that the treatment of fibers by alkalization helps in improving the mechanical interlocking and chemical bonding between the resin and the fiber, resulting in superior mechanical properties, with better performance by composites with long fibers than short fibers.

2.8.2 THEORETICAL VALUES OF ELASTIC MODULUS

The elastic modulus of a composite, E_c, can be theoretically calculated using the Rule of Mixture, whereby the elastic modulus of each component of the composite contributes to the final product based on its volume fraction. This is formulated as follows:

$$E_c = E_f V_f + E_m V_m \tag{2.2}$$

where E_f is the elastic modulus of the fiber, V_f is the volume fraction of the fiber, E_m is the elastic modulus of the matrix, and V_m is the volume fraction of the matrix. A Krenchel factor, n_θ, is introduced into the modified Rule of Mixture equation to account for the orientation of the fibers distributed throughout the composite [9,36–37]. This alters the original equation as follows:

$$E_c = n_\theta E_f V_f + E_m V_m \tag{2.3}$$

For a unidirectional fiber/polymer composite, $n_\theta = 1$, thus indicating that the fiber stiffness is entirely used by the composite. However, for a randomly oriented fiber/

TABLE 2.2

Theoretical Modulus of Elasticity of Bamboo Fiber/Polyester Composite

n_θ	BPC-0 (GPa)	BPC-4 (GPa)	BPC-6 (GPa)	BPC-8 (GPa)
1.00	4.07	4.28	4.73	3.67
0.25	3.03	3.08	3.19	2.93
$E_{experimental}$	3.18	3.31	3.81	3.65

polymer composite, $n_\theta = 0.25$. This means that the mechanical properties of the composite will depend more on the matrix, with fibers contributing only 25% to the overall stiffness. Table 2.2 represents the theoretical values for the modulus of elasticity of bamboo fiber/polyester composites used in this study by assuming the volume fraction of fibers as 20%.

The theoretical elastic modulus calculated for the randomly oriented fiber composite with 0% treatment is comparable with the experimental value obtained, with only a 4.7% difference between the two values. As the treatment concentration is increased, the gap difference between the theoretical and experimental values is enlarged. One possible explanation for this is that the equation of the Rule of Mixture did not consider the improvement of the interfacial adhesion between the fibers and the matrix as the result of the alkali treatment. The enhancement of the stiffness of the fibers due to treatment is reflected in the equation. Fibers treated at 8% NaOH exhibited the lowest stiffness. Thus, theoretically, they produced a composite with the lowest modulus of elasticity. However, the experimental value obtained for BPC-8 is higher than those for BPC-0 and BPC-4, indicating that the improved fiber–matrix adhesion plays an influential role in the improved stiffness of the composite.

It is often observed that the addition of fibers as reinforcement in a polymer matrix increases the strength and modulus of the composite since fibers have much higher strength and stiffness values than those of the matrices [38]. Theoretically, if the potential of the bamboo fibers were used to its fullest, whereby all fibers were aligned uniformly in the direction of the applied tensile load, the elastic modulus will be much higher than the NP resin, thus improving the mechanical properties of the polyester. This improvement is shown in Table 2.2 for an n_θ value equal to 1 for a unidirectional fiber orientation. However, due to the nature of the fibers extracted through the steam explosion process that are relatively shorter than the length of the composite, it is impossible to fabricate a continuous uniformly oriented fiber composite. A study by Wong et al. [36] showed that even when short bamboo fibers were arranged in a unidirectional manner, only a maximum of 25% increase in the tensile strength of polyester composites was achieved. The authors also highlighted that only the longitudinal fibers contribute in the energy dissipation and in crack retardation of the composite system.

2.9 MORPHOLOGY OF FRACTURED COMPOSITES

The fractured surfaces of the tensile tested samples are presented in Figure 2.29. The NP shows a rough surface profile with some samples experiencing multiple fracture

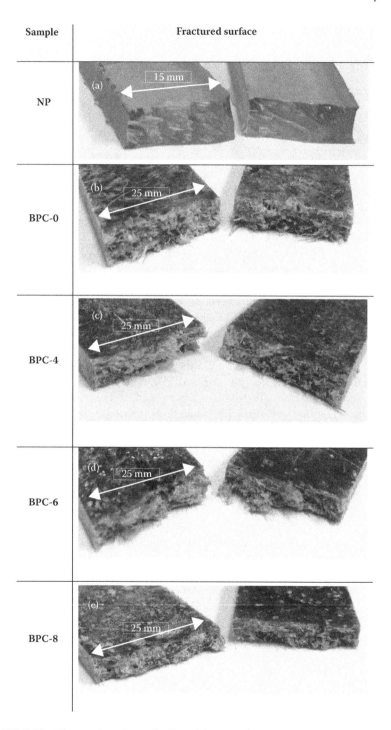

FIGURE 2.29 Fractured surface of NP and bamboo fiber/polyester composites. (a) NP, (b) BPC-0, (c) BPC-4, (d) BPC-6, (e) BPC-8.

points. Once the failure load was reached, failure occurred abruptly, demonstrating the brittleness of the material. With the addition of bamboo fibers, the composites did not fail in this manner. The samples were mainly intact with one fracture point within the gauge length. The presence of the bamboo fibers helped repress the abrupt separation of any fractured polyester particles. Cracking sounds were audible even before the maximum load was reached, indicating that the failure occurred gradually. At close inspection, it is interesting to note that the exposed fibers at the failure surface for treated composites had more resin particles attached to them compared with the untreated composite. This suggests better fiber–matrix adhesion for the treated composites, and a more thorough examination was performed using SEM.

The micrographs of the fractured surfaces of the samples are presented in Figure 2.30. Defragmentation of polyester particles was observed throughout the surface of the NP. The failed surface is uneven and rough due to the brittle nature of the failure condition. For the untreated composites, holes on the matrix surface were observed, left by debonded fibers that were pulled out during tensile loading. Higher magnification at the fiber–matrix interface shows that the fiber and matrix are separated by voids, proving that bonding between the two materials is poor. Therefore, with low interlocking strength for BPC-0, the fibers can easily slip through the polyester resin, leading to failure of composites through fiber pullouts. For 4% NaOH-treated fibers, a similar morphology to BPC-0 on the fiber–matrix adhesion was observed, as some areas showed poor resin bonding. However, resin particles were found to adhere along the fibers as observed in Figure 2.30e. The treatment had provided a better fiber surface than the raw fibers for matrix adhesion but may still be insufficient. Some surface impurities could still be present on the fibers. Thus, this has only slightly improved the strength of the composite for BPC-4. Polyester resin showed better adhesion to 6% NaOH-treated fibers, as some parts of the resin were still bonded on the fiber, although the surrounding matrix was fractured and removed. Resin particles were also found to adhere on the fibers, even in between the fibrils. The improved fiber–matrix interlocking was further confirmed by the mode of failure shown in Figure 2.30g whereby the breakage of a fiber bundle may imply that the fibers were damaged first before debonding could take place. As for BPC-8, the condition of the fibers was poor due to their vulnerability to the high alkali strength. Fibril damage and erosion were evident but the fiber–matrix adhesion was very good as no gaps were observed between the fiber and the matrix, as well as the apparent chunk of resin still adhered to a fibril. This supports the SFFT graph (Figure 2.17) where the displacement reading for the 8% NaOH-treated fibers was comparable to that for the 6% NaOH-treated fibers due to good fiber–matrix interlocking. The lower tensile strength for BPC-8, therefore, is influenced by the damaged fibers. It is interesting to note that the stiffness of BPC-8 is higher than that of NP, BPC-0, and BPC-4, which may imply that the elastic modulus is highly influenced by the fiber–matrix adhesion rather than by the individual components of the composite.

The morphologies of tensile fractured surfaces of other natural fiber polymer composites treated with NaOH were reported to have similar properties compared with those of the bamboo fiber/polyester composite in this study. Alkalized banana fiber/polypropylene composites were reported by Paul et al. [39] to have significant improvement in fiber–matrix adhesion as shown by the absence of holes and

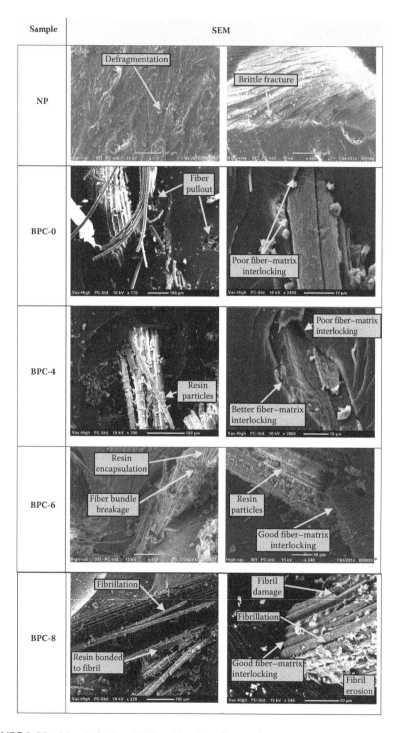

FIGURE 2.30 Morphologies of NP and bamboo fiber–polyester composites.

debonding of the fibers, which were apparent for untreated fibers. Instead, fiber breakage rather than pullout was detected, indicating better interfacial strength. This was also observed by Rout et al. [16] on 5% alkali-treated coir/polyester composites. Due to stronger adhesion between the treated coir fibers and the polyester, failure was found to occur on the fibers based on the cracks observed on the broken fiber ends. Another study on coir fibers by Nam et al. [13] showed that the gaps present in untreated fibers with poly(butylene succinate) resin were unobservable in the alkali-treated composites. A larger quantity of residual epoxy resin was observed by Rong et al. [19] on the surface of alkali-treated sisal/epoxy composites and even at the gaps between the ultimate cells, proving the penetration of epoxy into alkali-treated sisal fiber bundles. Overall, these authors agree that the SEM observations indicated that the increase in strength of these composites was influenced by the improvement in fiber–matrix adhesion due to the effect of alkalization on the fibers.

2.10 CONCLUSIONS

Bamboo fibers were treated with NaOH at a concentration of 4, 6, and 8% by weight. The characteristics of the fibers were observed and studied to determine the changes as a result of the alkalization process. These changes were correlated with the morphology of the fibers as observed through SEM. The results have shown that:

1. The fiber matrix was partially removed through alkalization. The amount of removal is higher as the concentration of treatment is increased, causing a rougher fiber surface with voids observed between the microfibrils. However, there is an optimum NaOH concentration for natural fiber treatment whereby at a concentration higher than the optimum level, degradation of the cellulosic component of the fiber may occur, which will affect the strength of the bamboo fiber. This was observed in the 8% NaOH-treated fiber whereby deterioration of the fibril surface was observed.
2. Alkalization in bamboo fibers has resulted in changes in appearance and texture. Treated fibers are darker than untreated fibers. The change in color may suggest that lignin was removed from the fiber, which alters its composition. In addition, hemicelluloses and other surface impurities that were also partially removed may contribute to the presence of voids within the fiber. This causes a slight reduction in the diameter and density of the treated fibers.
3. The tensile properties of bamboo fibers improved at 6% NaOH treatment, with an increment of 181 and 47% in tensile strength and elastic modulus, respectively. On the other hand, the tensile strength and stiffness decrease at 8% NaOH treatment due to the damaged fibrils, with 24 and 29% reduction observed in tensile strength and stiffness, respectively, compared with untreated fibers.
4. The interfacial properties of bamboo fibers with polyester matrix improved with alkali treatment. This correlates well with the fact that the alkalization roughens and removes surface impurities for a better interfacial bonding between the two components of the composite. Optimum interfacial shear

strength was observed for 6% NaOH-treated fibers with 22% improvement. However, the damaged fibrils at 8% NaOH treatment hinder the effect of the improved bonding, causing a reduction in the IFSS of the composite.

5. The tensile strength of the composite is lower than the NP due to the random orientation of the fibers, which hinders full usage of the strength of the fibers. Fiber treatment caused a slight enhancement in the tensile strength of the composite for 6% NaOH-treated fibers (about 6%), but reduced the strength as much as 18% at 8% NaOH due to fiber damage.

6. The elastic modulus of the composite is optimized at 6% NaOH treatment, where it is 14% higher than the NP resin. Apart from the increase in fiber stiffness due to alkalization, the improved interfacial adhesion plays a major role in providing a higher elastic modulus of the treated composites.

7. The SEM observations confirmed that fiber–matrix interlocking was improved with the increase in NaOH concentration. Untreated fibers were surrounded by voids at the interface with some fiber pullouts observed. BPC-6 and BPC-8 showed good fiber–matrix adhesion even though the 8% NaOH-treated fibers suffered some fibril damage.

REFERENCES

1. Zou, L., Jin, H., Lu, W.Y., and Li, X., Nanoscale structural and mechanical characterization of the cell wall of bamboo fibers. *Materials Science and Engineering: C*, 2009. 29(4): p. 1375–79.

2. Tong, J., Ma, Y., Chen, D., Sun, J., and Ren, L., Effects of vascular fiber content on abrasive wear of bamboo. *Wear*, 2005. 259(1–6): p. 78–83.

3. Yao, F., Wu, Q., Lei, Y., Guo, W., and Xu, Y., Thermal decomposition kinetics of natural fibers: Activation energy with dynamic thermogravimetric analysis. *Polymer Degradation and Stability*, 2008. 93(1): p. 90–8.

4. Shao, S., Wen, G., and Jin, Z., Changes in chemical characteristics of bamboo (Phyllostachys pubescens) components during steam explosion. *Wood Science and Technology*, 2008. 42(6): p. 439–51.

5. Methacanon, P., Weerawatsophon, U., Sumransin, N., Prahsarn, C., and Bergado, D.T., Properties and potential application of the selected natural fibers as limited life geotextiles. *Carbohydrate Polymers*, 2010. 82(4): p. 1090–6.

6. John, M. and Thomas, S., Biofibres and biocomposites. *Carbohydrate Polymers*, 2008. 71(3): p. 343–64.

7. Wong, K.J., Yousif, B.F., and Low, K.O., Effects of alkali treatment on the interfacial adhesion of bamboo fibres. *Journal of Materials: Design and Applications*, 2010. 224: p. 139–48.

8. Spinacé, M.A.S., Lambert, C.S., Fermoselli, K.K.G., and De Paoli, M.A., Characterization of lignocellulosic curaua fibres. *Carbohydrate Polymers*, 2009. 77(1): p. 47–53.

9. Aziz, S.H. and Ansell, M.P., The effect of alkalization and fibre alignment on the mechanical and thermal properties of kenaf and hemp bast fibre composites: Part 1—polyester resin matrix. *Composites Science and Technology*, 2004. 64(9): p. 1219–30.

10. Sydenstricker, T.H.D., Mochnaz, S., and Amico, S.C., Pull-out and other evaluations in sisal-reinforced polyester biocomposites. *Polymer Testing*, 2003. 22(4): p. 375–80.

11. Boopathi, L., Sampath, P.S., and Mylsamy, K., Investigation of physical, chemical and mechanical properties of raw and alkali treated Borassus fruit fiber. *Composites Part B: Engineering*, 2012. 43(8): p. 3044–52.

12. Elenga, R.G., Djemia, P., Tingaud, D., Chauveau, T., Maniongui, J.G., and Dirras, G., Effects of alkali treatment on the microstructure, composition, and properties of the Raffia textilis fiber. *BioResources*, 2013. 8(2): p. 2934–49.

13. Nam, T.H., Ogihara, S., Tung, N.H., and Kobayashi, S., Effect of alkali treatment on interfacial and mechanical properties of coir fiber reinforced poly(butylene succinate) biodegradable composites. *Composites Part B: Engineering*, 2011. 42(6): p. 1648–56.

14. Mohanty, A., Misra, M., and Hinrichsen, G., Biofibres, biodegradable polymers and biocomposites: An overview. *Macromolecular Materials and Engineering*, 2000. 276(1): p. 1–24.

15. Khan, F. and Ahmad, S., Chemical modification and spectroscopic analysis of jute fibre. *Polymer Degradation and Stability*, 1996. 52(3): p. 335–40.

16. Rout, J., Misra, M., Tripathy, S.S., Nayak, S.K., and Mohanty, A.K., The influence of fibre treatment on the performance of coir-polyester composites. *Composites Science and Technology*, 2001. 61: p. 1303–10.

17. Ratna Prasad, A.V. and Mohana Rao, K., Mechanical properties of natural fibre reinforced polyester composites: Jowar, sisal and bamboo. *Materials and Design*, 2011. 32(8–9): p. 4658–63.

18. Mwaikambo, L., *Sustainable Composite Materials: Exploitation of Plant Resourced Materials for Industrial Application*. 2009, VDM Publishing: Saarbrücken, Germany.

19. Rong, M.Z., Zhang, M.Q., Liu, Y.L., Yang, G.C., and Zeng, H.M., The effect of fiber treatment on the mechanical properties of unidirectional sisal-reinforced epoxy composites. *Composites Science and Technology*, 2001. 61: p. 1437–47.

20. ASTM, *ASTM D3379 - 75(1989)e1 Standard Test Method for Tensile Strength and Young's Modulus for High Modulus Single Filament Materials*. 1989, ASTM International: West Conshohocken, PA.

21. Bessadok, A., Marais, S., Roudesli, S., Lixon, C., and Métayer, M., Influence of chemical modifications on water-sorption and mechanical properties of Agave fibres. *Composites Part A: Applied Science and Manufacturing*, 2008. 39(1): p. 29–45.

22. Yousif, B.F., Orupabo, C., and Azwa, Z.N., Characteristics of kenaf fibre immersed in different solutions. *Journal of Natural Fibers*, 2012. 9(4): p. 207–218.

23. Kalia, S., Kaith, B.S., and Kaur, I., Pretreatments of natural fibers and their application as reinforcing material in polymer composites—A review. *Polymer Engineering & Science*, 2009. 49(7): p. 1253–72.

24. Sawpan, M.A., Pickering, K.L., and Fernyhough, A., Effect of various chemical treatments on the fibre structure and tensile properties of industrial hemp fibres. *Composites Part A: Applied Science and Manufacturing*, 2011. 42(8): p. 888–95.

25. Obi Reddy, K., Uma Maheswari, C., Shukla, M., Song, J.I., and Varada Rajulu, A., Tensile and structural characterization of alkali treated Borassus fruit fine fibers. *Composites Part B: Engineering*, 2013. 44(1): p. 433–8.

26. Defoirdt, N., Biswas, S., Vriese, L.D., Tran, L.Q.N., Acker, J.V., Ahsan, Q., et al., Assessment of the tensile properties of coir, bamboo and jute fibre. *Composites Part A: Applied Science and Manufacturing*, 2010. 41(5): p. 588–95.

27. Dhakal, H., Zhang, Z., and Richardson, M., Effect of water absorption on the mechanical properties of hemp fibre reinforced unsaturated polyester composites. *Composites Science and Technology*, 2007. 67(7–8): p. 1674–83.

28. Khedari, J., Watsanasathaporn, P., and Hirunlabh, J., Development of fibre-based soil–cement block with low thermal conductivity. *Cement and Concrete Composites*, 2005. 27(1): p. 111–6.

29. Kelly, A. and Tyson, W., Tensile properties of fibre-reinforced metals: Copper/tungsten and copper/molybdenum. *Journal of the Mechanics and Physics of Solids*, 1965. 13(6): p. 329–50.

30. Saravanakumar, S.S., Kumaravel, A., Nagarajan, T., Sudhakar, P., and Baskaran, R., Characterization of a novel natural cellulosic fiber from Prosopis juliflora bark. *Carbohydrate Polymers*, 2013. 92(2): p. 1928–33.

31. Wambua, P., Ivens, J., and Verpoest, I., Natural fibres: can they replace glass in fibre reinforced plastics? *Composites Science and Technology*, 2003. 63(9): p. 1259–64.

32. Akil, H., Omar, M., Mazuki, A., Safiee, S., Ishak, Z., and Abu Bakar, A., Kenaf fiber reinforced composites: A review. *Materials and Design*, 2011. 32(8): p. 4107–21.

33. ASTM, *ASTM D638 Standard Test Method for Tensile Properties of Plastics*. 2010, ASTM International: West Conshohocken, PA.

34. Manalo, A.C., Karunasena, W., and Lau, K.T., Mechanical properties of bamboo-fiber polyester composites, in 22nd Australasian Conference on the Mechanics of Structures and Materials. 2012, Sydney.

35. Lee, S.H. and Wang, S., Biodegradable polymers/bamboo fiber biocomposite with bio-based coupling agent. *Composites Part A: Applied Science and Manufacturing*, 2006. 37(1): p. 80–91.

36. Wong, K.J., Zahi, S., Low, K.O., and Lim, C.C., Fracture characterisation of short bamboo fibre reinforced polyester composites. *Materials and Design*, 2010. 31(9): p. 4147–54.

37. Krenchel, H., *Fibre Reinforcement Akademisk Forlag Copenhagen*. 1964, Denmark.

38. Ku, H., Wang, H., Pattarachaiyakoop, N., and Trada, M., A review on the tensile properties of natural fiber reinforced polymer composites. *Composites Part B: Engineering*, 2011. 42(4): p. 856–73.

39. Paul, A.S., Boudenne, A., Ibos, L., Candau, Y., Joseph, K., and Thomas, S., Effect of fiber loading and chemical treatments on thermophysical properties of banana fiber/polypropylene commingled composite materials. *Composites Part A: Applied Science and Manufacturing*, 2008. 39(9): p. 1582–8.

3 Alternative Solutions for Reinforcement of Thermoplastic Composites

Nadir Ayrilmis and Alireza Ashori

CONTENTS

3.1 INTRODUCTION

Over the past two decades, the thermoplastic industry has been attempting to decrease the dependence on petroleum-based fuels and products due to increased environmental consciousness. This leads to the need to investigate environmentally friendly, sustainable materials to replace existing ones. Currently, the most viable method of dealing with ecofriendly composites is the use of natural fibers as reinforcement. Natural fibers have received considerable attention as a substitute for synthetic fiber reinforcements in plastics. As replacements for conventional synthetic fibers such as aramid and glass fibers, natural fibers are increasingly used for reinforcement in thermoplastics due to their low density, good thermal insulation and mechanical properties, reduced tool wear, unlimited availability, low price, and problem-free disposal. Natural fibers also offer economical and environmental advantages compared with traditional inorganic reinforcements and fillers. As a result of these advantages, natural fiber-reinforced thermoplastic composites are gaining popularity in automotives, garden decking, fencing, railing, and nonstructural building applications, such as exterior window and door profiles, as well as siding.

The combination of interesting physical, mechanical, and thermal properties together with their sustainable nature has triggered various activities in the area of green composites. This chapter aims at providing a short review on developments in the area of natural fibers and their applications in fiber-based industries such as wood-plastic composites (WPCs).

3.2 CELLULOSE-BASED FIBERS

3.2.1 NATURAL FIBERS

Fibers can be classified into two main groups: man-made and natural. The term "natural fibers" is used to designate various types of fibers, which naturally originate from plants, minerals, and animals [1]. All plant fibers are composed of cellulose, hemicelluloses, and lignin whereas animal fibers consist of proteins (hair, silk, and wool). To clarify our case, the word "plant" might be cited as "vegetable," "cellulosic," or "lignocellulosic." Natural fibers offer the potential to deliver greater added value, sustainability, renewability, and lower costs compared with man-made fibers [2].

Plant fibers can be subdivided into nonwood fibers and wood fibers. Nonwood fibers can be classified according to which part of the plant they originate from. These include bast (stem or soft sclerenchyma) fibers, leaf, seed, fruit, root, grass, cereal straw, and wood [3]. Some of the important plant fibers are listed in Table 3.1 [4]. Agricultural residuals such as wheat straw, rice straw, bagasse, and corn stalks are also sources of natural fibers, although they have a lower cellulose content compared with wood [5]. Reddy and Yang [6] reported that velvet leaf (*Abutilon theophrasti*), which is currently considered a weed and an agricultural nuisance, could be used as a source for high-quality natural fibers. The fibers of the velvet leaf stem have properties similar to those of common bast fibers such as hemp and kenaf. The availability of large qualities of such fibers with well-defined mechanical properties is a general prerequisite for their successful use, namely in reinforcing plastics.

3.2.2 WOOD FIBERS

Wood is built up of cells, most of which are fibrous. Wood fiber is a composite material that is composed of a reinforcement of cellulose microfibril in a cementing matrix of hemicellulose and lignin [7]. Lignocellulosic materials contain cellulose, hemicellulose, lignin, and extractives in various amounts and chemical compositions. The chemical composition of different lignocellulosic fibers is displayed in Table 3.2 [8]. The properties of the wood flour are dependent not only on the main polymeric components but also on the structural arrangement of these components at the micro and macro scales. The chemical properties and behavior of wood or nonwood components

TABLE 3.1
List of Important Plant Fibers

Fiber Source	Species	Origin
Abaca	*Musa textilis*	Leaf
Bagasse	–	Grass
Bamboo	(>1250 species)	Grass
Banana	*Musa indica*	Leaf
Broom root	*Muhlenbergia macroura*	Root
Cantala	*Agave cantala*	Leaf
Caroa	*Neoglaziovia variegate*	Leaf
China jute	*Abutilon theophrasti*	Stem
Coir	*Cocos nucifera*	Fruit
Cotton	*Gossypium sp.*	Seed
Curaua	*Ananas erectifolius*	Leaf
Date palm	*Phoenix dactyifera*	Leaf
Flax	*Linum usitatissimum*	Leaf
Hemp	*Cannabis sativa*	Stem
Henequen	*Agave foourcrocydes*	Leaf
Isora	*Helicteres isora*	Stem
Istle	*Samuela carnerosana*	Leaf
Jute	*Corchorus capsularis*	Stem
Kapok	*Ceiba pentranda*	Fruit
Kenaf	*Hibiscus cannabinus*	Stem
Kudzu	*Pueraria thunbergiana*	Stem
Mauritius hemp	*Furcraea gigantea*	Leaf
Nettle	*Urtica dioica*	Stem
Oil palm	*Elaeis guineensis*	Fruit
Piassava	*Attalea funifera*	Leaf
Pineapple	*Ananas comosus*	Leaf
Phormium	*Phormium tenas*	Leaf
Roselle	*Hibiscus sabdariffa*	Stem
Ramie	*Boehmeria nivea*	Stem
Sansevieria	*Sansevieria*	Leaf
Sisal	*Agave sisilana*	Leaf
Sponge gourd	*Luffa cylinderica*	Fruit
Straw (cereal)	—	Stalk
Sunn hemp	*Crorolaria juncea*	Stem
Cadillo/urena	*Urena lobata*	Stem
Wood	(>10,000 species)	Stem

Source: John, M.J. and Thomas, S. 2008. Biofibers and biocomposites. *Carbohydrate Polymers* 71:343–64. With permission [4].

during pulping and bleaching are very important. The proportions for wood are, on average, 40%–50% cellulose, 20%–30% lignin, and 25%–35% hemicellulose [9].

It should be noted that natural fibers are cylindrical in shape with length to diameter ratios commonly in the range of 10%–60%. Natural fibers are usually hollow with walls having a composite, multilayered structure of commonly three main

TABLE 3.2

Chemical Composition of Common Lignocellulosic Fibers

Fiber	Cellulose (%)	Hemicellulose (%)	Lignin (%)	Extractives (%)	Ash Content (%)	Water Soluble (%)
Cotton	82.7	5.7	—	6.3	—	1.0
Jute	64.4	12	11.8	0.7	—	1.1
Flax	64.1	16.7	2.0	1.5–3.3	—	3.9
Ramie	68.6	13.1	0.6	1.9–2.2	—	5.5
Sisal	65.8	12.0	9.9	0.8–0.11	—	1.2
Oil palm EFB	65	—	19.0	—	2.0	—
Oil palm frond	56.0	27.5	20.5	4.4	2.4	—
Abaca	56–63	20–25	7–9	3	—	1.4
Hemp	74.4	17.9	3.7	0.9–1.7	—	—
Kenaf	53.4	33.9	21.2	—	4.0	—
Coir	32–43	0.15–0.25	40–45	—	—	—
Banana	60–65	19	5–10	4.6	—	—
PALF	81.5	—	12.7	—	—	—
Sun hemp	41–48	8.3–13.0	22.7	—	—	—
Bamboo	73.9	12.5	10.2	3.2	—	—
Hardwood	31–64	25–40	14–34	0.1–7.7	<1	—
Softwood	30–60	20–30	21–37	0.2–8.5	<1	—

Source: Jawaid, M. and Abdul Khalil, H.P.S. 2011. Cellulosic/synthetic fiber reinforced polymer hybrid composites: A review. *Carbohydrate Polymers* 86:1–18. With permission [8].

layers (Figure 3.1 [10]): the middle lamella (ML), the primary layer (P), and the secondary layer (S1, S2, and S3) [9]. In living, undamaged plant tissue, the walls taper at the ends of the fiber to form a sealed envelope around the central cavity, or lumen. Each layer of the fiber cell wall is made up of millions of microfibrils that are wound in a semi-structured helical fashion around the main fiber axis with varying quantities of lignin and amorphous hemicellulose binding the microfibrils into bundles, called "fibrils" [10].

Cellulose, the most abundant biopolymer on Earth, is also the main constituent of wood. It is located predominantly in the secondary wall of the wood fiber [11]. Approximately 45%–50% of extractive-free dry substance in most plant species is cellulose, and it is the single most important component in the fiber cell wall in terms of its volume and effect on the characteristics of wood [12]. Cellulose molecules consist of long linear chains of homo-polysaccharide, composed of β-D-glucopyranose units, which are linked by (1-4)-glycosidic bonds. Most of its chemical properties may be related to the hydroxyl groups in each monomer unit and the glycosidic bonds. The glycosidic bonds are not easily broken, thus causing cellulose to be stable under a wide range of conditions. However, the hydroxyl groups in cellulose can be readily oxidized, esterified, and converted to ethers. During compounding in an extruder, the thermal, chemical, and mechanical degradation of cellulosic fibers often deteriorate the mechanical properties of

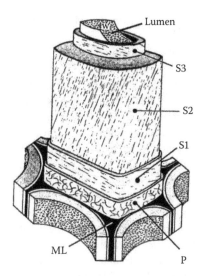

FIGURE 3.1 Diagram of cell wall organization. (From Smook, G.A. 1992. *Handbook for Pulp and Paper Technologists.* 2nd ed. Vancouver, BC: Angus Wilde. With permission [10].)

cellulosic fibers. Usually, cellulose is relatively insensitive to the effect of heating at moderate temperatures over short periods of time. However, thermal degradation begins to appear as the temperature and duration of heating are increased. At lower temperatures (below 300°C), thermal degradation of cellulosic fibers results in the decomposition of the glycosyl units of cellulose with the evolving of water, carbon dioxide, and carbon monoxide, when cellulosic fibers are exposed to the effects of heat, air, and moisture [13]. These reaction products can accelerate the degradation process.

Cellulose has a strong tendency to form intra- and intermolecular hydrogen bonds that are responsible for much of the super-molecular structure, as well as for the physical and chemical behavior of cellulose [11]. Within the microfibrils in the cell wall, cellulose is arranged into zones that display varying degrees of crystallinity and a degree of polymerization (DP) of 7,000–10,000 in native (or unpulped) wood. Cellulose of various plant origins (wood and nonwood) exhibits a wide range of crystallinity (20%–90%), which greatly affects its resistance to chemical attack and utility [14]. In addition, the tensile strength of wood flour comes primarily from the crystalline cellulose microfibrils.

The lignin contents in different woods range between 20% and 30%, typically 26%–32% in softwoods and 20%–28% in hardwoods. Simultaneously, nonwood fibers contain between 5% and 23% lignin [11]. Lignin is a highly complex noncrystalline, heterogeneous polymer with nonrepeating monomeric units and chemical linkages. Lignin acts as the plastic matrix that, in combination with hemicellulose, binds wood fibers together and provides wood with its structural rigidity and resistance to moisture and microbial attack [13]. Lignin concentration varies in different morphological regions of the plant and in different types of plant cells. In general, lignin concentration is high in the ML and P wall, and it is low in the S wall (Figure 3.1).

Noncellulosic carbohydrates or hemicelluloses are heteropolysaccharides that contain hexosan and pentosan monomer units. The amount of hemicellulose in wood is usually between 20% and 30%. Softwood hemicellulose consists of both pentosans and hexosans, whereas hardwood hemicellulose consists mainly of pentosans. Usually, hemicellulose can be characterized as a thermoplastic polymer and is similar to cellulose in possessing high-backbone rigidity through intermolecular hydrogen bonding. The pentosan content in softwoods is about 7%–10%, and in hardwoods it is about 19%–25% [14]. Most hemicelluloses have a DP of about 100–200 [11], are branched, and generally do not form crystalline regions.

All species of wood and other plant tissues contain small to moderate quantities of chemical substances in addition to the macromolecules of cellulose, hemicellulose, and lignin [9]. To distinguish them from the major cell wall components, these additional materials are known as the "extractive components," or simply "extractives." Extractive contents in most temperate and tropical wood species are 4%–10% and 20% of the dry weight, respectively. A wide range of different substances is included under the following extractive headings: flavonoids, lignans, stilbenes, tannins, inorganic salts, fats, waxes, alkaloids, proteins, simple and complex phenolics, simple sugars, pectins, mucilages, gums, terpenes, starch, glycosides, saponins, and essential oils. Extractives occupy certain morphological sites in the wood structure [11]. Many woods contain extractives that are toxic to bacteria, fungi, and termites; other extractives can add color and odor to wood.

Of the wood fiber cell wall components, cellulose and hemicellulose are essentially linear polysaccharides, whereas lignin is a three-dimensional phenolic component. Cellulose, a semicrystalline polymer with a crystallinity of about 60%–70%, is the primary component of wood flour. The tensile strength of wood flour comes primarily from the crystalline cellulose microfibril. Hemicellulose is composed of noncellulosic polysaccharides that serve as a matrix for wood flour and are probably amorphous in their naturally occurring state [9].

3.2.3 Paper Fibers

Pulp technology deals with the liberation of fibers fixed in the wood or plant matrix. Pulp consists of fibers, usually acquired from wood (Figure 3.2). The pulping processes aim first and foremost to liberate the fibers from the wood matrix. In principle, this can be achieved either mechanically or chemically.

By grinding wood or wood chips, the fibers in the wood are released and a mechanical pulp is obtained. Mechanical methods demand a substantial amount of electric power; on the other hand, they make use of practically all of the wood material, that is, the yield of the process is high (>90%) [15]. Since the fibers in wood and plants are glued together with lignin, the chemical way in which pulp is produced is by removing most of the lignin, thereby releasing the fibers. The delignification of wood is achieved through degrading the lignin molecules by introducing chemically charged groups, keeping the lignin fragments in solution, and eventually removing them by washing. No pulping chemicals are entirely selective toward

Pulp technology   Paper technology

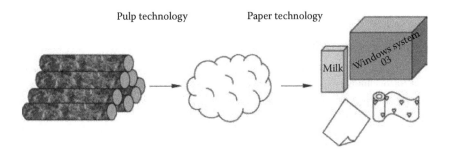

FIGURE 3.2 The making of pulp and paper products from wood.

lignin; the carbohydrates of the wood are also, to a varying extent, lost. In chemical pulping, only approximately half of the wood becomes pulp, and the other half is dissolved [16].

Paper fibers can be obtained from thermo-mechanical pulps (TMP) and chemically modified mechanical pulps. The TMP are produced by the mechanical defibration of wood chips at about 160°C under steam pressure in refiners. The grinding under steam pressure, which should have succeeded in softening the lignin-rich layer between the fibers before the wood structure is broken, results in a greater retention of fiber length than in conventional grinding. Chemically modified mechanical pulps are produced by treating wood chips with sodium sulfite. Then, the treated wood chips are refined either at atmospheric pressure to produce a chemi-mechanical pulp (CMP) or at higher pressure to produce a chemi-thermo-mechanical pulp (CTMP). The chemical treatment preserves the lengths of fibers and causes the fiber surfaces to be richer in hydrophilic polymers [9].

3.2.4 Nanocellulose

Over the past decade, and particularly in the past 5 years, a growing number of research groups worldwide have reported the formation and utilization of celluloses with widths of the fibrils or crystals in the nanometer range. Engineering fiber and design of lignocellulosics or rod-like cellulose nanoparticles and microfibrils to get high value-added products with special performance can reach new markets through nanotechnology [17]. It has been shown that the cellulose microfibrils present in wood can be liberated by high-pressure homogenizer procedures. The product, microfibrillated cellulose (MFC), exhibits gel-like characteristics. A second type of nanocellulose, nanocrystalline cellulose (NCC), is generated by the removal of amorphous sections of partially crystalline cellulose by acid hydrolysis. The NCC suspensions have liquid–crystalline properties. In contrast to MFC and NCC, which are prepared from already biosynthesized cellulose sources, a third nanocellulose variant, bacterial nanocellulose (BNC), is prepared from low-molecular-weight resources, such as sugars, by using acetic acid bacteria of the genus *luconacetobacter*. The in situ biofabrication of BNC opens up unique possibilities for the control of shape and the structure of the nanofiber network [18]. The family of nanocellulose materials is presented in Table 3.3 [18].

TABLE 3.3

The Family of Nanocellulose Materials

Type of Nanocellulose	Selected Reference and Synonyms	Typical Sources	Formation and Average Size
Microfibrillated cellulose (MFC)	Microfibrillated cellulose, nanofibrils and microfibrils, and nanofibrillated cellulose	Wood, sugar beet, potato tuber, hemp, and flax	Delamination of wood pulp by mechanical pressure before and/or after chemical or enzymatic treatment Diameter: 5–60 nm Length: several micrometers
Nanocrystalline cellulose (NCC)	Cellulose nanocrystals, crystallites, whiskers, and rodlike cellulose microcrystals	Wood, cotton, hemp, flax, wheat straw, mulberry bark, ramie, Avicel, tunicin, cellulose from algae, and bacteria	Acid hydrolysis of cellulose from many sources Diameter: 5–70 nm Length: 100–250 nm (from plant celluloses); 100 nm to several micrometers (from celluloses of tunicates, algae, and bacteria)
Bacterial nanocellulose (BNC)	Bacterial cellulose, microbial cellulose, and biocellulose	Low-molecular-weight sugars and alcohols	Bacterial synthesis Diameter: 20–100 nm; different types of nanofiber networks

Source: Klemm, D., Kramer, F., Moritz, S., Lindström, T., Ankerfors, M., Gray, D., and Dorris, A. 2011. Nanocelluloses: A new family of nature-based materials. *Angewandte Chemie International Edition* 50:5438–66. With permission [18].

Micro/nanofibrils isolated from natural fibers have garnered much attention for their use in composites, coatings, resins, and film because of their high-specific surface areas, renewability, and unique mechanical properties in the past two decades [19]. Nanocrystals are rice-like, needle-shaped, and strong with diameters in the 5 to 10 nm range and lengths on the order of 100 to 200 nm, depending on the source. In contrast, nanofibrils, which tend to have roughly 5 nm diameters, are spaghetti-like because they are longer (a micrometer or more), flexible, and easily entangled [20]. The potential uses of natural fiber composite (NFC) and NCC are presented in Figure 3.3 [21]. Some of the attributes (and benefits) of nanocellulose are as follows [22]:

- Abundant, sustainable, environmentally and biologically safe, and derived from plant material
- Long fiber structure, providing composite material strength enhancements, such as in plastics, glass, or concrete

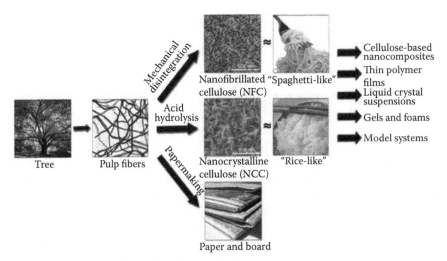

FIGURE 3.3 Potential use of natural fiber composite and nanocrystalline cellulose (NNC). (From Noticias de Nanotecnología. 2014. http://www.tecnologianano.com [Accessed 20 June 2014]. With permission [21].)

- High-tensile strength with a strength/weight ratio being eight times that of steel
- Clear or transparent in water but, with manipulation, it possesses very unique and interesting optic properties due to its structure such that with modifications it can screen ultraviolet (UV) rays or enhance color brilliance
- Benign to the human body and thus can be used as a unique carrier for drug delivery systems or to build tissue scaffolding, create microfilters for blood, or strengthen bones
- Safe for human consumption, offering properties that can enhance the food and beverage markets, where it can be used for delivery of taste, or enhancing food textures such as in yogurts or salad dressings or as an additive in low-calorie products
- Unique properties for aerogels or hydrogels
- Easily functionalized, meaning it can be manipulated or combined with other polymers for many applications

Composites, such as glass or plastics, can benefit greatly with the addition of nano-cellulose, with one study showing a 3,000 times strength gain by adding one ounce of nanocellulose to one pound of plastic. In this example, engineers can significantly reduce the weight of a plastic composite without sacrificing strength (benefiting auto manufacturing, for instance) or keep the mass of plastic the same while dramatically increasing strength. When properly aligned, nanocellulose offers an alternative to even stronger applications, such as replacing Kevlar® from DuPont. It is being investigated by the Department of Defense for its use in body armor and ballistic glass. Due to its transparency, strength, and optic properties, nanocellulose is being researched by Pioneer Electronics as a replacement for glass for flexible screens. Due to these optical and other properties, it is a unique additive for inks, paints, dyes, or glazing, thue adding strength while enhancing color brilliance. Combined with its safe attributes, it has applications in cosmetics and personal care products (for instance, as a sunscreen) [22].

3.2.4.1 Microfibrillated Cellulose

A cellulose fiber is composed of bundles of microfibrils where the cellulose chains are stabilized laterally by inter- and intramolecular hydrogen bonding. Cellulose fibers are turned into nanofibrous forms by chemical and mechanical treatments. One type of such nanofibers is called MFC, which can be obtained by a high-pressure homogenizing treatment [23–24]. The properties of MFC have been previously reviewed by Siro and Plackett [25]. The MFC consists of moderately degraded long fibrils that have a greatly expanded surface area. Typically, the traditional MFC consists of cellulose microfibril aggregates with a diameter ranging from 20 to even 100 nm and a length of several micrometers, rather than single nanoscale microfibrils [26]. Microfibrils comprise elementary fibrils where monocrystalline domains are linked by amorphous domains. A photograph showing MFC and an optical micrograph showing the microstructure of the cellulose fibrils are presented in Figure 3.4 [27].

Regarding the production of MFC, several mechanical treatments have been used, such as a two-step process, including a refining and a high-pressure homogenization step, cryocrushing, and grinding methods. Developed in 1983 by Turbak et al. [27], the homogenization technology allows the production of a network of interconnected cellulose microfibrils. Without any cellulose pretreatment, the two-step mechanical process has usually led to MFC with the smallest diameters. In general, MFC is obtained from cellulose fibers after a two-step mechanical disintegration process, consisting of an initial refining step followed by a high-pressure homogenization step. More recently, there has been a focus on energy-efficient production methods, whereby fibers are pretreated by various physical, chemical, and enzymatic methods before homogenization to decrease the energy consumption. Pretreatments

(a)   (b)

FIGURE 3.4   (a) Photograph showing microfibrillated cellulose (MFC) (5% MFC and 95% water) and (b) optical micrograph showing the microstructure of the cellulose fibrils. (From Turbak, A.F., Snyder, F.W., and Sandberg, K.R. 1983. Microfibrillated cellulose, a new cellulose product: Properties, uses and commercial potential. *Journal of Applied Applied Polymer Science* 37:815–27. With permission [27].)

are alkaline pretreatment, oxidative pretreatment, and enzymatic pretreatment [28]. The most important characteristics of MFC are the dimensions and distribution of the fibrillar material, and the rheology of the resulting dispersion [18]. The scheme of interaction between cellulose molecular chains within the crystalline region of cellulose microfibrils is presented in Figure 3.5 [29]. These microfibrils have disordered (amorphous) regions and highly ordered (crystalline) regions. In the crystalline regions, cellulose chains are closely packed together by a strong and highly intricate intra- and intermolecular hydrogen-bond network (Figure 3.5), whereas the amorphous domains are regularly distributed along the microfibrils [29].

The hydrophilic nature of MFC constitutes a major obstacle for its use in composite applications. To tackle this problem, one strategy involves the chemical modification of surface hydroxyl groups of the MFC, to prevent hornification phenomena and/or decrease the nanofiber surface hydrophilicity. In past decades, the chemical modification of MFC has received significant interest from the scientific community. Thus, many reactions have already been performed to permanently modify the surface properties of the MFC (i.e., surface polarity), involving the use of 2,2,6,6-tetramethylpiperidine-1-oxyl (TEMPO) oxidative agent, silane reagents, carboxymethylation, acetylation, isocyanates, poly(ε-caprolactone), or anhydrides [30].

3.2.4.2 Nanocrystalline Cellulose

NCC, also known as "whiskers," consists of rodlike cellulose crystals with widths and lengths of 5–70 nm and between 100 nm and several micrometers, respectively (Figure 3.6). They are generated by the removal of amorphous sections of a purified cellulose source by acid hydrolysis, often followed by ultrasonic treatment [18]. The commercialization of cellulose nanocrystals is still at an early stage, but appears very promising, as the strengthening effect and optical properties of NCC may find

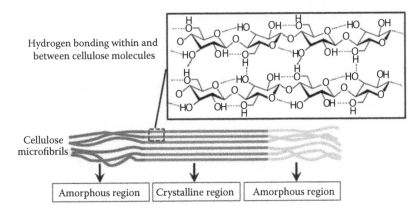

FIGURE 3.5 Scheme of interaction between cellulose molecular chains within the crystalline region of cellulose microfibrils. (From Zhou, C. and Wu, Q. 2012. Chapter 6: Recent development in applications of cellulose nanocrystals for advanced polymer-based nanocomposites by novel fabrication strategies. In: *Nanotechnology and Nanomaterials: Nanocrystals—Synthesis, Characterization and Applications*, ed. Nerella, S. Intec: Rijeka, Croatia. With permission [29].)

FIGURE 3.6 Example of NCC.

use in nanocomposites, paper making, coating additives, security papers, food packaging, and gas barriers. High aspect ratio, low density, low energy consumption,
inherent renewability, biodegradability, and biocompatibility are the advantages of
environmentally friendly crystalline nanocellulose (CNC) [29]. Bacterial cellulose,
microbial cellulose, or biocellulose is formed by aerobic bacteria, such as acetic acid
bacteria of the genus *Gluconacetobacter* and *Acetobacter xylinum*, which serves as
a pure component of their biofilms [18].

During the past decade, CNC has attracted considerable attention, which could be
attributed to its unique features. First, CNC has nanoscale dimensions and excellent
mechanical properties. The theoretical value of Young's modulus along the chain axis
for perfect native CNC is estimated to be 167.5 GPa, which is even theoretically stronger than steel and similar to Kevlar, whereas the Young's modulus of native CNC from
cotton and tunicate reaches 105 and 143 GPa, respectively [29].

An excellent review on cellulose whiskers summarized the dimensional characteristics with their respective sources, description of isolation processes, hydrolysis conditions, and techniques of determination and performance of this material
in suspension and in polymeric matrixes [17]. Comments on dispersity, related to
their tendency toward agglomeration, and their compatibility with commercial
hydrophobic polymers were also discussed. A recent review showed that NCC
exhibited intriguing scientific and engineering discoveries and advancements [31].
However, the authors pointed out that the field is still in its infancy and open to

opportunities for new advancements and discoveries. Other authors showed that cellulose nanocrystals are attractive material for incorporation into composites, because they can introduce additional strength gains with highly versatile chemical functionality [17].

3.3 NATURAL AND WOOD FIBER-REINFORCED PLASTIC COMPOSITES

WPCs refer to any composites that contain natural (including wood and nonwood) fibers and thermosets or thermoplastics [1]. Thermosets are plastics that, once cured, cannot be remelted and reshaped. These include resins such as epoxies and phenolics, and plastics that the forest products industry is most familiar with. Thermoplastics are polymers that can be repeatedly melted [15]. This property allows other materials, such as wood fibers, to be mixed with the plastic to form a composite product. Polypropylene (PP), polyethylene (PE), and polyvinyl chloride (PVC) are the most widely used thermoplastics for WPC and currently they are very common in buildings, construction, furniture, and automotive products. WPCs are used in nearly every field of application, but their main application is in construction engineering, with a main focus on decking for terraces and balconies. Cost effectiveness is very important for decking material. For this reason, mainly commodity polymers such as PE or PP are used [32].

Over the past two decades, natural fibers have received considerable attention as a substitute for synthetic fiber reinforcements in plastics. As replacements for conventional synthetic fibers such as aramid and glass fibers, natural fibers are increasingly used for reinforcement in thermoplastics due to their low density, good thermal insulation and mechanical properties, reduced tool wear, unlimited availability, low price, and problem-free disposal [33]. Several types of natural fibers such as kenaf [34], jute [35], sisal [36], flax [37], hemp [38], and coir fiber [39] were studied as reinforcement for thermoplastics such as PP and PE. Natural fibers also offer economical and environmental advantages compared with traditional inorganic reinforcements and fillers. As a result of these advantages, natural fiber-reinforced thermoplastic composites are gaining popularity in automotive and nonstructural construction applications. Natural fibers have attracted remarkable interest in the automotive industry owing to their hard-wearing quality and high hardness (not fragile such as glass fiber) and good acoustic resistance, in addition to, being moth-proof, nontoxic, resistant to microbial and fungi degradation, and not easily combustible [40]. The natural fiber-reinforced thermoplastic composites serve as a replacement for glass fiber in automotive components. They are used as trim parts in dashboards, door panels, parcel shelves, seat cushions, backrests, and cabin linings.

Forest product companies see WPCs as a way to increase the value-added utilization of waste wood and wood of low-commercial value. Plastic processors see wood as a readily available, relatively inexpensive filler that can lower resin costs, improve stiffness, increase profile extrusion rates, and act as an environmentally friendly method by which the use of petroleum-based plastics could be decreased. WPCs are resistant to moisture, insects, decay, and warping when compared with traditional pressure-treated lumber. They are stiffer, exhibit less creep, and are more dimensionally stable than unfilled plastic lumber. In addition, WPCs offer a "wood" look and feel with minimum maintenance [41].

WPCs are usually produced by mixing wood with polymers or by adding wood fiber as a filler in a polymer matrix, and then pressing or molding it under high pressure and suitable temperature. Additives such as colorants, coupling agents, stabilizers, blowing agents, reinforcing agents, fire retardants, biocides, foaming agents, and lubricants help tailor the end product to the target area of application [42]. The predominant manufacturing processes that are used to produce WPCs are usually extrusion and injection molding. In comparison to pure wood materials, more complex shapes can be formed by injection molding (Figure 3.7 [43]).

The characteristics and specifications of composites made from cellulose-based materials and thermoplastics (usually WPC or NFC) are given in the European Standard (EN)15534 [44] as follows:

- Part 1: Test methods for characterization of compounds and products
- Part 2: Load-bearing applications—Determination of modification factors for bending properties
- Part 3: Specifications of materials
- Part 4: Specifications for decking profiles and tiles
- Part 5: Specifications for cladding profiles and tiles
- Part 6: Specifications for fencing profiles
- Part 7: Specifications for general purpose profiles in external applications (outdoor)
- Part 8: Specifications for outdoor furniture

FIGURE 3.7 Injection-molded wood-plastic composite (WPC) samples. (From Ton-That, M.T. and Denault, J. 2007. *Development of Composites Based on Natural Fibers.* The Institute of Textile Science: Ottawa, ON. With permission [43].)

WPC lumber also tends to be quite dense compared with regular wood, and this means that several approaches have been adopted to reduce lumber weight (Figure 3.8 [45]). One approach deals with reducing section weight through the use of hollow profile cross-sections [46].

WPCs have been widely used for several years, and their market share is continuously growing. It is widely recognized that the use of a polymer and one or more solid fillers allows obtaining several advantages and, in particular, a combination of the main properties of the two (or more) solid phases. Among the fillers used, it is worth citing calcium carbonate, glass fibers, talc, kaolin, mica, wollastonite, dolomite, silica, graphite, synthetic fillers (e.g., polyethylene terephthalate (PET)- or polyvinyl alcohol (PVA)-based fibers), and high-performance fibers (carbon, aramidic, etc.) [47]. Major industries such as the aerospace, automotive, construction, or packaging have shown enormous interest in the development of new composite materials. One example of this is the replacement of inorganic fibers, such as glass or aramid fibers, by natural fibers that serve as fillers [48].

In general, making composites with large particles is difficult with the extrusion process. In addition, the range of thickness and the width of the natural fiber-reinforced plastic composites made using the extrusion process are lower than those of wood-based panels such as fiberboard and particleboard. Another little explored possibility is to produce natural fiber-reinforced thermoplastic composites on a flat-press such as the traditional wood-based panels. Flat-pressing technology can be considered a promising alternative for manufacturing large dimensioned panels, as slit-die extrusion is limited in width, thickness, and output rate (Figure 3.9 [49]). The advantage of this technology is that only a relatively low pressure level is required,

FIGURE 3.8 Extruded WPC samples. (From Web catalog. 2014. Greiner Tech. Profile GmbH, Austria, http://www.greiner-techprofile.com [Accessed 16 June 2014]. With permission [45].)

FIGURE 3.9 Flat-pressed WPC panel samples. (Benthien, J.T., Ohlmeyer, M., and Fruhwald, A. 2011. Wood plastic composites (WPC) flat pressed and large-dimensioned. Poster article, Department of Wood Science and Technology, Hamburg University. With permission [49].)

compared with extrusion and injection molding. The productivity of the pressing technology is much higher than that of extrusion and injection molding. Dimensions of flat-pressed WPC panels resemble more those of wood-based panels with a thermoset as an adhesive, such as particleboard and medium-density fiberboard (MDF), so that new application fields of WPC could be discovered in the future, particularly when elevated moisture resistance is required [50–52].

PVC provides the greatest strength and stiffness for WPC composites followed by PE and PP. It has an excellent cost/benefit ratio when compared with other polymer resins. WPC products made with PVC exhibit good impact strength, stiffness, and strength-to-weight ratio. PVC products offer good dimensional stability at ambient temperatures, resistance to chemicals and oils, durability, and a nonflammable nature [53]. However, PVC is the most toxic plastic for one's health. It is also the most environmentally harmful plastic.

3.3.1 MODIFIERS AND ADDITIVES FOR COMPOSITES

Natural fibers have a number of advantages and disadvantages compared with traditional synthetic fibers. Their ecological nature, biodegradability, low costs, nonabrasive nature, safe fiber handling, high possible filling levels, low energy consumption, high specific properties, low density, and wide variety of fiber types are very important factors for their acceptance in large volume markets, such as the automotive and construction industries. Furthermore, the public generally regards products made from renewable raw materials as environmentally friendly [2,54].

Wood flour is derived from various scrap wood from wood processors. High-quality wood flour must be of a specific species or species group and must be free from bark, dirt, and other foreign matter [55]. The most commonly used wood flours for plastic composites are made from pine, oak, maple, birch, spruce, fir, and poplar. Some species, such as red oak and chestnut, can contain low-molecular-weight phenolic compounds and tannins, which may cause stains if the composite is repeatedly wetted. Wood flour is typically about $0.10–0.30/kg. Most commercially manufactured wood flours used as fillers in thermoplastics are less than 425 μm (40 US standard mesh). Wood flour having a 40–60 mesh size is suitable for extrusion, whereas wood flour having a 80–100 mesh size is suitable for injection molding. Very fine wood flours can cost more and increase melt viscosity more than coarser wood flours, but composites made with them typically have more uniform appearance and a smoother finish [55].

However, certain drawbacks, such as a tendency to form aggregates during processing, low thermal stability, low resistance to moisture, and seasonal quality variations (even between individual plants in the same cultivation), greatly reduce the potential of natural fibers to be used as reinforcement for polymers [2,54]. The high moisture adsorption of natural fibers leads to swelling and the presence of voids at the interface, which results in poor mechanical properties and reduces the dimensional stability of composites [56]. It is, therefore, clear that chemical modification or use of adhesion promoters can be interesting paths to improve the overall mechanical properties.

The greater the wood content, the better the stiffness properties of the composite. However, there is a direct trade-off between wood content and the moisture-resistant properties of the WPC lumber. When the wood content of WPC increases beyond 65 wt.%, the resulting water absorption will increase accordingly because the wood is less likely to be fully encapsulated by the matrix polymer. With high-plastic percentages, WPCs are less likely to have absorbed much water from immersion tests [46].

Modification relies on physical and chemical techniques, which are mainly focused on grafting chemical groups that are capable of improving the interfacial interactions between filler particles and polymer matrix. The main techniques may be summarized as follows [57–59]:

- Thermal modification of wood fiber: Thermal modification is considered one of the most effective methods used to improve dimensional stability, durability, equilibrium moisture content, permeability, and surface quality of wood and wood-based composites [60–65]. It is the oldest, the least expensive, and the most eco-friendly modification method that has been popularly used during the past decade [66]. Various thermal modification methods are used in the world and some of those methods have been registered, such as Thermowood (Finland), Perdure (France), Plato (Netherlands), and Menz Holz (Germany) [67]. All these methods have some major differences such as process conditions, wet or dry process, steering schedules, process steps, atmosphere (oxygen or nitrogen), steaming, and use of oil [68]. Having lower equilibrium moisture content and density along with increased wettability are also important advantages of heat-treated wood [69]. However, the adverse influence of heat treatment on the mechanical properties of wood is inevitable [70]. The physical and chemical properties of wood under heat treatment change at a temperature near 150°C and this continues with increasing temperature [66]. A typical heat treatment is applied at temperature levels and exposure times ranging from 120 to 250°C and from 15 minutes to 24 hours, respectively, depending on the process, species, sample size, moisture content, and target utilization [71]. Recent efforts on thermal treatment of wood fibers lead to an improvement in the dimensional stability of WPC. Ayrilmis et al. [62] reported that water absorption and thickness swelling of WPC containing *Eucalyptus camaldulensis* wood fibers treated at three different temperatures (120, 150, or 180°C) for 20 or 40 minutes under saturated steam in a laboratory autoclave was significantly lower than those of the WPC containing untreated wood fibers. Hot water extraction of wood chips is also effective in improving the dimensional stability of WPC [72].

- Alkali treatment (also called mercerization): This is usually performed on short fibers, by heating at approximately 80°C in 10% NaOH aqueous solution for about 3–4 hours, washing, and drying in a ventilated oven. This process disrupts the formation of fiber clusters in order to obtain smaller and better-quality fibers. It should also improve fiber wetting.

- Acetylation: The fibers are usually first immersed in glacial acetic acid for 1 hour, then immersed in a mixture of acetic anhydride and a few drops of concentrated sulfuric acid for a few minutes, filtrated, washed, and dried in a ventilated oven. This is an esterification method that should stabilize the cell walls, especially in terms of humidity absorption and consequent dimensional variation.

- Treatment with stearic acid: The acid is added to an ethyl alcohol solution, up to 10% of the total weight of the fibers to be treated, and the obtained solution is thus added drop wise to the fibers, which are then dried in an oven. It is an esterification method as well.

- Benzylation: The fibers are immersed in 10% NaOH, then stirred with benzoyl chloride for 1 hour, filtrated, washed, dried, immersed in ethanol for 1 hour, rinsed, and dried in an oven. This method allows for decreasing of the hydrophilicity of the fibers.

- Toluene diisocyanate (TDI) treatment: The fibers are immersed in chloroform with a few drops of a catalyst (based on dibutyltin dilaurate) and stirred for 2 hours after adding toluene-2,4-diisocyanate. Finally, the fibers are rinsed in acetone and dried in an oven.

- Peroxide treatment: The fibers are immersed in a solution of dicumyl (or benzoyl) peroxide in acetone for about half an hour, then decanted, and dried. Recent studies have highlighted significant improvements with regard to mechanical properties with this treatment.

- Anhydride treatment: This is usually carried out by using maleic anhydride or maleated PP (or PE) in a toluene or xylene solution, where the fibers are immersed for impregnation and a reaction with the hydroxyl groups on the fiber surface. Literature reports a significant reduction of water absorption.

- Permanganate treatment: The fibers are immersed in a solution of $KMnO_4$ in acetone (typical concentrations may range between 0.005% and 0.205%) for 1 minute, then decanted and dried. Investigations have pointed out a decreased hydrophilic nature of the fibers on performing this treatment.

- Silane treatment: The fibers are immersed in a 3:2 alcohol–water solution, containing a silane-based adhesion promoter for 2 hours at pH ≈ 4, rinsed in water, and dried in an oven. Silanes should react with the hydroxyl groups of the fibers and improve their surface quality.
- Isocyanate treatment: The isocyanate group can react with the hydroxyl groups on the fiber surface, thus improving the interface adhesion with the polymer matrix. The treatment is typically performed with isocyanate compounds at intermediate temperatures (around 50°C) for approximately 1 hour.
- Plasma treatment: This recent method allows a significant modification of the fiber surface. However, chemical and morphological modification can be very heterogeneous depending on the treatment conditions, and therefore it is not easy to generalize; process control is a critical aspect, and the final surface modifications strongly depend on it. More specifically, TDI, dicumylperoxide, and silane treatment seem to guarantee the best results with regard to mechanical properties, whereas alkali treatment and acetylation seem to give better improvements in thermal and dimensional stability.

One of the major disadvantages of natural fibers is the poor compatibility exhibited between the fibers and the polymeric matrices, which results in the nonuniform dispersion of fibers within the matrix and poor mechanical properties. Wood and plastic are similar to oil and water, and they do not mix well [73]. Most polymers, especially thermoplastics, are nonpolar ("hydrophobic," repelling water) substances that are not compatible with polar ("hydrophilic," absorbing water) wood fibers and, therefore, this can result in poor adhesion between polymers and fibers in WPC. To improve the affinity and adhesion between fibers and thermoplastic matrices in production, chemical "coupling" or "compatibilizing" agents have been used [54]. Chemical coupling agents are substances, typically polymers, that are used in small quantities to treat a surface so that bonding occurs between it and other surfaces, for example, wood and thermoplastics. The coupling forms include covalent bonds, secondary bonding (such as hydrogen bonding and van der Waals' forces), polymer molecular entanglement, and mechanical interblocking [74]. Therefore, chemical treatments can be considered in modifying the properties of natural fibers. Some compounds are known to promote adhesion by chemically coupling the adhesive to the material, such as sodium hydroxide, silane, acetic acid, acrylic acid, isocyanates, potassium permanganate, peroxide, and so on. The mechanism of the compatibilizing agent is shown in Figure 3.10 [75].

The coupling agent chemically bonds with the hydrophilic fiber and blends by wetting in the polymer chain [76]. An example of the improvements that a maleated adhesion promoter can assure to a WPC based on PP and 30 wt.% wood flour is reported in Table 3.4 [77].

Another issue is the processing temperature, which restricts the choice of matrix material. Natural fibers are composed of various organic materials (primarily cellulose as well as hemicellulose and lignin), and, therefore, their thermal treatment leads to a variety of physical and chemical changes. Thermal degradation of those fibers leads to poor organoleptic properties, such as odor and colors, and moreover

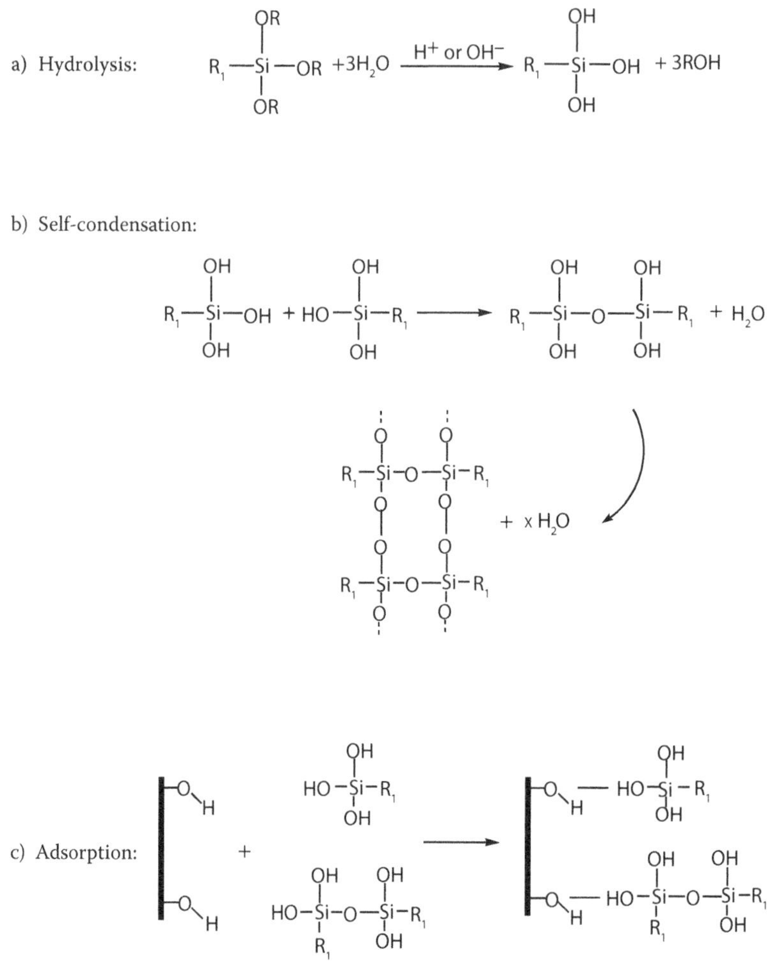

a) Hydrolysis:

b) Self-condensation:

c) Adsorption:

FIGURE 3.10   Interaction of silane with natural fibers by hydrolysis. (From Xie, Y., Hill, C.A.S., Xiao, Z., Militz, H., and Mai, C. 2010. Silane coupling agents used for natural fiber/polymer composites: A review. *Composites: Part A* 41:806–19. With permission [75].)

to the deterioration of their mechanical properties. It also results in the generation of gaseous products, when processing takes place at temperatures above 200°C, which can create high porosity, low density, and reduced mechanical properties. For the improvement of thermal stability, attempts have been made to coat and/or graft the fibers with monomers [4].

3.3.2   PROCESSING TECHNIQUES, PROPERTIES, AND APPLICATIONS OF COMPOSITES

The term "WPC" usually means compounding equal amounts of wood flour or particulates with thermoplastic polymers such as PP, PE, polystyrene (PS), PVC, or acrylonitrile-butadiene styrene (ABS). Since the wood content in the polymer matrix is high, the appearance and odor of WPC are similar to those of natural wood [15].

TABLE 3.4

Main Mechanical Properties of PP-Wood Flour/Fiber Composite with or without Maleated Adhesion Promoter

Property	Unit	PP	PP +40 wt.% Wood Flour	PP + 40% Wood Flour + 3% Coupling Agent	PP + 40% Wood Fiber	PP + 40% Wood Fiber + 3% Coupling Agent
Specific gravity		0.90	1.05	1.05	1.03	1.03
Tensile strength	MPa	28.5	25.4	32.3	28.2	52.3
Tensile modulus	GPa	1.53	3.87	4.10	4.20	4.23
Bending strength	MPa	38.30	44.20	53.10	47.90	72.40
Bending modulus	GPa	1.19	3.03	3.08	3.25	3.22

Source: Stark, N.M. and Rowlands, R.E. 2003. Effects of wood fiber characteristics on mechanical properties of wood/polypropylene composites. *Wood and Fiber Science* 35:167–74. With permission [77].

WPCs can be produced by different processes depending on the consumer's need. Since the major markets for WPC are decking, railing, fencing, and siding, 97% of WPCs are produced by profile extrusion [73]. Wood flour is well accepted by plastic processors as a filler, because it is cheap and readily available. The stiffness and strength of plastics can be increased by adding wood flour as a filler. Wood also causes less abrasion to an extruder than mineral fillers such as glass fiber and talc. Since wood cools faster than plastics, there is no need for "calibrating" to shape a part of WPC lumber [78]. The result was the introduction of WPC injection mold industries that produce construction-related products such as windows, door sills, railing spindles, railing post skirts, and caps, as well as nonconstruction-related products, including automotive parts and furniture components. The ability to use recycled plastics in WPCs is an economic incentive, and there is a general perception that WPCs are quite durable and resistant to decay since the wood particulates are expected to be completely encapsulated by plastic.

Wood flour usually contains at least 4% moisture when delivered, which must be removed before or during processing with thermoplastics. Though moisture could potentially be used as a foaming agent to reduce density, this approach is difficult to control and is not common industrial practice. Commercially, moisture is removed from the wood flour (1) before processing using a dryer, (2) by using the first part of an extruder as a dryer in some in-line process, or (3) during a separate compounding step (or in the first extruder in a tandem process). Once dried, wood flour can still absorb moisture quickly. Depending on the ambient conditions, wood flour can absorb several weight percent of moisture within hours. Even compounded material often needs to be dried before further processing, especially if high weight percentages of wood flour are used. The hygroscopicity of wood flour can also affect the end composite. Absorbed moisture interferes with and reduces hydrogen bonding between cell wall polymers and alters the mechanical performance of the product [55]. The wood flour should be dried in a dryer at 100°C for 24 hours to reach 0%–1% moisture content and then be stored in sealed plastic bags until blending with the polymer matrix.

The properties, ability to process, and rate of production of WPC can be improved drastically if proper additives are used. Plastics, such as high-density PE (HPPE) and PP, tend to absorb much less moisture than wood in the natural environment. Therefore, WPCs are less affected by moisture and possess better dimensional stability and fungus/termite resistance than solid wood, because wood particulates are encapsulated by the polymer matrix. Since thermoplastic polymers such as PP and HDPE are nonpolar (hydrophobic) materials whereas wood particulates are polar (hydrophilic) materials, there is a high probability of obtaining poor adhesion between wood and polymer, resulting in low tensile and flexural strengths of the WPC [74]. To improve the adhesion between wood and plastics, maleic anhydride grafted polymers are generally introduced as a compatibilizer to improve the strength of WPC. In addition, when the adhesion of wood flour and plastics is enhanced, the rate of water adsorption also decreases in the presence of the coupling agent. The production rate of WPC can be substantially increased when lubricants such as zinc stearate or fatty acids are added in the compounding process. Lubricants help suppress edge tearing and melt fracture phenomena happening in the extrusion process. Generally, the density of WPC is higher than that of solid wood and this limits the applications of WPC [79]. The density of WPC can be reduced by as much as 30% by adding blowing agents that make the density of WPC similar to that of real wood. Foaming also helps a manufacturer save on material cost.

Fungal and insect resistances of WPC are improved by adding boron compounds such as zinc borate, boric acid, borax, or disodium octaborate tetrahydrate (DOT) [80]. The boron compounds can be incorporated into the wood particles by conventional methods such as dipping-diffusion or vacuum-pressure treatments. The WPC manufacturers usually use the boron compounds as dry powder in a blender due to its easy application. However, the dipping-diffusion or vacuum-pressure treatments of boron compounds are more effective than dry blending of powders of wood and boron compounds against wood-destroying insects and fungi. This is because dipping-diffusion or vacuum-pressure treatments enable the chemicals to be more deeply impregnated into the wood particles or wood flour as compared with their powder form. The key issue to expand boron use for wood protection appears to be their fixation into wood while allowing for sufficient mobility so they remain fungicidal. In a recent study, boric acid was fixed into wood with condensed tannins and hexamine through a non-formaldehyde emission polymer network. This treatment greatly reduced boron leaching [81].

WPCs can be colored easily by adding colorants. Adding a UV stabilizer can improve the UV stability and weatherability of WPC. Light stabilizers can be classified into two types according to their action mode: UV absorbers act by shielding the material from UV light, and hindered amine light stabilizers (HALS) act by scavenging the radical intermediates formed in the photooxidation process. The WPC manufacturers offer specific light stabilizers for each polymer. The amount of protection can be influenced by both photostabilizer concentration and exposure variables [82]. Antioxidants are also applied here to prevent the polymer from degrading during the compounding process. Finally, some fillers such as glass fiber, talc, calcium carbonate, and nanoclay are recommended as the second filler in the WPC to improve mechanical properties and creep resistance. Indeed, the desired properties of a WPC can be achieved with a combination of different additives [1]. Overall, a coupling agent, lubricant, and blowing agent are the additives that attract more attention in the WPC industry.

The WPC manufacturers are introducing new applications for the furniture industry. Further expansion into the residential construction industry and development of applications for the furniture industry require an understanding of the fire performance of WPCs. For some applications, it may be necessary to improve the fire performance. Therefore, a knowledge of the effect of fire retardants in WPCs is also critical. Polymers used in WPCs burn and drip in case of fire, resulting in a very risky scenario. Burning plastics may produce hazards such as the evolution of toxic gases, loss of physical integrity, and melting and dripping, thereby providing other ignition sources [41]. Thus, fire retardant agents must be used to improve fire behavior [83]. The most effective and widely used fire retardant chemicals for WPCs are ammonium polyphosphate and expandable graphite, decabromodiphenyl oxide, magnesium hydroxide, melamine polyphosphate, aluminum hydroxide, and boron compounds (zinc borate) [51,84–86]. In a previous study, Ayrilmis et al. [51] reported that higher levels of wood flour content resulted in significantly improved fire resistance of the WPC panels with and without fire retardant as measured in the cone calorimeter test.

These days, the applications of WPCs are limited by material performance. The reason for this is that the flexural strength of plastics used in compounding WPC is approximately 40–80 MPa, whereas the flexural modulus is only about 1.5–2.5 GPa; the corresponding values for natural wood can be as high as 80 MPa for strength and 9 GPa for the modulus [79]. Thus, any combination of wood flour and plastic will lead to WPC flexural and tensile moduli that are significantly lower than those of natural wood. A deck constructed out of WPC, therefore, will flex much more than an identical wood deck for the same load, and this is undesirable. In addition, more research needs to be done on the long-term material behavior such as weather durability and creep. The flame retardancy of WPC is especially important in some states such as California, which have regulations in this regard. Material performance of WPCs these days has already met builder acceptance as a nonstructural material for decking. With more and more research being done in this area, it may be possible to use WPCs as structural materials in the future. Therefore, product evaluation standards of WPCs have to be established to meet the consumer's demand and safety requirements.

3.4   CONCLUSIONS

A new generation of composites is emerging as material behavior is better understood, processing and performance are improved, and new opportunities are identified. Recent trends such as the desire to decrease petroleum dependence, increase bio-content, commercial production of nanocellulose, and changing markets will play a major role in the future of these composites. For example, composite manufacturers are seeking to take advantage of the favorable balance of properties (e.g., low density, good mechanical properties) of bast fibers from plants in composite applications. Natural fiber-reinforced polymer composites have received much attention because of their low density, low cost, nonabrasive, combustible, nontoxic, and biodegradable properties. A new building material, WPC, has emerged.

One main challenge of natural fiber-reinforced thermoplastic composites is their inherent moisture sensitivity, a major cause for fungal decay, mold growth, and dimensional instability, resulting in decreased service life as well as costly maintenance. Another issue is to understand the critical issues of durability, color stability, and UV weathering of natural fiber-reinforced thermoplastic composites. Optimum content of additives such as flame retardants, antioxidants, UV stabilizers, absorbers, pigments, antioxidants, and biocidies significantly improves the properties of natural fiber-reinforced thermoplastic composites. A serious issue of natural fibers is their strong polar nature, which creates incompatibility with most polymer matrices. Surface treatments are potentially able to overcome the problem of incompatibility. Chemical treatments can increase the interface adhesion between the fiber and the matrix, and they can decrease the water absorption of fibers.

Nanocellulose-based reinforcements constitute another class of naturally sourced reinforcements of recent interest. Application of cellulose nanofibers in polymer reinforcement is a relatively new research field. The main reason for using cellulose nanofibers in composite materials is that one can potentially exploit the high stiffness of the cellulose crystal for reinforcement.

REFERENCES

1. Ashori, A. 2008. Wood-plastic composites as promising green-composites for automotive industries. *Bioresource Technology* 99:4661–7.
2. Bismarck, A., Baltazar, Y.J., and Sarlkakis, K. 2006. Green composites as Panacea? Socio-economic aspects of green materials. *Environment, Development and Sustainability* 8:445–63.
3. Peijs, T., Cabrera, N., Alcock, B., Schimanski, T., and Loos, J. 2002. Pure all-polypropylene composites for ultimate recyclability. Paper presented at the 9th International Conference on Fiber Reinforced Composites—FRC 2002, Newcastle.
4. John, M.J. and Thomas, S. 2008. Biofibers and biocomposites. *Carbohydrate Polymers* 71:343–64.
5. Panthapulakkal, S., Zereshkian, A., and Sain, M. 2006. Preparation and characterization of wheat straw fibers for reinforcing application in injection molded thermoplastic composites. *Bioresource Technology* 97:265–72.
6. Reddy, N. and Yan, Y. 2008. Characterizing natural cellulose fibers from velvet leaf (*Abutilon theophrasti*) stems. *Bioresource Technology* 99:2449–54.
7. Ashori, A., Jalaluddin, H., Raverty, W.D., and Mohd Nor, M.Y. 2006. Chemical and morphological characteristics of Malaysian cultivated kenaf (*Hibiscus cannabinus*) fiber. *Polymer-Plastics Technology & Engineering* 45:131–34.
8. Jawaid, M. and Abdul Khalil, H.P.S. 2011. Cellulosic/synthetic fiber reinforced polymer hybrid composites: A review. *Carbohydrate Polymers* 86:1–18.
9. Ashori, A. 2006. Non-wood fibers—A potential source of raw material in papermaking. *Polymer-Plastics Technology & Engineering* 45:1133–6.
10. Smook, G.A. 1992. *Handbook for Pulp and Paper Technologists*. 2nd ed. Vancouver, BC: Angus Wilde.

11. Sjöström, E. 1993. *Wood Chemistry: Fundamentals and Applications.* 2nd ed. San Diego, CA: Academic Publisher.
12. Rowell, R.M., Young, R.A., and Rowell, J.K. 1997. *Paper and Composites from Agrobased Resources.* Boca Raton, FL: CRC Press.
13. Rowell, R.M. and Clemons, C.M. 1992. Chemical modification of wood fiber for thermoplasticity, compatibilization with plastics and dimensional stability. Paper presented at the International Particleboard/Composite Materials Symposium, Pullman, WA.
14. TAPPI Test Methods. 2002. TAPPI Press: Atlanta, GA.
15. Nourbakhsh, A., Ashori, A., Ziaei Tabari, H., and Rezaei, F. 2010. Mechanical and thermo-chemical properties of wood-flour polypropylene blends. *Polymer Bulletin* 65:691–700.
16. Ek, M., Gellerstedt, G., and Henriksson, G. 2009. *Paper Chemistry and Technology.* Walter de Gruyter GmbH and Co. KG: Germany.
17. Durán, N., Lemes, A.P., Durán, M., Freer, J., and Baeza, J. 2011. A mini review of cellulose nanocrystals and its potential integration as co-product in bioethanol production. *Journal of the Chilean Chemical Society* 56:672–77.
18. Klemm, D., Kramer, F., Moritz, S., Lindström, T., Ankerfors, M., Gray, D., and Dorris, A. 2011. Nanocelluloses: A new family of nature-based materials. *Angewandte Chemie International Edition* 50:5438–66.
19. Spence, K.L., Venditti, R.A., Rojas, O.J., Habibi, Y., and Pawlak, J.J. 2011. A comparative study of energy consumption and physical properties of microfibrillated cellulose produced by different processing methods. *Cellulose* 18:1097–111.
20. Jacoby, M. 2014. Nano from the forest. *Chemical & Engineering News* 92:9–12.
21. Noticias de Nanotecnología. 2014. http://www.tecnologianano.com [Accessed 20 June 2014].
22. Anonymous. 2014. *Silvantris: Primer on NanoFibers and NanoCellulose.* Silvantris, LLC: Orem, UT, http://www.silvantris.com [Accessed 15 June 2014].
23. Iwamoto, S., Yamamoto, S., Lee, S.H., and Endo, T. 2014. Mechanical properties of polypropylene composites reinforced by surface-coated microfibrillated cellulose. *Composites: Part A* 59: 26–9.
24. Jang, J.H., Lee, S.H., Endo, T., and Kim, N.H. 2013. Characteristics of microfibrillated cellulosic fibers and paper sheets from Korean white pine. *Wood Science and Technology* 47:925–37.
25. Siro, I. and Plackett, D. 2010. Microfibrillated cellulose and new nanocomposite materials: A review. *Cellulose* 17:459–94.
26. Kettunen, M. 2013. Cellulose nanofibrils as a functional material. PhD diss., Aalto University, Helsinki.
27. Turbak, A.F., Snyder, F.W., and Sandberg, K.R. 1983. Microfibrillated cellulose, a new cellulose product: Properties, uses and commercial potential. *Journal of Applied Applied Polymer Science* 37:815–27.
28. Kwon, J.H., Lee, S.H., Ayrilmis, N., and Han, T.H. 2014. Effect of microfibrillated cellulose content on the bonding performance of urea-formaldehyde resin. Paper presented at the 57th International Convention of Society of Wood Science and Technology, Zvolen.
29. Zhou, C. and Wu, Q. 2012. Chapter 6: Recent development in applications of cellulose nanocrystals for advanced polymer-based nanocomposites by novel fabrication strategies. In: *Nanotechnology and Nanomaterials: Nanocrystals—Synthesis, Characterization and Applications,* ed. Nerella, S. Intech: Rijeka, Croatia.
30. Tingaut, P., Eyholzer, C., and Zimmermann, T. 2011. Chapter 14: Functional polymer nanocomposite materials from microfibrillated cellulose. In: *Advances in Nanocomposite Technology,* ed. Hashim, A. Intech: Rijeka, Croatia.
31. Habibi, Y., Lucia, L.A., and Rojas, O.J. 2010. Cellulose nanocrystals: Chemistry, self-assembly, and applications. *Chemical Reviews* 110:3479–500.

32. Seefeldt, H. 2012. Flame retardancy of wood-plastic composites. PhD diss., Technischen Universität Berlin.

33. Bledzki, A.K. and Gassan, J. 1999. Composites reinforced with cellulose based fibers. *Progress in Polymer Science* 24:221–74.

34. Chow, P., Lambert, R.J., Bowers, C.T., McKenzie, N., Youngquist, J.A., Muehl, J.H., and Kryzsik, A.M. 2000. Physical and mechanical properties of composite panels made from kenaf plant fibers and plastics. Paper presented at the 2000 International Kenaf Symposium, Hiroshima.

35. Rahman, R., Hasan, M., Huque, M., and Islam, N. 2010. Physico-mechanical properties of jute fiber reinforced polypropylene composites. *Journal Reinforced Plastics and Composites* 29:445–55.

36. Joseph, P.V., Mathew, G., Joseph, K., Thomas, S., and Pradeep, P. 2003. Mechanical properties of short sisal fiber-reinforced polypropylene composites: comparison of experimental data with theoretical predictions. *Journal of Applied Polymer Science* 88:602–11.

37. Arbelaiz, A., Cantero, B.G., Llano-Ponte, R., Valea, A., and Mondragon, I. 2005. Mechanical properties of flax fiber/polypropylene composites. Influence of fiber/matrix modification and glass fiber hybridization. *Composites Part A* 36:1637–44.

38. Schirp, A. and Stender, J. 2010. Properties of extruded wood-plastic composites based on refiner wood fibers (TMP fibers) and hemp fibers. *European Journal of Wood and Wood Products* 68:219–31.

39. Ayrilmis, N., Jarusombuti, S., Fueangvivat, V., Bauchongkol, P., and White, R.H. 2011. Coir fiber reinforced polypropylene composite panel for automotive interior applications. *Fibers and Polymers* 12:919–26.

40. Mohanty, A.K., Misra, M., and Drzal, L.T. 2005. *Natural Fibers, Biopolymers and Biocomposites.* Boca Ranton, FL:Taylor & Francis.

41. Stark, N.M., White, R.H., Mueller, S.A., and Osswald, T.A. 2010. Evaluation of various fire retardants for use in wood flourepolyethylene composites. *Polymer Degradation and Stability* 95:1903–10.

42. Tabari, H.Z., Nourbakhsh, A., and Ashori, A. 2011. Effects of nanoclay and coupling agent on the mechanical, morphological, and thermal properties of wood flour/polypropylene composites. *Polymer Engineering & Science* 51:272–7.

43. Ton-That, M.T. and Denault, J. 2007. *Development of Composites Based on Natural Fibers.* The Institute of Textile Science: Ottawa, ON.

44. EN 15534. 2014. Wood plastic composites (WPCs). Test methods for characterization of WPC materials and products. European Committee for Standardization, Brussels, Belgium.

45. Web catalog. 2014. Greiner Tech. Profile GmbH, Austria, http://www.greiner-techprofile.com [Accessed 16 June 2014].

46. Gardner, D.J. and Murdock, D. 2010. Extrusion of wood plastic composites. University of Maine, Orono.

47. La Mantia, F.P. and Morreale, M. 2008. Accelerated weathering of polypropylene/wood flour composites. *Polymer Degradation and Stability* 93:1252–8.

48. Alemdar, A. and Sain, M. 2008. Isolation and characterization of nanofibers from agricultural residues—Wheat straw and soy hulls. *Bioresource Technology* 99:1664–71.

49. Benthien, J.T., Ohlmeyer, M., and Frühwald, A. 2011. Wood plastic composites (WPC) flat pressed and large-dimensioned. Poster article, Department of Wood Science and Technology, Hamburg University.

50. Benthien, J.T. and Thoemen, H. 2013. Effects of agglomeration and pressing process on the properties of flat pressed WPC panels. *Journal of Applied Polymer Science* 129:3710–7.

51. Ayrilmis, N., Benthien, J.T., Thoemen, H., and White, R.H. 2012. Effects of fire retardants on physical, mechanical, and fire properties of flat-pressed WPCs. *European Journal of Wood and Wood Products* 70:215–24.
52. Ayrilmis, N. and Jarusombuti, S. 2011. Flat-pressed wood plastic composite as an alternative to conventional wood-based panels. *Journal of Composite Materials* 45:103–12.
53. Gomez, J.A. 2014. *Application Profiles*. Plastic Technology, Gardner Business Media, Inc: Cincinnati, OH.
54. Kim, J.P., Yoon, T.H., Mun, S.P., Rhee, J.M., and Lee, J.S. 2006. Wood–polyethylene composites using ethylene–vinyl alcohol copolymer as adhesion promoter. *Bioresource Technology* 97:494–9.
55. Clemons, C.M. and Caufield, D.F. 2005. Chapter 15: Wood flour. In: *Functional Fillers for Plastics*, ed. Xanthos, M. Weinheim: Wiley. VCH Verlag GmbH & Co.
56. Gassan, J. and Gutowski, V.S. 2000. Effects of corona discharge and UV treatment on the properties of jute–fiber epoxy composites. *Composites Science and Technology* 60:2857–63.
57. Dipa, R., Sarkar, B.K., Rana, A.K., and Bose, N.R. 2001. Effect of alkali treated jute fibers on composite properties. *Bulletin of Materials Science* 24:129–35.
58. Dominkovics, Z., Dányádi, L., and Pukánszky, B. 2007. Surface modification of wood flour and its effect on the properties of PP/wood composites. *Composites: Part A* 38:1893–901.
59. Kalia, S., Kaith, B.S., and Kaura, I. 2009. Pretreatments of natural fibers and their application as reinforcing material in polymer composites—A review. *Polymer Engineering and Science* 49:1253–72.
60. Unsal, O. and Ayrilmis, N. 2005. Variations in compression strength and surface roughness of heat-treated Turkish river red gum (*Eucalyptus camaldulensis* Dehn.) wood. *Journal of Wood Science* 51:405–9.
61. Jarusombuti, J., Ayrilmis, N., Bauchongkol, P., and Fueangvivat, V. 2010. Surface characteristics and overlaying properties of MDF panels made from thermally treated rubberwood fibers. *Bioresources* 5:968–78.
62. Ayrilmis, N., Jarusombuti, S., Fueangvivat, V., and Bauchongkol, P. 2011. Effect of thermal-treatment of wood fibers on properties of flat-pressed wood plastic composites. *Polymer Degradation and Stability* 96:818–22.
63. Kamdem, D.P., Pizzi, A., and Jermannaud, A. 2002. Durability of heat treated wood. *Holz als Roh- und Werkstoff* 60:1–6.
64. Nguyen, C.T., Wagenfuhr, A., Phuong, L.X., Dai, V.H., Bremer, M., and Fischer, S. 2012. The effects of thermal modification on the properties of two Vietnamese bamboo species, Part I: Effects on physical properties. *Bioresources* 7:5355–66.
65. Chen, Y., Tshabalala, M.A., Gao, J., Stark, N.M., Fan, Y., and Ibach, R.E. 2014. Thermal behavior of extracted and delignified pine wood flour. *Thermochimica Acta* 591:40–4.
66. Salca, E.A. and Hiziroglu, S. 2014. Evaluation of hardness and surface quality of different wood species as function of heat treatment. *Materials and Design* 62:416–23.
67. Esteves, B. and Pereira, H.M. 2009. Wood modification by heat treatment. A review. *Bioresources* 4:370–404.
68. Militz, H. 2002. Thermal Treatment of Wood: European Processes and Their Background. IRG/WP 02-40241.
69. Candan, Z., Buyuksari, U., Korkut, S., Unsal, O., and Cakicier, N. 2012. Wettability and surface roughness of thermally modified plywood panels. *Industrial Crops Products* 36:434–6.
70. Kocaefe, D., Poncsak, S., Tang, J., and Bouazara, M. 2010. Effect of heat treatment on the mechanical properties of North American jack pine: thermogravimetric study. *Journal of Materials Science* 45:681–7.

71. Priadi, T. and Hiziroglu, S. 2013. Characterization of heat treated wood species. *Materials and Design* 49:575–82.
72. Ozdemir, F., Ayrilmis, N., Kaymakci, A., and Kwon, J.H. 2014. Improving dimensional stability of injection molded wood plastic composites using cold and hot water extraction methods. *Maderas Ciencia y Tecnología* 16:365–72.
73. Ziaei Tabari, H., Nourbakhsh, A., Ashori, A. 2011. Effects of nanoclay and coupling agent on the mechanical, morphological, and thermal properties of wood flour/polypropylene composites. *Polymer Engineering and Science* 51:272–7.
74. Lu, J.Z., Wu, Q., and McNabb, H.S. 2000. Chemical coupling in wood fiber and polymer composites: A review of coupling agents and treatments. *Wood and Fiber Science* 32:88–104.
75. Xie, Y., Hill, C.A.S., Xiao, Z., Militz, H., and Mai, C. 2010. Silane coupling agents used for natural fiber/polymer composites: A review. *Composites: Part A* 41:806–19.
76. Yan, S., Yin, J., Yang, Y., Dai, Z., Ma, J., and Chen, X. 2007. Surface-grafted silica linked with L-lactic acid oligomer: A novel nanofiller to improve the performance of biodegradable poly(L-lactide). *Polymer* 48:1688–94.
77. Stark, N.M. and Rowlands, R.E. 2003. Effects of wood fiber characteristics on mechanical properties of wood/polypropylene composites. *Wood and Fiber Science* 35:167–74.
78. Yeh, S.K. and Gupta, R.K. 2010. Nanoclay-reinforced, polypropylene-based wood-plastic composites. *Polymer Engineering and Science* 50:2013–20.
79. Yeh, S.K., Kim, K.J., and Gupta, R.K. 2013. Synergistic effect of coupling agents on polypropylene-based wood-plastic composites. *Journal of Applied Polymer Science* 127:1047–53.
80. Manning, M.J. 2008. Chapter 26: Borate wood preservatives: The current landscape. In: *Development of Commercial Wood Preservatives: Efficacy, Environmental, and Health Issues*, eds. Schultz, T.P. et al. ACS Symposium Series, Washington, DC.
81. Thévenon, M.F., Tondi, G., and Pizzi, A. 2010. Environmentally friendly wood preservative system based on polymerized tannin resin-boric acid for outdoor applications. *Maderas Ciencia y Tecnología* 12:253–7.
82. García, M., Hidalgo, J., Garmendia, I., García-Jaca, J. 2009. Wood–plastics composites with better fire retardancy and durability performance. *Composites: Part A:* 40:1772–6.
83. White, R.H., Stark, N., and Ayrilmis, N. 2011. Recent activities in flame retardancy of wood-plastic composites at the Forest Products Laboratory. Paper presented at the 22nd Annual Conference on Recent Advances in Flame Retardancy of Polymeric Materials, Stamford.
84. Ayrilmis, N., Benthien, J.T., Thoemen, H., and White, R.H. 2011. Properties of flat-pressed wood plastic composites containing fire retardants. *Journal of Applied Polymer Science* 122:3201–10.
85. Nikolaeva, M. and Karki, T. 2011. A review of fire retardant processes and chemistry, with discussion of the case of wood-plastic composites. *Baltic Forestry* 17:314–26.
86. Seefeldt, H. and Braun, U. 2012. A new flame retardant for wood materials tested in wood plastic composites. *Macromolecular Materials and Engineering* 297:814–20.

4 Matrices for Natural Fiber Composites

Juan Carlos Domínguez, Mercedes Oliet,
María Virginia Alonso, and Bo Madsen

CONTENTS

4.1   INTRODUCTION

In this chapter, we are introduced to the use of matrices together with natural fibers for the manufacture of reinforced composites. In particular, this chapter is focused on the use of the most important polymer matrices in the manufacture of composite materials. The properties of the final natural fiber-reinforced composites have been described for these most used conventional and bio-based polymeric matrices.

4.2 OVERVIEW OF MATRICES FOR NATURAL FIBER COMPOSITES

In the design of composites, the choice of a suitable matrix is one of the key factors for determining the final properties of the material. Thus, both the external aspect of materials (color, texture, durability, stability, environmental tolerance, etc.) and their structural properties, essentially provided by the fibrous reinforcement, should be taken into account because of the effects exerted on them. For instance, the compatibility between the fibers used and the compound selected must be considered as the matrix of the composite. Various types of materials can be used as matrices in the manufacture of composites: metal, polymer, and ceramic matrices. Polymeric matrices are the most studied in the production of composite materials with natural fibers serving as reinforcement. This chapter is focused on this type of matrix. Their main advantages and drawbacks, according to Verma et al. [1], are shown in Table 4.1.

Using conventional polymer matrices (petroleum based) together with natural fibers for the manufacture of composite materials has been widely studied in recent years. The effect caused by the replacement of traditional fibers such as glass or carbon fibers by different available types of natural fibers and the possibility of their use as a substitute of traditional fibers has been the goal of most of these studies. To really consider this substitution, the environmental benefits of using natural fibers should be taken into account together with their competitiveness: economically and in terms of the properties of the final material. Furthermore, the use of natural fibers as reinforcement in the manufacture of composite materials involves a study of the compounds used as matrices for these materials and, therefore, of the characteristics such as compatibility with fibrous reinforcements and properties of the obtained materials (thermal, mechanical, acoustic, etc.). Fibrous reinforcements from biological sources, natural fibers, have contributed to the use and development of new matrices and, complying with the mentioned characteristics required for composite materials, have increased the environmentally friendly nature of the obtained materials. Therefore, new biocomposites with bio matrices (considering a bio matrix as one that is biodegradable, either natural or oil based) have been developed [2]. Attending to all these considerations, a possible classification of the matrices used for composite manufacturing using natural fibers as reinforcements is shown in Figure 4.1.

TABLE 4.1
Advantages and Drawbacks of Polymer Matrices

Advantages	Limitations
Low densities	Low transverse strength
Good corrosion resistance	Low operational temperature limits
Low thermal conductivities	
Low electrical conductivities	
Translucence	
Aesthetic color effects	

Source: Verma, D. et al., *Journal of Materials and Environmental Science*, 3(6), 1079–1092, 2012 [1].

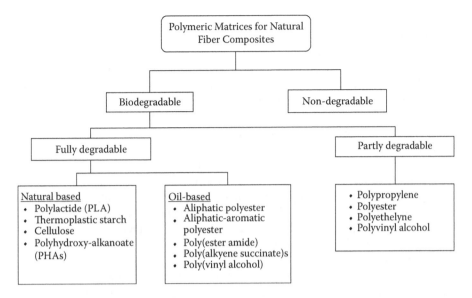

FIGURE 4.1 Classification of natural fiber composites.

A more traditional classification of the matrices used in the production of composite materials using natural fibers as reinforcement can be based on the nature of the polymer used as matrix: thermoplastic or thermosetting polymer [3–4]. In Sections 4.3 through 4.5, the most used matrices for the manufacturing of natural fiber-reinforced composite materials will be described according to this classification.

4.3 CONVENTIONAL THERMOPLASTIC MATRICES

All the thermoplastic polymers (TPS) used in the manufacture of composite materials employing natural fibers as reinforcement standout polypropylene (PP), polyethylene (PE), polystyrene (PS), and polyvinyl chloride (PVC) are petroleum-based polymers. The first two polymers are the most commonly used in the manufacturing of biocomposites. The main characteristics of both matrices together with natural fibers are described in Sections 4.3.1 and 4.3.2.

4.3.1 POLYPROPYLENE

One of the most studied properties in the case of the use of PP as a matrix together with natural fibers are the interfacial properties of the material and possible treatments that could improve it. In the manufacturing of flax-PP composites, Cantero et al. [5] applied chemical treatments to the fibers to enhance their adhesion to the PP matrix. They used maleic anhydride, maleic anhydride-PP copolymer, and vinyl trimethoxy silane as compatibilizers, and they produced composite materials by extrusion mixing and injection molding. These treatments increased the contact angle of the fibers with water. For hemp fiber-PP composite materials, Pickering et al. [6] applied an alkali treatment and a coupling agent to fibers. They found that the strongest fibers were produced using a 10-wt% sodium hydroxide (NaOH) solution at 160°C for 45 minutes and that the maleic anhydride-PP copolymer added in 3-wt% improved the interfacial bonding between the matrix and the fibers.

Park et al. [7] used treatments for fiber surface improvement of hemp and jute fiber-PP composites (NaOH and silane coupling agent) and studied the modified maleic anhydride-PP copolymer for the improvement of the PP matrix. Single fiber pullout and acoustic emission tests, and dynamic contact angle measurements, were performed on the materials. They found a decrease in the contact angle by acting on either the fibers (by the alkali treatment) or the matrix, by increasing the content of the maleic anhydride-PP copolymer. Another technique that was used to increase the compatibility of jute fiber with a PP matrix was the oxidation of the raw jute fiber and a posttreatment of the composites with urea after extruder and injection molding [8]; posttreated materials showed increased mechanical properties compared with oxidized and raw specimens. Other works showed improvements of the properties of the materials obtained for cellulose fiber and wood fiber-PP composites caused by surface treatments [9–10].

Along with the interfacial properties of the natural fiber-reinforced composite materials, their mechanical properties, closely related to what has been mentioned earlier, are probably the most studied in the literature. Some of the aspects and tensile properties of composite materials manufactured using PP as a matrix and natural fibers as reinforcement are described as follows. Keener et al. [11] used flax fibers as reinforcement (a fixed weight fraction of 30%) of maleated PP to manufacture composites by extrusion and injection molding using coupling agents (Eastman G-3003, Eastman G-3015, and Epolene E-43) in three different amounts (1%, 3%, and 5%). They obtained a maximum tensile strength of 53.4 MPa for a 3% optimal amount of the Eastman G-3015 coupling agent. The effect of different fiber weight fractions on porosity and tensile properties of flax/PP composites produced by filament winding, film stacking, and a vacuum heating process was studied by Madsen and Lilholt [12]. They found tensile strength—axial properties—to be in the range of 251–321 MPa and proved their rule-of-mixtures model as a useful tool to determine the optimal fiber weight fraction for a given matrix. Pickering et al. [6] found for hemp/PP-treated composites an increase in the tensile strength and elastic modulus, in comparison with the untreated material, from 22.8 to 47.2 MPa, and from 1.07 to 4.88 GPa, respectively. For wood fiber/PP composites, an increase of the tensile properties was found by an electron beam process [13]. In general, the use of coupling agents and previous treatments increases the tensile properties of natural fiber-reinforced PP composites.

4.3.2 POLYETHYLENE

Polyethylene is another thermoplastic polymer extensively studied in the literature, together with PP, in manufacturing biocomposites with natural fibers as reinforcement. PS and PVC have been studied as matrices together with natural fibers in a much-reduced mode. This section includes only the information corresponding to the use of high-density polyethylene (HDPE) as a matrix, since the low-density polyethylene (LDPE) has been studied to a lower extent. Tensional and flexural properties for natural fiber/HDPE composites have been studied using a wide range of fibers: flax fibers [14], sisal fibers [15–17], henequen fibers [18], and wood fibers obtained from a Kerbside collection waste [19]. Similar to the composites with a PP matrix, the interfacial properties of composites with an HDPE matrix have been studied in the literature, in many cases using various treatments to improve these properties: sisal fibers [16,20] and henequen fibers [18]. Finally, in the case of HDPE used as a matrix, impact properties have also been significantly studied in the literature. Some studies carried out for different types of natural fibers are as follows: sisal fiber (manufactured by rotomolding) reduced the ability to absorb energy of the neat matrix as the fiber content of the composite was increased [15], and rice straw fiber reduced the impact strength of the composite material, as proved by Yao et al. [21] for composites produced by compression molding (as an example, a reduction from 13.19 kJ·m^{-2} for a recycled unreinforced HDPE to 7.56 and 4.85 kJ·m^{-2} for a 30 and 50 wt% reinforced composite, respectively). In short, natural fiber reinforcement increases the brittleness of the material compared with the unreinforced polymer, and the increase is more intense as the fiber weight fraction of the composite is increased.

4.4 CONVENTIONAL THERMOSETTING MATRICES

Similar to conventional reinforcing fibers (glass and carbon fibers), in the case of natural fibers, thermosets are the other large group, together with thermoplastics, of polymeric matrices that are used in the manufacture of composite materials. Most resins used as matrices in the literature are epoxy, phenolic, and polyester resins.

4.4.1 EPOXY RESINS

This section presents the reported work on various aspects of natural fiber-reinforced epoxy composites and addresses some of the basic issues in the development of such composites. Epoxy resins are one of the most important resins used for fiber-reinforced plastic materials. Different types of natural fibers, matrix, and processing techniques are used for composite fabrication. Natural fibers are grouped into three types: seed hair, bast fibers, and leaf fibers, depending on the source. Some examples are cotton (seed hairs); ramie, jute, and flax (bast fibers); and sisal and abaca (leaf fibers). Of these fibers, jute, ramie, flax, and sisal are the most commonly used fibers for polymer composites. Natural fibers in the form of wood flour have also been used for preparation of natural fiber composites.

4.4.1.1 Synthesis, Main Properties, Processing, and Applications

Epoxy resins are thermosetting materials for which the resin precursors contain at least one epoxy function [22]. The epoxy functions are highly reactive toward diverse functions, such as hydroxyl, leading to extremely versatile materials that range from laminated circuit board, structural carbon fiber composites, electronic component encapsulations, and adhesives [23]. These thermosetting polymers have special advantages such as good properties of tensile strength and low shrinkage after curing, and they are usually used as matrix materials for composites of carbon fiber and aramid. Epoxy resins are usually prepared from compounds containing two or more epoxy groups that have been reacted with amines, anhydrides, or other groups that are capable of opening the epoxy ring and forming thermosetting products [24].

Castan [25] and Schlack [26] are credited with the earliest U.S. patents describing epoxy resin technology. Sylvan [27] further emphasized the use of bisphenols and their reaction with epichlorohydrin to yield diepoxides that are capable of a reaction with crude tall oil resin acids to yield useful resins for coatings. Fred and Lawn reported the use of diepoxide resins that were cured with amines in a U.S. patent in 1956 [28]. The introduction of epoxidation techniques for polyunsaturated natural oils led to industrial interest in the preparation of epoxy compounds that are useful for resin production. Nowadays, almost 90% of the world production of epoxy resins is based on the reaction between Bisphenol A (2,2-bis(4-hydroxyphenyl)propane and epichlorohydrin, yielding diglycidyl ether of Bisphenol A (DGEBA) as is shown in Figure 4.2 [29].

The reaction involves a chlorohydrin intermediate, which is then treated with a base to give the resulting epoxy compound. DGEBA contains one or more Bisphenol A moieties in terms of the applications. In thermosetting materials, epoxies are currently combined with a large range of co-reactants, so-called curing agents, such as amines, anhydrides, and amides [30]. Curing time and temperature to complete the polymerization reaction depend on the type and amount of the curing agent. The curing reaction can also involve homopolymerization catalyzed by Lewis acids or tertiary amines. In most cases, reactions catalyzed by Lewis acids are too fast for practical systems, whereas reactions catalyzed by tertiary amines are too slow and too highly temperature and concentration dependent to make them reliable as the sole curing agent. Accelerators are sometimes added to the liquid mix to speed up a slow reaction and shorten the curing time. Curing reactions result in the formation of tridimensional networks with a broad spectrum of performances, depending on the nature of the curing agent, and the extent and density of cross-linking [29]. The curing reaction that is used to transform liquid resin to the solid state is initiated by adding small amounts of curing agent just before incorporating fibers into the liquid mix.

The use of natural vegetable fibers as reinforcements in polymer composites to replace synthetic fibers such as glass is currently receiving increasing attention

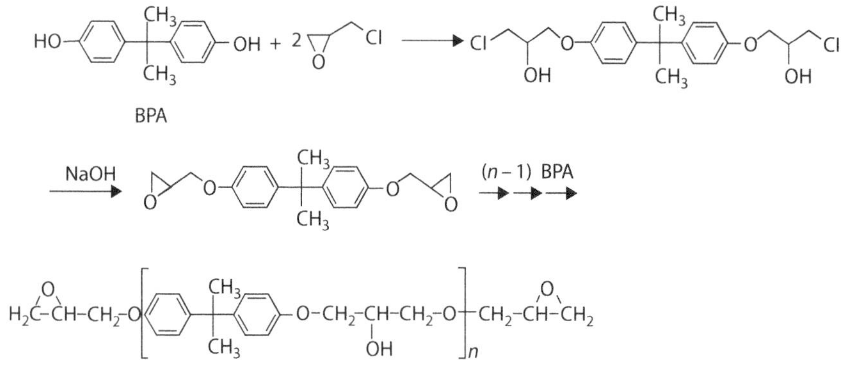

FIGURE 4.2. Synthesis of diglycidyl ether of Bisphenol A.

because of the advantages, including cost effectiveness, low density, high specific strength, as well as their availability as renewable resources [31–32]. The main disadvantages of natural fibers in respective composites are the poor compatibility between the fiber and the matrix and their relatively high moisture absorption [33]. Therefore, natural fiber modifications are considered as leading to a change in the fiber surface properties to improve their adhesion with different matrices. Additional advantages of the use of natural fibers in composites are their renewability, biodegradability, nontoxicity, and good insulation properties. A broad variety of fibers with different thermal and mechanical properties are abundantly synthesized in nature and are available for the development of high-performance composites.

Considering the advantages of natural fiber composite properties compared with glass fiber composites, such as being nonabrasive to equipment and freedom from health problems due to skin irritation during handling and processing, the use of natural fiber composites is preferred. Unfortunately, some drawbacks of natural fibers such as thermal and mechanical degradation during processing make them undesirable for certain applications. Natural fiber-reinforced composites also have several drawbacks such as poor wettability, incompatibility with some polymeric matrices, and high moisture absorption by the fibers.

The role of the matrix in a fiber-reinforced composite is to transfer stresses between the fibers, provide a barrier in adverse environments, and protect the surface of the fibers from mechanical abrasion. The properties of a cured epoxy resin depend principally on the cross-link density. In general, the tensile modulus, glass transition temperature, and thermal stability, as well as chemical resistance, are improved with increasing cross-link density, but the strain-to-failure and fracture toughness are reduced. Factors that control the cross-link density are the chemical structure of the starting liquid resin (e.g., number of epoxide groups per molecule and spacing between epoxide groups), functionality of the curing agent (e.g., number of active hydrogen atoms in diethylenediamine [DETA]), and the reaction conditions. The heat resistance of an epoxy is improved if it contains more aromatic rings in its basic chain [34].

The excellent properties of epoxy resins compared with other thermosets make them one of the best matrix materials for composites: good adhesion; high mechanical properties; low moisture content; little shrinkage after curing; processing ease; absence of volatile matters during cure; excellent resistance to chemicals and solvents; good adhesion to a wide variety of fillers, fibers, and other substrates; and a wide variety of properties (since a large number of starting materials, curing agents, and modifiers are available). The main drawbacks of epoxy resins for industrial use are their brittleness, long cure time, and high cost [34]. The ideal reinforcing material should be able to improve tensile strength, increase fracture toughness, and reduce cost of the composite compared with that for the neat epoxy resin. Properties of the fibers, the aspect ratio of the fibers, and the fiber–matrix interface govern the properties of the composites. The surface adhesion between the fiber and the polymer plays an important role in the transmission of stress from the matrix to the fiber and, thus, contributes toward the performance of the composite.

Numerous reports are available on natural fiber composites, which are composed of natural or synthetic resins that are reinforced with natural fibers. In their review article, Bledzki and Gassan reviewed the reinforcement of the most readily used natural fibers in polymer composites to the year 1999 [35]. More current reinforcement of natural fibers in polymer composites from the years 2000 to 2010 is reported by Faruk et al. [3]. The greatest challenge in working with natural fiber-reinforced plastic composites is their large variation in properties and characteristics. Natural fibers exhibit many advantageous properties; they are a low-density material yielding relatively light-weight composites with high specific properties. These fibers also have significant cost advantages and ease of processing along with being a highly renewable resource.

The interaction between fibers and matrix is important in designing a damage-tolerant structure. The processability and defects in composite materials strongly depend on the physical and thermal characteristics, such as viscosity, melting point, and curing temperature of the matrix [34]. Processing techniques of natural fiber composites are similar to those utilized in processing synthetic fibers. Some factors such as fiber type, fiber content, fiber orientation, and moisture content of fiber significantly influence the processing of natural fiber composites, as well as the properties of the final product [36]. For thermoset composites, the basic and most commonly used method for the manufacture of both small and large reinforced products is the hand layup technique [37]. Fiber reinforcements and resin are placed manually against the mold surface. The layers of materials are placed against the mold to control thickness. In this process, the uniformity of the composite in terms of thickness, fiber-to-matrix ratio, and void content throughout the sample depends on the workmanship skill.

The bag molding process is one of the most versatile processes used in the manufacturing of composite parts. The lamina is laid up in a mold and the resin is spread or coated, covered with a flexible diaphragm or bag, and cured with heat and pressure. After the required curing cycle, the materials become an integrated molded part, which is shaped to the desired configuration. The three basic molding methods involved are pressure bag, vacuum bag, and autoclave [38]. A large number of reinforced thermosetting resin products are made by matched die molding processes such as hot-press compression molding, injection molding, and transfer molding. Molding compounds can be divided into three broad categories: dough molding, sheet molding, and prepregs.

Resin transfer molding (RTM) has the potential of becoming a dominant low-cost process for the fabrication of large, integrated, and high-performance products. One of the major advantages of RTM is the modest requirement on the mold, since relatively low pressures and temperatures are encountered. When excess resin begins to flow from the vent areas of the mold, the resin flow is stopped and the molded component starts being cured. Once the composite develops sufficient strength, it can be removed from the tool and postcured [39]. Oksman [40] used high-quality natural fibers as reinforcements using the RTM processing technique. The fibers were unidirectional (UD) high-quality Arctic Flax, and the matrix was an epoxy resin. The mechanical properties of the composites were compared with those of conventional RTM-manufactured glass fiber composites, traditionally retted UD-flax fiber composites, and the pure epoxy. The results from mechanical testing showed that the (50/50) high-quality ArcticFlax/epoxy composite has a stiffness of about 40 GPa compared with a stiffness of 3.2 GPa for pure epoxy. The same composite has a tensile strength of 280 MPa compared with a strength of 80 MPa for the epoxy. The RTM proves to be a suitable processing technique for natural fiber composites when high-quality laminates are preferred [40].

A few other methods such as centrifugal casting, cold press molding, continuous laminating, encapsulation, filament winding, pultrusion, reinforced reaction injection molding, rotational molding, and vacuum forming are being used for composites, but the use of these methods for natural fiber composites is hardly reported [4]. In thermoset polymers, the fibers are used as UD tapes or mats. These are impregnated with the thermosetting resins and then exposed to high temperature for curing to take place. Pultrusion is an automated process that is used for manufacturing composite materials into continuous, constant cross-section profiles. The composite profile is produced by pulling the reinforcement through a heated dye, which is then mixed with the matrix [37]. Profiles may have high strength and stiffness in the length direction, with fiber content as high as 60%–65% by volume [38]. In vacuum-assisted resin transfer molding (VARTM), the resin is pulled inside under vacuum pressure and mixed with the fibers/fiber mats. Under this condition, the resin impregnation quality in a composite is much better than that fabricated by the hand layup technique and the void content can be kept as minimal as possible [37].

These days, natural fiber-reinforced composites are extensively found in automotive sectors. Natural fiber composite materials and associated design methods are also sufficiently mature to allow their widespread use, for example, as construction materials. A significant research effort is underway to develop natural fiber composite materials and explore their use as construction materials, especially for load-bearing applications [36].

4.4.1.2 Characterization of Mechanical Properties

Tensile, flexural, and impact properties are the most commonly investigated mechanical properties of natural fiber-reinforced composites [36]. Both the matrix and fiber properties are important in improving mechanical properties of the composites. The tensile strength is more sensitive to the matrix properties, whereas the modulus is dependent on the fiber properties [4]. Compared with composites from glass fiber reinforcements, natural fiber composites are of low tensile and low impact strength, which has been a disadvantage in structural applications. This is due to poor adhesion between the fibers and the matrix, and due to the flaws introduced by fiber bundles. There have been efforts to individualize fiber bundles, but the problem of adhesion has remained a key issue and has proved to be a major challenge in the quest for extended utilization of natural fibers in reinforcement [41]. The hydrophilic nature of natural fibers adversely affects adhesion to a hydrophobic matrix and results in loss of strength. In the past decade, numerous studies have analyzed the effect of different natural fibers on the mechanical and thermal properties of their epoxy composites. Other works have evaluated the effect of fiber treatment on the processing methods, the water absorption, and the fracture toughness, as described in this Section.

The effect of the fiber treatment on the mechanical properties of UD sisal-reinforced epoxy composites was reported by Rong et al. [42]. Treatments including alkalization, acetylation, cyanoethylation, the use of silane coupling agent, and heating were carried out to modify the fiber surface and its internal structure.

When the treated fibers were incorporated into an epoxy matrix, mechanical characterization of the laminates revealed the importance of two types of interface: one between fiber bundles and the matrix and the other between the ultimate cells. Ganan et al. [43] evaluated the mechanical and thermal properties of sisal fiber-reinforced epoxy matrix composites as a function of fiber modification by using mercerization and silane treatments. Both treatments clearly enhanced thermal performance and also mechanical properties of fibers, with other physical properties also being modified. Mercerization, above all, when combined with silanization, led to a significant enhancement of the mechanical properties of composites as a consequence of increasing mechanical properties of fibers and improving fiber–matrix adhesion.

Bachtiar et al. [44] studied the effect of alkaline treatment on tensile properties of sugar palm fiber-reinforced epoxy composites. The treatment was carried out using NaOH solutions at two different concentrations and three different soaking times. The composite specimens were tested for tensile property determination. Some fractured specimens were examined under a scanning electron microscope (SEM) to study the microstructure of the materials. From this study, it was found that the alkaline treatment significantly improved the tensile properties of sugar palm fiber-reinforced epoxy composites, particularly the tensile modulus. The hydrophilic nature of the sugar palm fibers was reduced due to this treatment, and, therefore, the interfacial bonding between the matrix and fibers was increased. However, at higher soaking times and alkaline concentrations, the effect of these parameters on tensile strength was not so pronounced because, under these conditions, fiber damages may have been dominant [44].

Bakar et al. [45] studied the mechanical properties of treated and untreated kenaf fiber-reinforced epoxy composites. The properties of kenaf fibers were improved through the alkali treatment process. The composites were prepared in both untreated and treated forms of the fiber with varying concentrations of weight, by using the hot-press method. The results of this work showed that the flexural modulus and strength were improved by 79% and 24.7%, respectively, and the impact strength was improved by 14.7%. Hybrid kenaf/glass fiber-reinforced epoxy composites were manufactured by Davoodi et al. [46] to enhance the desired mechanical properties for car bumper beams as automotive structural components. The tensile strength and modulus exhibited higher values than a typical car bumper beam, which led to the potential utilization of hybrid natural fibers in some car structural components such as bumper beams [46].

Hepworth et al. [47] studied the manufacture of hemp and flax fiber epoxy composites and tested their mechanical properties. By using an 80% volume fraction of flax fibers in epoxy resin (low viscosity), composites with a mean stiffness of 26 GPa and a mean strength of 378 MPa were produced. By reducing processing damage of the plant fibers, the mechanical properties could be increased by 40%. Strips of retted fiber tissue were found to be just as effective for reinforcement as fiber bundles and individual fibers. Two fiber pretreatments were devised to improve adhesion with resins. The first, 6 M urea, was used in natural fiber-epoxy composites, where it increased the stiffness but not the strength. The second pretreatment was a 50% polyvinyl alcohol (PVA) solution, which was cured before the addition of space-filling resin. The PVA treatment improved the stiffness and strength of natural fiber-epoxy composites [47].

Most polymer matrix composites absorb moisture from humid atmospheres. This is usually confined to the resin matrix, but some fibers also absorb moisture. Epoxy resins absorb between 1% and 10% by weight of moisture, so a typical composite with 60% by volume of fibers will absorb around 0.3%–3% moisture, but the systems in widespread use tend to absorb around 1%–2% by weight moisture [48]. In the work by Deo and Acharya [49], the moisture absorption behavior and its effect on mechanical properties of *Lantana camara* fiber-reinforced epoxy composite was studied. These authors prepared composite samples reinforced with different wt% by the hand layup technique. An increase in tensile and flexural strength was observed with an increase up to 30% in fiber content. The moisture diffusivity constant and equilibrium moisture uptake were calculated. Moisture absorption of the studied composites was proved to follow the kinetics of a Fickian diffusion process. Tensile and flexural strength of the composites was found to decrease with moisture absorption. After the mechanical tests, fracture characteristics analysis of the tested specimens was carried out to reveal a reasonable interaction between the reinforcement and the matrix.

Leman et al. [50] calculated the value of Fickian diffusivity constant, the moisture equilibrium content, and the correction factor for a combination of sugar palm fiber and epoxy resins. Tests were carried out on composite plates and two different fiber compositions were chosen, which were 10% and 20% by weight. Pure epoxy plates have been used for the control measures. From this study, plates with 20% fiber loading possessed the highest amount of moisture before the moisture absorption behavior test, which is 0.93%. In the moisture absorption behavior test, the corrected value of Fickian diffusivity constant for the 20% fiber loading was 3.76×10^{-7} mm^2/s, which is the highest among other composites. It is shown that, for composite plates that contain a higher fiber composition, the moisture absorption rate is even higher [50].

Woven sisal textile fiber-reinforced epoxy composites were used by Kim and Seo [51] to evaluate tensile and fracture toughness properties. These authors studied the durability of sisal fiber composites that were aged under distilled water and were exposed to controlled cycles of wetting and drying. The durability of sisal composites was measured as the loss of strength in cyclic time. In general, the tensile strengths of these materials were decreased after the moisture uptake, due to the effect of the water molecules, which change the structure and properties of the fiber, the matrix, and the interface between them. The water absorption characteristics of sisal composites were studied with cycled and aged conditions. Water uptake of the epoxy composites was found to increase with cyclic times. The mechanical properties were dramatically affected by the water absorption cycles. Water-saturated samples present poor mechanical properties such as lower values of maximum strength and extreme elongation [51].

Gonzalez-Murillo and Ansell [52] studied the effect of joint geometry on the strength of natural fiber composite joints. Epoxy-bonded single lap shear joints (SLJ) between henequen and sisal fiber composite elements were manufactured and tested in tension to assess the shear strength of the structural bonds [52].

Ku et al. [53] reviewed the reported works with regard to the effects of fiber loading, chemical treatments, manufacturing techniques, and process parameters on tensile properties of natural fiber-reinforced composites. Mathematical models were also found to be an effective tool to predict the tensile properties of natural fiber-reinforced composites.

4.4.2 Phenolic Resins

In this section, we review recent developments in the area of natural fiber-reinforced phenolic composites. In addition, the mechanical properties of the reinforced phenolic composites are commented on and discussed.

4.4.2.1 Synthesis, Main Properties, and Applications

The materials based on reinforced plastics with fibers began being used in 1908 with cellulose fibers in phenolic matrices and consisted of textile fabrics and "mats" or papers impregnated with the resin that were later cured. Currently, the glass fiber is the reinforcement that is dominant in thermoplastic and thermosetting composites. Doubtlessly, the most important factor that has led to the fast development of the composites is the reduction of weight that can be obtained from the use of low-density fiber, with high elastic module and strength. After decades of using artificial, high-cost, and low ecological fibers, natural fibers have been studied in recent years by to the scientific community, motivated by the search of materials that are compatible with the environment [54–57].

The present investigations show that, in certain composite applications, competitiveness of the natural fibers has demonstrated to be superior to glass fibers [18,58–62]. The advantages that are presented with these fibers are based on their availability, low cost, low density, strength, facility of being superficially modified, being renewable, and the biodegradable CO_2 kidnapping [58,63]. On the other hand, it has been verified that the presence of these fibers reduces the abrasion of the injection and extrusion molds, as well as distributes them in fused polymers. The development and commercialization of materials based on constituents derived from renewable resources would have a great impact, leading to both environmental and economic benefits [64–67]. In this way, some companies, mainly of the automotive sector, have begun using composites reinforced with vegetal fibers, since these are efficient in the absorption of noise and are more resistant and of lower cost than those of glass fibers [62,68–71]. In addition, composites reinforced with natural fibers are lighter, biodegradable, and can be obtained using 68% of the energy used to create glass fiber composites [66]. Currently, a large number of research groups exist, preparing composites with reinforced polymeric materials with vegetable fibers, which improve mechanical properties [35,72–77].

A great amount of natural fibers, such as jute, hemp, flax, kenaf, sisal, pineapple, cotton, oil of palm, and so on, has already been tested as reinforcement of composites [18,59–61,63,74,75,78–82]. Some authors have used leafy coniferous cellulose wood fibers [76,83–86] and even fibers of recycled paper [58]. Other authors used lignin Kraft for the natural fiber treatment [87]. Hybrid composites have also been studied, which contain some synthetic fibers next to natural fibers to increase the strength and reduce the cost [63,88]. The use of another natural compound, the lignin, is also being investigated, for instance, as stuffed in biocomposites and as a coupling agent between natural fibers and the thermosetting matrix [89–90].

The most common thermosetting polymers for natural fiber composites are polyester resins [60,91–98], which are used more than epoxy [42,60,78,99–100]. Few studies still exist on thermosetting polymeric matrices, and less exist with phenolic resins [61–62,74–76,101–102]. The phenolic resins are formulated from phenol and formaldehyde under acid or alkaline conditions. More detailed information is provided elsewhere [103–107]. Nowadays, this polymer is highly toxic due to the emissions. However, despite the emergence of new-generation materials, phenolic resins retain their industrial and commercial interest. What is the intersest of phenolic resins after a century? The main reasons are the excellent stability to high temperatures and, when exposed to fire, flame retardation and low smoke and toxic emissions. If they are reinforced with glass or carbon fibers, their mechanical properties are lower than when these materials are mixed with epoxy resins, but their behavior in the presence of fire is superior, besides their toxic emissions being reduced [108]. For that reason, the systems based on phenolic resins are generally focused on applications with critical requirements of behavior in front of fire, for example, the interiors of the civil airplanes and military.

An important disadvantage in the use of natural fibers in composites is the hydrophilic nature of the fibers in relation to the hydrophobic nature of the matrix, causing a poor adhesion between the fibers and the matrix [60]. In order to improve the adhesion, it is necessary to limit the water absorption of fibers before incorporating them into the matrix [86,109–111], for which some methods are based on the modification of the surface of the fibers [35,62].

The alkaline treatment obtains the maceration of natural fibers. Its effectiveness will depend on the type and concentration of the alkaline dissolution, the time of processing, and the temperature [61,79–81,112–114]. In general, this treatment allows reducing the material cemented in the surface of the fiber and the volatile ones, improving the tension properties. The methods of impregnating the compatible polymers with the matrix also improve the interaction between the fibers and the matrix. Another form that is used to superficially modify fibers consists of the use of coupling agents between the matrix and the fibers, which generally facilitate the transference of "stress" in the interphase and allow improving the properties of adhesion [115–117]. The general mechanism that takes place in the interphase is based on hydrolysis of alcoxisilane, followed by the condensation and formation of connections between the matrix and the fibers or particles. Finally, there is the possibility of other types of methods such as acetylation, the use of potassium permanganate or sodium chloride dissolutions, and the use of acrylic or maleic acid [79,80,118]. This practice, which gives rise to an improvement of the mechanical properties, besides reducing the humidity absorption, is important in certain applications, for example, when the materials are destined to exteriors [63]. Therefore, the superficial modification of fibers can enhance the performance of reinforced phenolic composites to a greater extent than unreinforced phenolic composites, as shown in the next section.

4.4.2.2 Characterization of Mechanical Properties

The properties of a reinforced composite are related to the fiber distribution, voids in the matrix, and good adhesion between the fiber and the matrix. Initial evaluation of structural composites often consists of tensile and flexure tests, and sometimes impact tests are also used, depending on the application. The measured parameters in composites include the Young's modulus and strength. In general, the values obtained for reinforced composites are compared in relation to unreinforced materials. Sreekala et al. [119] studied the tensile properties of 40% oil palm fiber-reinforced phenolic composites and compared the results with reinforced samples, which had been previously modified on the fiber surfaces. The results obtained are shown in Table 4.2.

TABLE 4.2
Mechanical Properties in Tensile Test for Unreinforced and Reinforced Phenolic Composites

Composite	Modulus (GPa)	Strength (MPa)	Ref.
40% oil palm fiber-reinforced phenolic composite			Sreekala et al. [119]
Untreated fiber	1.150	37	
Mercerized fiber	1.300	35	
Acetylated fiber	0.800	19	
Tri-ethoxy silane fiber	0.700	15	
Phenolic composite			Joseph et al. [122]
45% banana fiber reinforced			
Phenolic matrix	0.175	7	
10 mm fiber	0.268	23	
30 mm fiber	0.556	26	
40 mm fiber	0.398	23	
Phenolic composite			Kalia and Kaith [120]
Phenolic matrix	-	35	
5% flax fiber reinforced	-	29	
5% flax-g-poly(MMA) reinforced	-	64	
Phenolic composite			Milanese et al. [121]
Phenolic matrix	-	4.9	
33% woven sisal fiber ($h = 16.5\%$)	-	25.2	
33% woven sisal fiber ($h = 6.7\%$)	-	24.9	
Phenolic composite			Rojo et al. [76]
Phenolic matrix	0.835	13.3	
3% cellulose fiber reinforced	0.888	15.1	
Treated, 1% NaOH	0.868	13.3	
Treated, 5% NaOH	0.661	7.5	
APS silane[a]	0.872	14.7	
AAPS silane[b]	0.870	16.7	

[a] 3-(Amino-propyl) trimethoxysilane.

[b] 3-(2-Aminoethylamino) propyltrimethoxysilane.

The tensile modulus value for mercerized oil palm fiber-reinforced phenolic composite was higher (1.3 GPa) than that for unreinforced phenolic composite due to an improvement in the interfacial properties, although the strength value was similar to that of untreated oil palm fiber-reinforced phenolic composite. This fact is a result of energy absorption after partial failure of the material.

Phenolic composites reinforced with 3% cellulose fibers, untreated and treated, were studied by Rojo et al. [76]. The untreated cellulose fiber-reinforced phenolic composite exhibited better tension properties in relation to phenolic composite due to fibers supporting part of the material stress. The cellulose fibers treated at 1% NaOH (mercerized method) presented a tensile modulus similar to untreated cellulose fiber-reinforced composites. However, cellulose fibers treated at 5% NaOH produced materials with poor properties due to high alkali concentrations. Rojo et al. [76] observed damaged or weak fibers by SEM when these fibers were mercerized with 5% NaOH. In addition, voids were observed around the cellulose fibers, reducing the interfacial adhesion between the fiber and the matrix. When using silanes as coupling agent, the reinforced materials improved the fiber–matrix adhesion (Figure 4.3). It should be noted that a part of the phenolic matrix is adhered to the surface of cellulose fibers due to the coupling agent, that is, the fibers treated with the 3-(amino-propyl) trimethoxysilane (APS) and the 3-(2-aminoethylamino) propyltrimethoxysilane (AAPS) enhance the interface properties.

Kalia and Kaith [120] also found that the compressive strength was improved with the incorporation of flax fibers in the phenolic composite. In the case of tensile strength, 5% flax-g-poly methyl methacrylate (MMA) fiber-reinforced phenolic composite was significantly enhanced as compared to the phenolic composite (Table 4.2). The cause is the incorporation of monomer chains, poly(MMA), to the backbone of flax fiber; in this manner, the grafting decreased the crystallinity of flax fibers, with a resulting reduction in their stiffness and hardness. Milanese et al. [121] observed that the addition of woven sisal fibers to the phenolic matrix also improved the tensile properties. This was attributed to good adhesion between the fibers and the matrix.

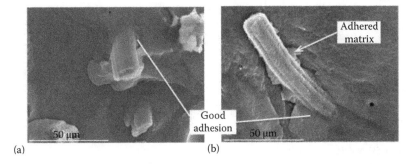

FIGURE 4.3 Scanning electron micrographs of the fiber–matrix interfaces of the phenolic composites: (a) cellulose fibers with APS and (b) cellulose fibers with AAPS. (From Rojo, E. et al., *Polymer Engineering & Science,* 54(10), 2228–2238, 2014 [76]. With permission from Wiley.)

Fiber length also influences the tensile modulus and strength values (Table 4.2). For instance, the modulus of phenolic composite reinforced with 45% banana fibers increased ~52% when the fiber length was increased from 10 to 30 mm [122]. A similar increase in strength was noted when the fiber length was increased. However, a higher fiber length (40 mm) reduces the tensile properties of banana fiber-reinforced phenolic composites, which was attributed to the fiber entanglements formed at higher lengths. The flexural modulus and strength of different phenolic composites are shown in Table 4.3. Sreekala et al. [119] observed that the incorporation of mercerized fibers in the phenolic composite improved the flexion properties similar to the results obtained in tension (Table 4.2). Rojo et al. [76] obtained similar behavior in their samples. It should be noted that reinforced composites with 3% cellulose fiber mercerized at 1% NaOH presented a ~30% superior flexural modulus for unreinforced composites. However, the composites with fiber treated with APS or AAPS exhibit poor mechanical properties due to large voids on the interface. Megiatto et al. [69] observed similar behavior.

TABLE 4.3

Mechanical Properties in Flexural Test for Unreinforced and Reinforced Phenolic Composites

Composite	Modulus (GPa)	Strength (MPa)	Ref.
40% oil palm fiber reinforced phenolic composite			Sreekala et al. [119]
Untreated fiber	3.050	49	
Mercerized fiber	2.950	75	
Acetylated fiber	1.900	36	
Tri-ethoxy silane fiber	1.200	8	
Phenolic composite 45% banana fiber reinforced			Joseph et al. [122]
Phenolic matrix	1.973	10	
10 mm fiber	0.516	25	
30 mm fiber	2.285	50	
40 mm fiber	2.481	50	
Phenolic composite			Rojo et al. [76]
Phenolic matrix	0.6895	24.2	
3% cellulose fiber reinforced	0.6543	23.4	
Treated, 1% NaOH	1.0004	26.9	
Treated, 5% NaOH	0.6566	19.1	
APS silane[a]	0.5963	23.4	
AAPS silane[b]	0.7269	25.1	

[a] 3-(Amino-propyl) trimethoxysilane.

[b] 3-(2-Aminoethylamino) propyltrimethoxysilane.

Joseph et al. [122] studied the influence of fiber length on phenolic composites reinforced with banana fiber. In the reinforced composites, the flexural modulus increases by increasing the length of the fiber. The flexural strength values for banana fiber-reinforced composites are higher than the tensile strength values, as shown in Tables 4.2 and 4.3. The main reason is that banana fiber introduces a plasticizing effect on the phenolic matrix and, as a result, the composite presents more toughness.

Other authors considered that thermally treated composites improve the adhesion fiber matrix, and the interface is stronger in relation to chemical modifications [123]. We believe that most authors are agreeing to modify the fiber surface structure in order to enhance the final properties of the material. However, we cannot forget the need to reduce water absorption of fibers, since this drastically reduces the mechanical properties of composites.

Finally, it should be noted that the incorporation of natural fibers in phenolic composites has beneficial effects on both the environment and the composite characteristics. However, a key factor limiting the manufacture and utilization of natural fiber-reinforced phenolic composites is the challenge in the production of uniform materials. Thus, techniques and methods must be developed for effectively dispersing the reinforcement into the matrix. This necessity is important, because natural fiber loading on the material is low in relation to commercial composites. Another outstanding challenge is the importance of the interface; thus, studies of surface modifications of fibers are required for the fabrication and performance of biocomposites in the future.

4.4.3 Polyester Resins

Polyester resins are, together with those previously described phenolic and epoxy resins, as well as polyurethane resins (which are outside the scope of this chapter), the most commonly used polymeric matrices in the manufacture of composite materials. In this section, some of the methods of synthesis, conventional and novel, main properties, new bio-polyester resins, and most important applications are described. Once the polymeric matrix is properly introduced, the mechanical properties of composite materials obtained when using some of the different types of natural fibers as reinforcement will be exhibited and commented on.

4.4.3.1 Synthesis, Main Properties, and Applications

Polyester resins are a class of unsaturated polyester that originate from the polycondensation reaction of a polyol and either a polyvalent acid or an acid anhydride [29]. Regarding different processes of polymerization of polyesters, both conventional and novel are included the literature review conducted by Pang et al. [124], where they are extensively detailed. Throughout this review, the different raw materials used in the manufacture of polyesters are described: conventional polymerization methods; the first step of prepolymerization and the second step of polycondensation; the polymerization kinetics of the process under study; and the use of catalysts and additives. It also includes a description of novel synthetic methods: cyclic oligoesters and the ring-opening polymerization of cyclic oligoesters.

The curing reactions of unsaturated polyesters, performed by radical or thermal processes with other comonomers, such as unsaturated styrene, result in the production of polyester resins. This curing process has been extensively described by Penczek et al. [125] in the review carried out in his book chapter. Furthermore, in addition to the extensive work done on the use of conventional polyester resins in the manufacture of composites with natural fibers as reinforcement, there is a wide research field for the development and application of new bio-polyester resins as substitutes of conventional polyesters. In these resins, the usual curing sites existing in the backbone of the polyester resin, commonly acid anhydrides, are replaced by renewable-based analogous building blocks (partially or totally); for instance, fatty acids or oils (e.g., soybean) can be used as polyacids and rigid carbohydrates can be used as polyols [126].

The most common current applications for polyester resins are as decorative coatings and in the manufacture of composite materials as a matrix. The produced composites are generally used as industrial equipment: tanks, boat building, production of blades for wind turbines, and so on. [29].

4.4.3.2 Characterization of Mechanical Properties

The polyester resins have been used in the manufacture of composite materials as a matrix with a large number of natural fibers (either modified or not) as reinforcement: flax, sisal, hemp, jute, banana, kenaf, and so on. The mechanical properties of these materials have been extensively studied in the literature. Here, some of the most relevant results are shown.

Burgueño et al. [127] produced cellular sandwich beams and plates using green flax fiber as reinforcement. They used a fiber weight fraction of 33% and cured the resin at two temperatures (100°C and 150°C) for a fixed time of 2 hours each. The elastic modulus and the tensile strength of the composites were 3.37 and 13.31 MPa, respectively. The experimental results were compared with the predictions for the elastic modulus and the strength of the Halpin-Pagano and the Lees models, respectively, obtaining good agreements.

Composite materials using polyester resin as a matrix and UD sisal fiber bundles as reinforcement either after treatment with 0.06 M NaOH solution (64.4% volume fraction) or without treatment (68.2% volume fraction) were produced by Towo and Ansell [128]. The composites were manufactured by using a 50°C, 60-bar, and 20-minute cure cycle followed by postcuring at 80°C overnight. The fatigue properties of the resulting composites were studied. Those materials containing alkali-treated fiber bundles have better mechanical properties than those with untreated fiber bundles.

Singh et al. [129] studied the effect of treated fibers on the moisture resistance. Fibers were treated with silane, titanate, and zirconate coupling agents and N-substituted methacrylamide. The resulting composites presented better properties in humid as well as dry conditions. Hemp fiber-reinforced polyester composite materials were produced by RTM [130]. Tensile and flexural properties of the resulting composite were found to increase when the fiber content was increased (reaching a maximum of 35% fiber volume fraction). They did not increase the fiber volume fraction until the maximum predicted by the Madsen et al. model [131].

Dhakal et al. [132] studied the tensile and flexural properties of composite materials manufactured using polyester resin as a matrix and nonwoven hemp fiber with a volume fraction from 0% to 26% as reinforcement. Composites were produced from preimpregnated hemp mats by conducting a pre-curing stage at 22°C, 10 bars, for 1.5 hours, and a postcuring stage at 80°C for 3 hours. They investigated the effect of water absorption on mechanical properties at 25°C and 100°C for different immersion times. In general, they found that higher cellulose content (higher fiber volume fraction) increased the moisture uptake of the composites. The tensile and flexural properties were decreased as the percentage of moisture uptake increased. The effect of moisture at high temperatures induced degradation of composite samples.

Tensile, flexural, impact, and shear properties of a jute/polyester composite were investigated by Gowda et al. [133]. A 45% volume fraction of hand-laden jute fabric was used as reinforcement. Tensile strength, modulus, and Poisson's ratio for the composite were 60 MPa, 7 GPa, and 0.8 GPa, respectively; 92.5 MPa flexural strength and 5.1 GPa flexural modulus were found for the resulting composite; and the average impact energy per unit area and the interlaminar shear strength of the composites were 29 kJ·m^{-2} and 10 MPa, respectively.

Tobias and Ibarra [134] studied the effect of cure temperature on flexural strength of different polyester-based composites using natural fiber reinforcement. Regarding temperature influence on mechanical properties, they found that the flexural strength increased with increasing cure temperature. Abaca fibers showed the highest strength when compared with banana and rice hull fibers for polyester composites.

4.5 BIO-BASED MATRICES

The manufacture of composite materials using bio-based polymers (from renewable resources) is of particular interest due to causes such as concern about the environment, climate change, and the scarcity of fossil fuels. This interest is increased if the used matrices are biodegradable, and has increased the scientific effort in their study in recent years for either thermoset or thermoplastic bio-based polymers [2]. The polymers most commonly used as bio-matrices together with natural fibers as reinforcements are starch (a naturally occurring polymer); polylactic acid (PLA), using raw materials for its manufacturing of lactic acid, the monomer obtained from the fermentation of sugar, and polyhydroxy-alkanoate (PHA), produced from vegetable oils. Other bio-based polymers that have been studied as matrices are soy-based resin (from soy oil), polycaprolactone (PCL), and polybutylene (PBS), to a lesser extent than the first ones [3,36].

4.5.1 SUSTAINABILITY OF BIO-BASED MATRICES

Bio-based polymeric matrices have a number of advantages and disadvantages beside those that are characteristics of the use of natural fibers as reinforcement in the manufacturing of composite materials. In Table 4.4, the main advantages and disadvantages of using bio-based polymers are shown according to Faruk et al. [36]. As can be observed in Table 4.4, some of the advantages presented for the bio-based polymer matrices are highly related to the disadvantages; for instance, the statement that bio-based polymer matrices are more sustainable and eco-friendly than conventional petroleum-based matrices is refuted by opposite statements included in the list of disadvantages. Recently, this has been a hot topic and an intense debate

TABLE 4.4

Advantages and Disadvantages of Bio-Based Polymer Matrices in Relation to Conventional Matrices

Advantages	Disadvantages
Bio-based matrices are more sustainable due to their ability to help in reducing dependency on oil as substitutes of conventional oil-based matrices	Usually, bio-based polymer matrices present lesser performance properties when compared with conventional matrices (petroleum based)
Bio-based matrices are made from renewable sources	Some of them present lower lifetime/durability than conventional matrices
Many bio-based matrices are fully biodegradable; produce more environmentally friendly materials	The manufacturing process significantly relies on energy obtained from petroleum or other nonrenewable energy sources
Bio-based matrices are used and promoted for recycled products	The environmental impact of bio-based polymer matrices has to be individually studied and compared with other matrices; their renewable sources do not necessarily imply a lower environmental impact than other matrices
The rising prices of oil have increased their competitiveness	There are concerns that bio-based matrices will upset existing recycling methods
The intense research on these polymers is leading to improved production techniques, and the manufactured products are more environmentally friendly	Their costs are still high compared with fossil-based matrices. Its market is fairly new; in the future, their costs will probably be reduced due to high-scale production economy

Source: Faruk, O. et al., *Macromolecular Materials and Engineering*, 299(1), 9–26, 2014 [36].

and research has been conducted to clarify this fact [135–137]. The most used tool in the literature to reach arguments that positively prove the higher sustainability of bio-based matrices is probably life cycle assessments (LCA).

Surprisingly—because of the source of these matrices—in the literature, some studies show that there is no clear advantage of bio-based matrices compared with petroleum-based polymers. In their LCA study of a bio-based matrix (in this case a bio-epoxy) used to manufacture a sandwich composite with natural fibers as reinforcement versus an epoxy/glass-fiber sandwich, La Rosa et al. [136] concluded that bio-based polymers show favorable results in terms of environmental impacts and energy use. However, they clearly specify that the obtained results are highly dependent on the system and boundary conditions considered.

Furthermore, the literature review conducted by Hottle et al. [135] for the most common three types of bio-based polymer matrices, PLA, PHA, and thermoplastic starch (TPS), compared with five petroleum-based matrices, shows similar results for the estimated impacts for both. This review reveals that most of the LCAs carried out are focused on the impact on the global warming potential (GWP) and fossil resource depletion, ignoring other impacts that can be important. As an improvement, Hottle et al. [135] suggested the inclusion of end-of-life (EOL) analysis in LCA to obtain more accurate results and therefore clarify the comparison between the sustainability of bio-based and petroleum-based polymer matrices.

Finally, due to the difficulty and the particularities of bio-based matrices, existing methodologies for conducting LCA have been reviewed and a number of recommendations and considerations proposed that, always taking into account the peculiarities of the matrix under analysis, obtain more accurate results as well as a better understanding of the environmental impacts from the production, use, and disposal of bio-based matrices [137]. However, the current limited availability of reliable data and the complexity of the system under study remain key issues, which make the sustainability of these matrices a subject of ongoing study along with the continuous review of the methodologies.

4.5.2 POLY(LACTIC ACID) OR POLYLACTIDE (PLA)

PLA is probably the most used bio-based polymer matrix and also one of the most promising as a substitute for conventional polymeric matrices. It is a bio-based thermoplastic polymer obtained from a renewable and biodegradable resource that has been utilized as a matrix in manufacturing biocomposites using natural fibers as reinforcement.

For man-made cellulose and abaca used as reinforcement in the manufacture of composite materials, Bledzki et al. [138] compared the mechanical properties obtained using PLA and PP as matrices. To produce the composites, they used a combined molding technology: two-step extrusion coating followed by injection molding. The results showed that man-made cellulose (30 wt%) increased the tensile strength and modulus of the composite material compared with the neat polymer, and abaca fibers, in the same weight fraction; it also enhanced the final properties of the material (man-made cellulose showed a better modulus performance than abaca, and so did abaca for the tensile strength compared with man-made cellulose reinforcement). PLA showed better mechanical performance than PP; however, impact properties of PLA composites were lower in the case of PLA/abaca composites than when PP was used as a matrix. In a later work, Bledzki and Jaszkiewicz [139] compared the mechanical properties of PLA biocomposites (they also studied poly(3-hydroxybutyrate-co-3-hydroxyvalerate, PHBV, for the comparison in this work) with PP conventional composites using three different natural fibers: man-made cellulose, jute, and abaca. The processing of composites was by injection molding. The PLA/jute composites showed the best mechanical properties of the tested composites. However, at room temperature, PP exhibited a slightly better impact strength performance than PLA.

PLA has also been studied as a matrix with flax fibers as reinforcement. Oksman et al. [140] produced composites using 30 and 40 wt% flax fiber reinforcement by extrusion. They studied the mechanical properties of the manufactured composites and also analyzed the effect of different amounts of a plasticizer (triacetin glycerol triacetate ester) while looking for an increase in the impact properties of the material. The flax/PLA composite showed enhanced mechanical behavior (50% better strength) compared with similar flax/PP composites. The plasticizer did not show any positive effect on the impact properties of the material, but it reduced the tensile stress and modulus of the composite as its weight fraction was increased. Bax and Müssig [141] found similar results for flax/PLA composites produced by web laying, shredding, and injection molding processing, and they proved that the impact strength increased with increasing weight fiber fractions.

In the literature, PLA has been used to produce biocomposites reinforced with other natural fiber such as jute [142–144], kenaf [145–147], coir [148–149], and bamboo [150]. Moreover, the influence on the final properties of the material of chemical treatments, manufacturing techniques, coupling agents, and so on has been an object of deep research [3].

4.5.3 STARCH

Starch is a bio-based TPS commonly used in the manufacture of biocomposites. The properties of these composites have been widely studied and, as in the case of PLA, they have been used to a great extent together with natural fibers used as reinforcement. Some of these fibers are as follows: kenaf, producing composites by press-forming [151] and together with PCL by injection molding [152]; and sisal, manufacturing biocomposites by mixing starch with the fibers and evaluating their melt rheological behavior, thermal stability/degradation, and creep properties [153–155]. Alvarez et al. [156] also studied the tensile and fracture properties of TPS/sisal short fibers processed by injection molding. Kuciel and Liber-Knec [157] tested the mechanical properties and the effect of water absorption in sisal fiber-reinforced composites. The tensile and flexural strengths of the composites reinforced with hemp fiber increased with increasing fiber content up to 70%: 365 and 223 MPa, respectively, according to Ochi [158].

4.5.4 FURAN RESINS

Polyfurfuryl alcohol resins, also called "furan resins," have a long history for their use in the production of metal-casting cores and molds, corrosion-resistant coatings, and wood adhesives [159]. Furan resins are produced using furfural derived from agricultural by-products such as corncobs and sugar cane bagasse, and an acid catalyst is used to carry out the curing of the resin. Recently, furan resins for composites have been under development in a number of European research and development projects (BIOCOMP, NATEX, and WOODY). These are a new type of bio resin that is intended for composites with an attractive profile of being eco-friendly, low cost, and highly fire resistant. Furan resins, however, present a key issue with respect to composite manufacturing. The existence of solvent water in the resins, in addition to the water generated by condensation during the curing reactions [159], potentially results in materials with low dimensional stability and nonnegligible porosity content. This needs to be taken into account in the selection of manufacturing techniques and conditions.

The curing process of furan resins has been studied from a chemical point of view to determine the involved cross-linking reactions. The curing mechanism for furan resin was originally proposed by Dunlop and Peters [160] and involves two reactions: first, condensation reactions of −OH groups to obtain methylene linkages, and second, condensation reactions of pairs of −OH groups to obtain dimethyl ether linkages. Choura et al. [161] proposed a more complex mechanism that could explain the observed color change of the resin during curing, by the formation of linear compounds with a high degree of conjugation. Several possible mechanisms were also suggested for branching of the polymer chains such as Diels-Alder reactions between furan rings (dienes) in oligomeric molecules and conjugated dihydrofuranic sequences.

The study of the curing kinetics of resins is important for the development and optimization of the cure cycles used in the manufacturing of composites. The curing kinetics of both the monomer furfuryl alcohol and the oligomer polyfurfuryl alcohol have been studied [162–163], as well as some blends of furan resins [162], using different thermal analysis techniques. Different catalysts, such as maleic anhydride, ferric chloride hexahydrate, phosphoric acid, and p-toluenesulfonic acid, have been used for the curing process [162–168]. In the study by Domínguez et al. [165], the amount of catalyst was used as an experimental variable, and the curing kinetics of the resin was studied by differential scanning calorimetry (DSC) nonisothermal measurements. The maximum activation energy was found to be about 115, 95, and 80 kJ·mol^{-1} before the gelation point for 2%, 4%, and 6% w/w amounts of catalyst, respectively. Based on a purely kinetic criterion, the most suitable amount of catalyst was assessed to be 4% w/w. Figure 4.4a shows an example of the determined curing degree of a furan resin as a function of temperature. The flow behavior and the viscosity dependence on temperature of furan resin have been studied by rheology. In the study by Domínguez and Madsen [169], an Arrhenius chemorheological model was applied to analyze the evolution of the complex viscosity of the resin during the pre-gel curing stage. Figure 4.4b shows an example of the determined complex viscosity of a furan resin as a function of time, under different isothermal conditions. In general, the good agreement between the model predictions and the experimental data in the two diagrams in Figure 4.4 indicates the usefulness of the models in the design of cure cycles for the manufacturing of composite materials with a furan matrix.

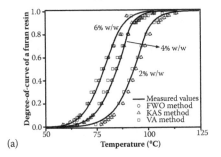
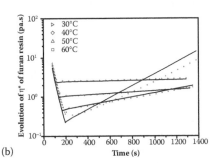

(a)   (b)

FIGURE 4.4 (a) Experimental data and model predictions of curing degree of a furan resin with different amounts of catalyst, as a function of temperature. Adapted from Domínguez et al. [165]. (b) Experimental data and Arrhenius model predictions of complex viscosity of a furan resin under different isothermal conditions. Adapted from Domínguez and Madsen. (From Domínguez, J. C. and Madsen, B., *Industrial Crops and Products, 52*, 321–328, 2014 [169].)

The mechanical properties of fiber-reinforced furan composites have been examined in a number of studies. In the study by Crossley et al. [170], flax and glass fiber composites were manufactured by the vacuum infusion technique using furan, phenolic, polyester, and epoxy resins as a matrix. Neat resin materials were manufactured to demonstrate that the furan has a comparable modulus, and a slightly lower strength than the conventional resins of polyester and epoxy, with values of 2.8 versus 3.1 GPa, and 30 versus 50 MPa, respectively (see Table 4.5). In the study by Pohl et al. [171], flax and glass fiber-reinforced furan composites were manufactured by a special adapted compression molding technique allowing water vapor to be removed from the furan resin during cyclic opening of the mold. The biaxial flax fiber composites were tested in two directions showing the expected larger modulus and strength in the 0°/90° testing direction compared with the ±45° testing direction, with values of 13.0 versus 7.7 GPa, and 70 versus 46 MPa, respectively (see Table 4.5). In the study by Domínguez and Madsen [172], glass fiber-reinforced furan composites were manufactured by the double-vacuum-bag technique using a two-stage cure cycle, and controlling the pressure difference between the two vacuum bags to allow removal of water from the resin. The fiber content of the composites was varied, showing the expected larger modulus at a higher fiber content, with values of 8.4 and 7.5 GPa for 31 and 41 vol%, respectively (see Table 4.5).

The fire resistance of furan resin is excellent compared with conventional resins. In the study by Crossley et al. [170], the fire resistance of flax composites with matrices of furan, phenolic, polyester, and epoxy was tested by the UL94 standard. In the horizontal test, the furan and phenolic composites achieved the highest C class rating. In the vertical test, the furan composite, however, failed, presumably due to the presence of microvoids running parallel to the fibers. In the study by Pohl et al. [171], the fire resistance of flax-reinforced furan composites was tested by the FMVSS 302 standard and passed the test successfully.

TABLE 4.5
Mechanical Properties for Unreinforced and Reinforced Furan Composites

Composite	Modulus (GPa)	Strength (MPa)	Ref.
Neat resin			Crossley et al. [170]
Furan	2.8	29	
Phenol	2.5	33	
Polyester	3.1	46	
Epoxy	3.1	53	
Biaxial flax fiber-reinforced furan composites, 53 vol%			Pohl et al. [171]
±45° testing direction	7.7	46	
0°/90° testing direction	13.0	70	
Biaxial glass fiber-reinforced furan composites, ±45° testing direction			Domínguez and Madsen [172]
31 vol% fibers	7.5	55	
41 vol% fibers	8.4	47	

4.6 CONCLUSIONS

A variety of matrices has been used in the manufacture of composite materials reinforced with natural fibers. Polymeric matrices, with thermoplastic and thermosetting polymers, are the most commonly used. Among these matrices, both conventional oil-based polymers and new bio polymers, with most of them still under development, have been used in the manufacture of fiber-reinforced composites.

The main drawback of natural fibers when used as reinforcement together with conventional polymeric matrices to produce composites lies in their hydrophilic nature, a high moisture content that might result in composites with a higher void fraction than synthetic fiber composites and therefore lower mechanical properties. Hence, a major part of recent research efforts on natural fiber composites has been directed toward fiber and/or matrix modification to improve their mutual compatibility, decrease moisture absorption, and enhance mechanical properties of the resultant composite materials.

Bio polymer matrices pose an even greater challenge, since in this particular case the use of both the reinforcement and the matrix increases the difficulties of manufacturing composites. However, in many cases, when considering the real requirements of end users, for example the required mechanical properties, these might be lower than those offered by conventional matrix composites. Thus, composites manufactured using natural fibers and bio matrices (fully biocomposites or biocomposites) may be used as substitutes for conventional materials as they have a lower environmental impact according to the environmental awareness of modern society.

REFERENCES

1. Verma, D., P. C. Gope, M. K. Maheshwari, and R. K. Sharma. 2012. Bagasse fiber composites: A review. *Journal of Materials and Environmental Science* 3 (6):1079–1092.
2. Sahari, J. and S. M. Sapuan. 2011. Natural fiber reinforced biodegradable polymer composites. *Reviews on Advanced Materials Science* 30:166–174.
3. Faruk, O., A. K. Bledzki, H. P. Fink, and M. Sain. 2012. Biocomposites reinforced with natural fibers: 2000–2010. *Progress in Polymer Science* 37 (11):1552–1596.
4. Saheb, D. N. and J. P. Jog. 1999. Natural fiber polymer composites: A review. *Advances in Polymer Technology* 18 (4):351–363.
5. Cantero, G., A. Arbelaiz, R. Llano-Ponte, and I. Mondragon. 2003. Effects of fiber treatment on wettability and mechanical behaviour of flax/polypropylene composites. *Composites Science and Technology* 63 (9):1247–1254.
6. Pickering, K. L., G. W. Beckermann, S. N. Alam, and N. J. Foreman. 2007. Optimising industrial hemp fiber for composites. *Composites Part A: Applied Science and Manufacturing* 38 (2):461–468.
7. Park, J. M., S. T. Quang, B. S. Hwang, and K. L. DeVries. 2006. Interfacial evaluation of modified Jute and Hemp fibers/polypropylene (PP)-maleic anhydride polypropylene copolymers (PP-MAPP) composites using micromechanical technique and nondestructive acoustic emission. *Composites Science and Technology* 66 (15):2686–2699.
8. Rahman, M. R., M. Hasan, M. M. Huque, and M. N. Islam. 2010. Physico-mechanical properties of jute fiber reinforced polypropylene composites. *Journal of Reinforced Plastics and Composites* 29 (3):445–455.
9. Bataille, P., L. Ricard, and S. Sapieha. 1989. Effects of cellulose fibers in polypropylene composites. *Polymer Composites* 10 (2):103–108.

10. Sain, M. M., B. V. Kokta, and C. Imbert. 1994. Structure-property relationships of wood fiber-filled polypropylene composite. *Polymer-Plastics Technology and Engineering* 33 (1):89–104.

11. Keener, T. J., R. K. Stuart, and T. K. Brown. 2004. Maleated coupling agents for natural fiber composites. *Composites Part A: Applied Science and Manufacturing* 35 (3):357–362.

12. Madsen, B. and H. Lilholt. 2003. Physical and mechanical properties of unidirectional plant fiber composites: An evaluation of the influence of porosity. *Composites Science and Technology* 63 (9):1265–1272.

13. Czvikovszky, T. 1996. Electron-beam processing of wood fiber reinforced polypropylene. *Radiation Physics and Chemistry* 47 (3):425–430.

14. Singleton, A. C. N., C. A. Baillie, P. W. R. Beaumont, and T. Peijs. 2003. On the mechanical properties, deformation and fracture of a natural fiber/recycled polymer composite. *Composites Part B: Engineering* 34 (6):519–526.

15. Torres, F. G. and C. L. Aragon. 2006. Final product testing of rotational moulded natural fiber-reinforced polyethylene. *Polymer Testing* 25 (4):568–577.

16. Choudhury, A. 2008. Isothermal crystallization and mechanical behavior of ionomer treated sisal/HDPE composites. *Materials Science and Engineering: A* 491 (1–2):492–500.

17. Joseph, K., S. Thomas, and C. Pavithran. 1996. Effect of chemical treatment on the tensile properties of short sisal fiber-reinforced polyethylene composites. *Polymer* 37 (23):5139–5149.

18. Herrera-Franco, P. J. and A. Valadez-González. 2005. A study of the mechanical properties of short natural-fiber reinforced composites. *Composites Part B: Engineering* 36 (8):597–608.

19. Jayaraman, K. and D. Bhattacharyya. 2004. Mechanical performance of woodfiber–waste plastic composite materials. *Resources, Conservation and Recycling* 41 (4):307–319.

20. Li, Y., C. Hu, and Y. Yu. 2008. Interfacial studies of sisal fiber reinforced high density polyethylene (HDPE) composites. *Composites Part A: Applied Science and Manufacturing* 39 (4):570–578.

21. Yao, F., Q. Wu, Y. Lei, and Y. Xu. 2008. Rice straw fiber-reinforced high-density polyethylene composite: Effect of fiber type and loading. *Industrial Crops and Products* 28 (1):63–72.

22. Ashcroft, W. R. and B. Ellis. 1993. *Chemistry and Technology of Epoxy Resins.* Glasgow, UK: Blackie Academic & Professional.

23. Puglisi, J. S. and M. A. Chaudhari. 1988. "Epoxies (EP)." In *Engineered Materials Handbook*, ed. J. N. Epel, J. M. Margolis, S. Newman, and R. B. Seymour, 240–242. Metals Park, OH: CRC.

24. Sandler, S. R., W. Karo, J. Bonesteel, and E. M. Pearce. 1998. *Polymer Synthesis and Characterization: A Laboratory Manual.* San Diego, CA: Academic Press.

25. Castan, P. 1943. Process of preparing synthetic resins. US2324483 A.

26. Schlack, P. 1938. Manufacture of organic compounds containing basic substituents. US2131120 A.

27. Sylvan, O. G. 1950. Tall oil esters. US2493486 A.

28. Fred, W. and R. J. Lawn. 1956. Coating compositions containing coal tar pitch and an epoxy ether resin. US2765288 A.

29. Raquez, J. M., M. Deléglise, M. F. Lacrampe, and P. Krawczak. 2010. Thermosetting (bio)materials derived from renewable resources: A critical review. *Progress in Polymer Science* 35 (4):487–509.

30. Carfagna, C., E. Amendola, and M. Giamberini. 1997. Liquid crystalline epoxy based thermosetting polymers. *Progress in Polymer Science* 22 (8):1607–1647.

31. Satyanarayana, K. G., G. G. C. Arizaga, and F. Wypych. 2009. Biodegradable composites based on lignocellulosic fibers: An overview. *Progress in Polymer Science* 34 (9):982–1021.

32. Summerscales, J., N. P. J. Dissanayake, A. S. Virk, and W. Hall. 2010. A review of bast fibers and their composites. Part 1–Fibers as reinforcements. *Composites Part A: Applied Science and Manufacturing* 41 (10):1329–1335.

33. Ghali, L. H., M. Aloui, M. Zidi, H. B. Daly, S. Msahli, and F. Sakli. 2011. Effect of chemical modification of *Luffa cylindrica* fibers on the mechanical and hygrothermal behaviours of polyester/luffa composites. *BioResources* 6 (4):3836–3849.

34. Mallick, P. K. 1993. *Fiber-Reinforced Composites: Materials, Manufacturing, and Design.* Boca Raton, FL: CRC press.

35. Bledzki, A. K. and J. Gassan. 1999. Composites reinforced with cellulose based fibers. *Progress in Polymer Science* 24 (2):221–274.

36. Faruk, O., A. K. Bledzki, H. P. Fink, and M. Sain. 2014. Progress report on natural fiber reinforced composites. *Macromolecular Materials and Engineering* 299 (1):9–26.

37. Kabir, M. M., H. Wang, K. T. Lau, and F. Cardona. 2012. Chemical treatments on plant-based natural fiber reinforced polymer composites: An overview. *Composites Part B: Engineering* 43 (7):2883–2892.

38. Thomas, S. and L. A. Pothan. 2009. *Natural Fiber Reinforced Polymer Composites: From Macro to Nanoscale.* Paris: Editions des archives contemporaines.

39. Wallenberger, F. T. and N. E. Weston. 2004. *Natural Fibers, Plastics and Composites.* New York, NY: Springer.

40. Oksman, K. 2001. High quality flax fiber composites manufactured by the resin transfer moulding process. *Journal of Reinforced Plastics and Composites* 20 (7):621–627.

41. Bhowmick, M., S. Mukhopadhyay, and R. Alagirusamy. 2012. Mechanical properties of natural fiber-reinforced composites. *Textile Progress* 44 (2):85–140.

42. Rong, M. Z., M. Q. Zhang, Y. Liu, G. C. Yang, and H. M. Zeng. 2001. The effect of fiber treatment on the mechanical properties of unidirectional sisal-reinforced epoxy composites. *Composites Science and Technology* 61 (10):1437–1447.

43. Ganan, P., S. Garbizu, R. Llano-Ponte, and I. Mondragon. 2005. Surface modification of sisal fibers: Effects on the mechanical and thermal properties of their epoxy composites. *Polymer Composites* 26 (2):121–127.

44. Bachtiar, D., S. M. Sapuan, and M. M. Hamdan. 2008. The effect of alkaline treatment on tensile properties of sugar palm fiber reinforced epoxy composites. *Materials & Design* 29 (7):1285–1290.

45. Bakar, A., S. Ahmad, and W. Kuntjoro. 2010. The mechanical properties of treated and untreated kenaf fiber reinforced epoxy composite. *Journal of Biobased Materials and Bioenergy* 4 (2):159–163.

46. Davoodi, M. M., S. M. Sapuan, D. Ahmad, A. Ali, A. Khalina, and M. Jonoobi. 2010. Mechanical properties of hybrid kenaf/glass reinforced epoxy composite for passenger car bumper beam. *Materials & Design* 31 (10):4927–4932.

47. Hepworth, D. G., D. M. Bruce, J. F. V. Vincent, and G. Jeronimidis. 2000. The manufacture and mechanical testing of thermosetting natural fiber composites. *Journal of Materials Science* 35 (2):293–298.

48. Matthews, F. L., and R. D. Rawlings. 1994. *Composite Materials: Engineering and Science.* London: Chapman & Hall.

49. Deo, C. and S. K. Acharya. 2010. Effect of moisture absorption on mechanical properties of chopped natural fiber reinforced epoxy composite. *Journal of Reinforced Plastics and Composites* 29 (16):2513–2521.

50. Leman, Z., S. M. Sapuan, A. M. Saifol, M. A. Maleque, and M. M. H. M. Ahmad. 2008. Moisture absorption behavior of sugar palm fiber reinforced epoxy composites. *Materials & Design* 29 (8):1666–1670.

51. Kim, H. J. and D. W. Seo. 2006. Effect of water absorption fatigue on mechanical properties of sisal textile-reinforced composites. *International Journal of Fatigue* 28 (10):1307–1314.

52. Gonzalez-Murillo, C., and M. P. Ansell. 2010. Co-cured in-line joints for natural fiber composites. *Composites Science and Technology* 70 (3):442–449.

53. Ku, H., H. Wang, N. Pattarachaiyakoop, and M. Trada. 2011. A review on the tensile properties of natural fiber reinforced polymer composites. *Composites Part B: Engineering* 42 (4):856–873.

54. Athijayamani, A., M. Thiruchitrambalam, U. Natarajan, and B. Pazhanivel. 2010. Influence of alkali-treated fibers on the mechanical properties and machinability of roselle and sisal fiber hybrid polyester composite. *Polymer Composites* 31 (4):723–731.

55. Idicula, M., P. A. Sreekumar, K. Joseph, and S. Thomas. 2009. Natural fiber hybrid composites: A comparison between compression molding and resin transfer molding. *Polymer Composites* 30 (10):1417–1425.

56. Young, R. A. 1997. "Utilization of natural fibers: characterization, modification and applications." In *Lignocellulosic-Plastic Composites*. 1–21. São Paulo: USP & UNESP.

57. Frollini, E., J. M. F. Paiva, W. G. Trindade, I. A. T. Razer, and S. P. Tita. 2004. "Plastics and composites from lignin phenols." In *Natural Fibers, Plastics and Composites*, ed. F. T. Wallenberger and N. Weston, 193–225. New York, NY: Kluwer Academic Publishers.

58. Huda, M. S., L. T. Drzal, A. K. Mohanty, and M. Misra. 2006. Chopped glass and recycled newspaper as reinforcement fibers in injection molded poly(lactic acid) (PLA) composites: A comparative study. *Composites Science and Technology* 66 (11–12):1813–1824.

59. Idicula, M., A. Boudenne, L. Umadevi, L. Ibos, Y. Candau, and S. Thomas. 2006. Thermophysical properties of natural fiber reinforced polyester composites. *Composites Science and Technology* 66 (15):2719–2725.

60. Kaddami, H., A. Dufresne, B. Khelifi, A. Bendahou, M. Taourirte, M. Raihane, N. Issartel, H. Sautereau, J. F. Gérard, and N. Sami. 2006. Short palm tree fibers: Thermoset matrices composites. *Composites Part A: Applied Science and Manufacturing* 37 (9):1413–1422.

61. Razera, I. A. T. and E. Frollini. 2004. Composites based on jute fibers and phenolic matrices: Properties of fibers and composites. *Journal of Applied Polymer Science* 91 (2):1077–1085.

62. Trindade, W. G., W. Hoareau, J. D. Megiatto, I. A. T. Razera, A. Castellan, and E. Frollini. 2005. Thermoset phenolic matrices reinforced with unmodified and surface-grafted furfuryl alcohol sugar cane bagasse and curaua fibers: Properties of fibers and composites. *Biomacromolecules* 6 (5):2485–2496.

63. Mohanty, A. K., M. Misra, and L. T. Drzal. 2005. *Natural Fibers, Biopolymers, and Biocomposites*. Boca Raton, FL: CRC Press.

64. El-Saied, H., A. I. El-Diwany, A. H. Basta, N. A. Atwa, and D. E. El-Ghwas. 2008. Production and characterization of economical bacterial cellulose. *Bioresources* 3 (4):1196–1217.

65. Joshi, S. V., L. T. Drzal, A. K. Mohanty, and S. Arora. 2004. Are natural fiber composites environmentally superior to glass fiber reinforced composites? *Composites Part A: Applied Science and Manufacturing* 35 (3):371–376.

66. Mohanty, A. K., A. Wibowo, M. Misra, and L. T. Drzal. 2004. Effect of process engineering on the performance of natural fiber reinforced cellulose acetate biocomposites. *Composites Part A: Applied Science and Manufacturing* 35 (3):363–370.

67. Singha, A. S. and V. K. Thakur. 2010. Mechanical, morphological, and thermal characterization of compression-molded polymer biocomposites. *International Journal of Polymer Analysis and Characterization* 15 (2):87–97.

68. Gu, H. 2009. Tensile behaviours of the coir fiber and related composites after NaOH treatment. *Materials & Design* 30 (9):3931–3934.
69. Megiatto, J. D., F. B. Oliveira, D. S. Rosa, C. Gardrat, A. Castellan, and E. Frollini. 2007. Renewable resources as reinforcement of polymeric matrices: Composites based on phenolic thermosets and chemically modified sisal fibers. *Macromolecular Bioscience* 7 (9–10):1121–1131.
70. Riedel, U., and J. Nickel. 1999. Natural fiber-reinforced biopolymers as construction materials: New discoveries. *Die Angewandte Makromolekulare Chemie* 272 (1):34–40.
71. Wötzel, K., R. Wirth, and M. Flake. 1999. Life cycle studies on hemp fiber reinforced components and ABS for automotive parts. *Die Angewandte Makromolekulare Chemie* 272 (1):121–127.
72. Harikumar, K. R., K. Joseph, and S. Thomas. 1999. Jute sack cloth reinforced polypropylene composites: Mechanical and sorption studies. *Journal of Reinforced Plastics and Composites* 18 (4):346–372.
73. Mohanty, A. K., M. Misra, and L. T. Drzal. 2002. Sustainable bio-composites from renewable resources: Opportunities and challenges in the green materials world. *Journal of Polymers and the Environment* 10 (1–2):19–26.
74. Ramires, E. C. and E. Frollini. 2012. Tannin–phenolic resins: Synthesis, characterization, and application as matrix in biobased composites reinforced with sisal fibers. *Composites Part B: Engineering* 43 (7):2851–2860.
75. Ramires, E. C., J. D. Megiatto Jr, C. Gardrat, A. Castellan, and E. Frollini. 2010. Biobased composites from glyoxal–phenolic resins and sisal fibers. *Bioresource Technology* 101 (6):1998–2006.
76. Rojo, E., M. Oliet, M. V. Alonso, B. D. Saz-Orozco, and F. Rodriguez. 2014. Mechanical and interfacial properties of phenolic composites reinforced with treated cellulose fibers. *Polymer Engineering & Science* 54 (10):2228–2238.
77. Sreekala, M. S., S. Thomas, and G. Groeninckx. 2005. Dynamic mechanical properties of oil palm fiber/phenol formaldehyde and oil palm fiber/glass hybrid phenol formaldehyde composites. *Polymer Composites* 26 (3):388–400.
78. Bledzki, A. K., H. P. Fink, and K. Specht. 2004. Unidirectional hemp and flax EP- and PP-composites: Influence of defined fiber treatments. *Journal of Applied Polymer Science* 93 (5):2150–2156.
79. Li, X., S. A. Panigrahi, L. G. Tabil, and W. J. Crerar. 2004. "Flax fiber-reinforced composites and the effect of chemical treatments on their properties." North central ASAE/CSAE Conference.
80. Wielage, B., T. Lampke, H. Utschick, and F. Soergel. 2003. Processing of natural-fiber reinforced polymers and the resulting dynamic–mechanical properties. *Journal of Materials Processing Technology* 139 (1–3):140–146.
81. Aziz, S. H., M. P. Ansell, S. J. Clarke, and S. R. Panteny. 2005. Modified polyester resins for natural fiber composites. *Composites Science and Technology* 65 (3):525–535.
82. Doan, T. T. L., S. L. Gao, and E. Mäder. 2006. Jute/polypropylene composites I. Effect of matrix modification. *Composites Science and Technology* 66 (7):952–963.
83. Abdelmouleh, M., S. Boufi, M. N. Belgacem, A. P. Duarte, A. B. Salah, and A. Gandini. 2004. Modification of cellulosic fibers with functionalised silanes: Development of surface properties. *International Journal of Adhesion and Adhesives* 24 (1):43–54.
84. Bastidas, J. C., R. Venditti, J. Pawlak, R. Gilbert, S. Zauscher, and J. F. Kadla. 2005. Chemical force microscopy of cellulosic fibers. *Carbohydrate Polymers* 62 (4):369–378.
85. Vander Wielen, L. C., M. Östenson, P. Gatenholm, and A. J. Ragauskas. 2006. Surface modification of cellulosic fibers using dielectric-barrier discharge. *Carbohydrate Polymers* 65 (2):179–184.

86. Zhang, W., S. Okubayashi, and T. Bechtold. 2005. Fibrillation tendency of cellulosic fibers: Part 4. Effects of alkali pretreatment of various cellulosic fibers. *Carbohydrate Polymers* 61 (4):427–433.

87. Thielemans, W. and R. P. Wool. 2005. Kraft lignin as fiber treatment for natural fiber-reinforced composites. *Polymer Composites* 26 (5):695–705.

88. Joseph, S., M. S. Sreekala, P. Koshy, and S. Thomas. 2008. Mechanical properties and water sorption behavior of phenol–formaldehyde hybrid composites reinforced with banana fiber and glass fiber. *Journal of Applied Polymer Science* 109 (3):1439–1446.

89. Avérous, L. and F. Le Digabel. 2006. Properties of biocomposites based on lignocellulosic fillers. *Carbohydrate Polymers* 66 (4):480–493.

90. Thielemans, W. and R. P. Wool. 2004. Butyrated kraft lignin as compatibilizing agent for natural fiber reinforced thermoset composites. *Composites Part A: Applied Science and Manufacturing* 35 (3):327–338.

91. Devi, L. U., S. S. Bhagawan, and S. Thomas. 1997. Mechanical properties of pineapple leaf fiber-reinforced polyester composites. *Journal of Applied Polymer Science* 64 (9):1739–1748.

92. Fernandes Jr, V. J., A. S. Araujo, V. M. Fonseca, N. S. Fernandes, and D. R. Silva. 2002. Thermogravimetric evaluation of polyester/sisal flame retarded composite. *Thermochimica acta* 392:71–77.

93. Hassan, M. L. and A. A. Nada. 2003. Utilization of lignocellulosic fibers in molded polyester composites. *Journal of Applied Polymer Science* 87 (4):653–660.

94. Mahdi, E., A. S. M. Hamouda, and A. C. Sen. 2004. Quasi-static crushing behaviour of hybrid and non-hybrid natural fiber composite solid cones. *Composite structures* 66 (1):647–663.

95. Misra, S., M. Misra, S. S. Tripathy, S. K. Nayak, and A. K. Mohanty. 2002. The influence of chemical surface modification on the performance of sisal-polyester biocomposites. *Polymer Composites* 23 (2):164–170.

96. Paiva Junior, C. Z., L. H. De Carvalho, V. M. Fonseca, S. N. Monteiro, and J. R. M. d'Almeida. 2004. Analysis of the tensile strength of polyester/hybrid ramie–cotton fabric composites. *Polymer Testing* 23 (2):131–135.

97. Rout, J., S. S. Tripathy, M. Misra, A. K. Mohanty, and S. K. Nayak. 2001. The influence of fiber surface modification on the mechanical properties of coir-polyester composites. *Polymer Composites* 22 (4):468–476.

98. Sydenstricker, T. H. D., S. Mochnaz, and S. C. Amico. 2003. Pull-out and other evaluations in sisal-reinforced polyester biocomposites. *Polymer Testing* 22 (4):375–380.

99. Gassan, J., and A. K. Bledzki. 1999. Effect of cyclic moisture absorption desorption on the mechanical properties of silanized jute-epoxy composites. *Polymer Composites* 20 (4):604–611.

100. Gassan, J. and A. K. Bledzki. 1999. Possibilities for improving the mechanical properties of jute/epoxy composites by alkali treatment of fibers. *Composites Science and Technology* 59 (9):1303–1309.

101. Barbosa Jr, V., E. C. Ramires, I. A. T. Razera, and E. Frollini. 2010. Biobased composites from tannin–phenolic polymers reinforced with coir fibers. *Industrial Crops and Products* 32 (3):305–312.

102. Milanese, A. C., M. O. H. Cioffi, and H. J. C. Voorwald. 2012. Thermal and mechanical behaviour of sisal/phenolic composites. *Composites Part B: Engineering* 43 (7):2843–2850.

103. Alonso, M. V., M. Oliet, F. Rodríguez, G. Astarloa, and J. M. Echeverría. 2004. Use of a methylolated softwood ammonium lignosulfonate as partial substitute of phenol in resol resins manufacture. *Journal of Applied Polymer Science* 94 (2):643–650.

104. Astarloa Aierbe, G., J. M. Echeverría, M. D. Martin, A. M. Etxeberria, and I. Mondragon. 2000. Influence of the initial formaldehyde to phenol molar ratio (F/P) on the formation of a phenolic resol resin catalyzed with amine. *Polymer* 41 (18):6797–6802.

105. Astarloa-Aierbe, G., J. M. Echeverría, A. Vázquez, and I. Mondragon. 2000. Influence of the amount of catalyst and initial pH on the phenolic resol resin formation. *Polymer* 41 (9):3311–3315.

106. Gonçalves, A. R. and P. Benar. 2001. Hydroxymethylation and oxidation of Organosolv lignins and utilization of the products. *Bioresource Technology* 79 (2):103–111.

107. Knop, A. and L. A. Pilato. 1985. *Phenolic Resins, Chemistry, Applications and Performance: Future Directions*. Berlin, Germany: Springer.

108. Miravete, A., J. Cuartero, and Compuestos Asociación Española de Materiales. 2003. *Materiales compuestos*. Antonio Miravete.

109. de Lange, P. J., E. Mäder, K. Mai, R. J. Young, and I. Ahmad. 2001. Characterization and micromechanical testing of the interphase of aramid-reinforced epoxy composites. *Composites Part A: Applied Science and Manufacturing* 32 (3–4):331–342.

110. Komai, K., K. Minoshima, K. Tanaka, and T. Tokura. 2002. Effects of stress waveform and water absorption on the fatigue strength of angle-ply aramid fiber/epoxy composites. *International Journal of Fatigue* 24 (2-4):339–348.

111. Tanaka, K., K. Minoshima, W. Grela, and K. Komai. 2002. Characterization of the aramid/epoxy interfacial properties by means of pull-out test and influence of water absorption. *Composites Science and Technology* 62 (16):2169–2177.

112. Rojo, E., M. Virginia Alonso, J. C. Domínguez, B. D. Saz-Orozco, M. Oliet, and F. Rodriguez. 2013. Alkali treatment of viscose cellulosic fibers from eucalyptus wood: Structural, morphological, and thermal analysis. *Journal of Applied Polymer Science* 130 (3):2198–2204.

113. Van de Weyenberg, I., T. Chi Truong, B. Vangrimde, and I. Verpoest. 2006. Improving the properties of UD flax fiber reinforced composites by applying an alkaline fiber treatment. *Composites Part A: Applied Science and Manufacturing* 37 (9):1368–1376.

114. Yan, L., N. Chouw, and X. Yuan. 2012. Improving the mechanical properties of natural fiber fabric reinforced epoxy composites by alkali treatment. *Journal of Reinforced Plastics and Composites* 31 (6):425–437.

115. Alix, S., L. Lebrun, C. Morvan, and S. Marais. 2011. Study of water behaviour of chemically treated flax fibers-based composites: A way to approach the hydric interface. *Composites Science and Technology* 71 (6):893–899.

116. Rojo, E., M. Virginia Alonso, M. Oliet, B. D. Saz-Orozco, and F. Rodriguez. 2015. Effect of fiber loading on the properties of treated cellulose fiber-reinforced phenolic composites. *Composites Part B: Engineering* 68 (0):185–192.

117. Xie, Y., C. A. S. Hill, Z. Xiao, H. Militz, and C. Mai. 2010. Silane coupling agents used for natural fiber/polymer composites: A review. *Composites Part A: Applied Science and Manufacturing* 41 (7):806–819.

118. John, M. J. and R. D. Anandjiwala. 2008. Recent developments in chemical modification and characterization of natural fiber-reinforced composites. *Polymer Composites* 29 (2):187–207.

119. Sreekala, M. S., M. G. Kumaran, S. Joseph, M. Jacob, and S. Thomas. 2000. Oil palm fiber reinforced phenol formaldehyde composites: Influence of fiber surface modifications on the mechanical performance. *Applied Composite Materials* 7 (5–6):295–329.

120. Kalia, S. and B. S. Kaith. 2008. Mechanical properties of phenolic composites reinforced with flax-g-copolymers prepared under different reaction conditions: A comparative study. *Journal of Chemistry* 5 (1):177–184.

121. Milanese, A. C., M. O. H. Cioffi, and H. J. C. Voorwald. 2011. Mechanical behavior of natural fiber composites. *Procedia Engineering* 10 (0):2022–2027.

122. Joseph, S., M. S. Sreekala, Z. Oommen, P. Koshy, and S. Thomas. 2002. A comparison of the mechanical properties of phenol formaldehyde composites reinforced with banana fibers and glass fibers. *Composites Science and Technology* 62 (14):1857–1868.

123. Jacob, M., S. Thomas, and K. T. Varughese. 2006. Novel woven sisal fabric reinforced natural rubber composites: Tensile and swelling characteristics. *Journal of Composite Materials* 40 (16):1471–1485.

124. Pang, K., R. Kotek, and A. Tonelli. 2006. Review of conventional and novel polymerization processes for polyesters. *Progress in Polymer Science* 31 (11):1009–1037.

125. Penczek, P., P. Czub, and J. Pielichowski. 2005. Unsaturated polyester resins: Chemistry and technology. *Crosslinking in Materials Science* 184:1–95.

126. Miyagawa, H., A. K. Mohanty, R. Burgueño, L. T. Drzal, and M. Misra. 2007. Novel biobased resins from blends of functionalized soybean oil and unsaturated polyester resin. *Journal of Polymer Science Part B: Polymer Physics* 45 (6):698–704.

127. Burgueño, R., M. J. Quagliata, A. K. Mohanty, G. Mehta, L. T. Drzal, and M. Misra. 2004. Load-bearing natural fiber composite cellular beams and panels. *Composites Part A: Applied Science and Manufacturing* 35 (6):645–656.

128. Towo, A. N. and M. P. Ansell. 2008. Fatigue of sisal fiber reinforced composites: Constant-life diagrams and hysteresis loop capture. *Composites Science and Technology* 68 (3-4):915–924.

129. Singh, B., M. Gupta, and A. Verma. 1996. Influence of fiber surface treatment on the properties of sisal-polyester composites. *Polymer Composites* 17 (6):910–918.

130. Rouison, D., M. Sain, and M. Couturier. 2006. Resin transfer molding of hemp fiber composites: Optimization of the process and mechanical properties of the materials. *Composites Science and Technology* 66 (7-8):895–906.

131. Madsen, B., A. Thygesen, and H. Lilholt. 2009. Plant fiber composites: Porosity and stiffness. *Composites Science and Technology* 69 (7–8):1057–1069.

132. Dhakal, H. N., Z. Y. Zhang, and M. O. W. Richardson. 2007. Effect of water absorption on the mechanical properties of hemp fiber reinforced unsaturated polyester composites. *Composites Science and Technology* 67 (7-8):1674–1683.

133. Gowda, T. M., A. C. B. Naidu, and R. Chhaya. 1999. Some mechanical properties of untreated jute fabric-reinforced polyester composites. *Composites Part A: Applied Science and Manufacturing* 30 (3):277–284.

134. Tobias, B. C., and E. Ibarra. 1997. Influence of cure temperature on the flexural strength of natural based composites. 42nd International SAMPE Symposium and Exhibition, Anaheim, CA.

135. Hottle, T. A., M. M. Bilec, and A. E. Landis. 2013. Sustainability assessments of biobased polymers. *Polymer Degradation and Stability* 98 (9):1898–1907.

136. La Rosa, A. D., G. Recca, J. Summerscales, A. Latteri, G. Cozzo, and G. Cicala. 2014. Bio-based versus traditional polymer composites. A life cycle assessment perspective. *Journal of Cleaner Production* 74:135–144.

137. Pawelzik, P., M. Carus, J. Hotchkiss, R. Narayan, S. Selke, M. Wellisch, M. Weiss, B. Wicke, and M. K. Patel. 2013. Critical aspects in the life cycle assessment (LCA) of bio-based materials: Reviewing methodologies and deriving recommendations. *Resources, Conservation and Recycling* 73 (0):211–228.

138. Bledzki, A. K., A. Jaszkiewicz, and D. Scherzer. 2009. Mechanical properties of PLA composites with man-made cellulose and abaca fibers. *Composites Part A: Applied Science and Manufacturing* 40 (4):404–412.

139. Bledzki, A. K., and A. Jaszkiewicz. 2010. Mechanical performance of biocomposites based on PLA and PHBV reinforced with natural fibers: A comparative study to PP. *Composites Science and Technology* 70 (12):1687–1696.

140. Oksman, K., M. Skrifvars, and J. F. Selin. 2003. Natural fibers as reinforcement in polylactic acid (PLA) composites. *Composites Science and Technology* 63 (9):1317–1324.

141. Bax, B., and J. Müssig. 2008. Impact and tensile properties of PLA/Cordenka and PLA/flax composites. *Composites Science and Technology* 68 (7–8):1601–1607.

142. Hu, R. H., M. H. Jang, Y. J. Kim, Y. J. Piao, and J. K. Lim. 2010. Fully degradable jute fiber reinforced polylactide composites applicable to car interior panel. *Advanced Materials Research* 123:1151–1154.

143. Plackett, D., T. L. Andersen, W. B. Pedersen, and L. Nielsen. 2003. Biodegradable composites based on L-polylactide and jute fibers. *Composites Science and Technology* 63 (9):1287–1296.

144. Cho, D., J. M. Seo, W. H. Park, S. K. Han, T. W. Hwang, C. H. Choi, and S. J. Jung. 2007. Improvement of the interfacial, flexural, and thermal properties of jute/poly (lactic acid) biocomposites by fiber surface treatments. *Journal of Biobased Materials and Bioenergy* 1 (3):331–340.

145. Dobreva, T., R. Benavente, J. M. Perena, E. Perez, M. Avella, M. Garcia, and G. Bogoeva-Gaceva. 2010. Effect of different thermal treatments on the mechanical performance of poly (L-lactic acid) based eco-composites. *Journal of Applied Polymer Science* 116 (2):1088–1098.

146. Yussuf, A. A., I. Massoumi, and A. Hassan. 2010. Comparison of polylactic acid/kenaf and polylactic acid/rise husk composites: The influence of the natural fibers on the mechanical, thermal and biodegradability properties. *Journal of Polymers and the Environment* 18 (3):422–429.

147. Ochi, S.. 2008. Mechanical properties of kenaf fibers and kenaf/PLA composites. *Mechanics of Materials* 40 (4–5):446–452.

148. Iovino, R., R. Zullo, M. A. Rao, L. Cassar, and L. Gianfreda. 2008. Biodegradation of poly(lactic acid)/starch/coir biocomposites under controlled composting conditions. *Polymer Degradation and Stability* 93 (1):147–157.

149. Wu, C. S. 2009. Renewable resource-based composites of recycled natural fibers and maleated polylactide bioplastic: Characterization and biodegradability. *Polymer Degradation and Stability* 94 (7):1076–1084.

150. Okubo, K., T. Fujii, and E. T. Thostenson. 2009. Multi-scale hybrid biocomposite: Processing and mechanical characterization of bamboo fiber reinforced PLA with microfibrillated cellulose. *Composites Part A: Applied Science and Manufacturing* 40 (4):469–475.

151. Shibata, S., Y. Cao, and I. Fukumoto. 2005. Press forming of short natural fiber-reinforced biodegradable resin: Effects of fiber volume and length on flexural properties. *Polymer Testing* 24 (8):1005–1011.

152. Di Franco, C. R., V. P. Cyras, J. P. Busalmen, R. A. Ruseckaite, and A. Vázquez. 2004. Degradation of polycaprolactone/starch blends and composites with sisal fiber. *Polymer Degradation and Stability* 86 (1):95–103.

153. Alvarez, V. A., J. M. Kenny, and A. Vázquez. 2004. Creep behavior of biocomposites based on sisal fiber reinforced cellulose derivatives/starch blends. *Polymer Composites* 25 (3):280–288.

154. Alvarez, V. A., A. Terenzi, J. M. Kenny, and A. Vazquez. 2004. Melt rheological behavior of starch-based matrix composites reinforced with short sisal fibers. *Polymer Engineering & Science* 44 (10):1907–1914.

155. Alvarez, V. A. and A. Vázquez. 2004. Thermal degradation of cellulose derivatives/starch blends and sisal fiber biocomposites. *Polymer Degradation and Stability* 84 (1):13–21.

156. Alvarez, V., A. Vazquez, and C. Bernal. 2006. Effect of microstructure on the tensile and fracture properties of sisal fiber/starch-based composites. *Journal of Composite Materials* 40 (1):21–35.

157. Kuciel, S., and A. Liber-Knec. 2009. Biocomposites on the base of thermoplastic starch filled by wood and kenaf fiber. *Journal of Biobased Materials and Bioenergy* 3 (3):269–274.

158. Ochi, Sh.. 2006. Development of high strength biodegradable composites using Manila hemp fiber and starch-based biodegradable resin. *Composites Part A: Applied Science and Manufacturing* 37 (11):1879–1883.

159. Gandini, A. and M. N. Belgacem. 1997. Furans in polymer chemistry. *Progress in Polymer Science* 22 (6):1203–1379.

160. Dunlop, A. P., and F. N. Peters. 1953. *The Furans.* New York, NY: Reinhold Publishing Co.

161. Choura, M., N. M. Belgacem, and A. Gandini. 1996. Acid-catalyzed polycondensation of furfuryl alcohol mechanisms of chromophore formation and cross-linking. *Macromolecules* 29 (11):3839–3850.

162. Guigo, N., A. Mija, L. Vincent, and N. Sbirrazzuoli. 2010. Eco-friendly composite resins based on renewable biomass resources: Polyfurfuryl alcohol/lignin thermosets. *European Polymer Journal* 46 (5):1016–1023.

163. Wewerka, E. M., K. L. Walters, and R. H. Moore. 1969. Differential thermal analysis of furfuryl alcohol resin binders. *Carbon* 7 (1):129–141.

164. Botelho, E. C., N. Scherbakoff, and M. C. Rezende. 2000. Rheological analysis of the phenolic and furfuryl resins used in the carbon materials processing. *Materials Research (Sao Carlos, Brazil)* 3 (2):1923.

165. Domínguez, J. C., J. C. Grivel, and B. Madsen. 2012. Study on the non-isothermal curing kinetics of a polyfurfuryl alcohol bioresin by DSC using different amounts of catalyst. *Thermochimica Acta* 529 (1):29–35.

166. Gaefke, C. B., E. C. Botelho, N. G. Ferreira, and M. C. Rezende. 2007. Effect of furfuryl alcohol addition on the cure of furfuryl alcohol resin used in the glassy carbon manufacture. *Journal of Applied Polymer Science* 106 (4):2274–2281.

167. Ma, C. C. M., M. S. Yn, J. L. Han, C. J. Chang, and H. D. Wu. 1995. Pultruded fiber-reinforced furfuryl alcohol resin composites: 1. Process feasibility study. *Composites Manufacturing* 6 (1):45–52.

168. Ma, C. C. M., M. S. Yn, J. L. Han, C. J. Chang, and H. D. Wu. 1995. Pultruded fiber-reinforced furfuryl alcohol resin composites: 2. Static, dynamic mechanical and thermal properties. *Composites Manufacturing* 6 (1):53–58.

169. Domínguez, J. C. and B. Madsen. 2014. Chemorheological study of a polyfurfuryl alcohol resin system: Pre-gel curing stage. *Industrial Crops and Products* 52:321–328.

170. Crossley, R., P. Schubel, and A. Stevenson. 2014. Furan matrix and flax fiber as a sustainable renewable composite: Mechanical and fire-resistant properties in comparison to phenol, epoxy and polyester. *Journal of Reinforced Plastics and Composites* 33 (1):58–68.

171. Pohl, T., M. Bierer, E. Natter, B. Madsen, H. Hoydonckx, and R. Schledjewski. 2011. Properties of compression moulded new fully biobased thermoset composites with aligned flax textiles. *Plastics, Rubber and Composites* 40 (6/7):249–299.

172. Domínguez, J. C. and B. Madsen. 2014. Development of new biomass-based furan/glass composites manufactured by the double-vacuum-bag technique. *Journal of Composite Materials* DOI: 10.1177/0021998314559060.

5 Interfacial Compatibility and Adhesion in Natural Fiber Composites

Le Quan Ngoc Tran, Carlos A. Fuentes, Ignace Verpoest, and Aart Willem Van Vuure

CONTENTS

5.1 INTRODUCTION

Fiber polymer composites exhibit advantages based on the combination of the properties of the constituents, in which the fibers act as a high strength and stiffness component and the surrounding matrix keeps them in a desired location and orientation, providing structural integrity. In such a system, the mechanical properties of the composites cannot be achieved with either fiber or matrix playing alone. The composites work through the presence of the fiber–matrix interface. The interface is a surface layer formed by the boundary of the fiber and matrix that maintains the bonds between the materials and is responsible for the load transfer; in this way, the fiber reinforcement can fulfill its role.

The knowledge of the interface has been developed for existing synthetic fiber (e.g., glass and carbon fiber) composites, but it is not yet fully understood for natural fiber composites. Unlike synthetic fibers, technical natural fibers are usually extracted from different parts of the plant, which typically have different surface chemical compositions, leading to different properties in terms of surface energy and potential for chemical reactions. Figure 5.1 displays the technical fibers of various plants. In the case of coconut, the technical coir fiber is naturally present as such in the husk and it is surrounded by organic tissues, whereas in the case of flax or bamboo, its technical fiber has a configuration depending on the fiber extraction method, which separates a bundle of elementary fibers from the stem or culm and forms a technical fiber. In general, most natural fibers are relatively hydrophilic, have a rough surface, and are physicochemically heterogeneous, which affects the interfacial compatibility and adhesion when used in composite materials.

In this chapter, the interface in natural fiber composites is presented by a systematic understanding of fiber–matrix wettability, compatibility, and interfacial adhesion. Focus will mainly be on three natural fibers for use in composites, namely flax, coir, and bamboo. The fiber surface properties in terms of surface energy and surface chemistry are studied to predict the fiber–matrix wettability and compatibility based on wetting analysis. The fiber–matrix interfacial adhesion can be evaluated

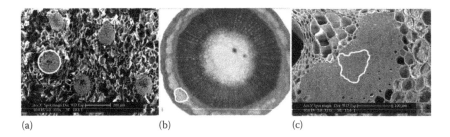

(a)   (b)   (c)

FIGURE 5.1 Technical natural fiber (circles) from different plants: (a) coir from coconut shell, (b) flax fiber from the stem [57], and (c) bamboo fiber from the culm [58]. (Data from Charlet, K., C. Baley, C. Morvan, J. P. Jernot, M. Gomina, and J. Bréard. 2007. Characteristics of Hermès flax fibres as a function of their location in the stem and properties of the derived unidirectional composites. *Compos. Part A Appl. S.* 38 (8):1912–1921 [57]; Osorio, L., E. Trujillo, A. W. Van Vuure, and I. Verpoest. 2011. Morphological aspects and mechanical properties of single bamboo fibers and flexural characterization of bamboo/epoxy composites. *J. Reinf. Plast. Compos.* 30 (5):396–408 [58]. With permission.)

by different methods at different levels, including single-fiber pullout tests and other interface mechanical tests. The combined knowledge of interfacial compatibility and adhesion is important for optimization of the interface of natural fiber composites by intelligently selecting a fiber treatment or matrix modification.

5.2   MAIN TYPES OF BONDING AT THE COMPOSITE INTERFACE

Generally, the adhesion at the composite interface can be described by the following main interactions: (1) physicochemical interactions, related to wettability and compatibility of the fiber and the matrix plus physical adhesion (e.g., van der Waals forces, acid–base interactions, and hydrogen bonds); (2) chemical bonding (covalent bonds); and (3) mechanical interlocking created on rough fiber surfaces. There are also other interactions such as molecular entanglement, interdiffusion, and so on [1]. Good interfacial adhesion initially requires a good wetting between the fiber and the matrix, to achieve an extensive and proper interfacial contact. The wettability mainly depends on the surface energies of the two materials. The surface energy of fibers generally should be higher than that of the liquid resin for a good wetting to take place during composite processing. Moreover, the surface energies will play an important role in maintaining stable contact after consolidation of the composite, through physical adhesion mechanisms [2]. Essentially the fiber–matrix interactions are controlled by the functional groups on the surface of the fiber and the matrix in the interfacial contacting area. These functional groups determine whether physical or chemical adhesion mechanisms can be used.

5.2.1   PHYSICOCHEMICAL INTERACTIONS

When the composite interface is characterized at the molecular level, the interaction between the fiber and the matrix is determined by chemical groups present on the surface of both phases. The physicochemical interaction is quantitatively characterized by the work of adhesion, which is defined in terms of surface energy components of the fiber and the matrix. The physicochemical interactions will be discussed in greater detail in Section 5.3 of this chapter.

5.2.2   CHEMICAL BONDING

Physical interactions mainly depend on van der Waals forces and polar interactions such as acid–base interactions, whereas chemical bonding mechanisms are based on primary covalent bonds at the interface. Chemical bonds are formed across the interface when atoms of the fiber surface molecules and the matrix swap or share electrons, producing typically high bond strength as compared with the physical intermolecular forces. This type of interaction can be promoted and predicted beforehand, assuming that the chemical composition of the fiber surface and the matrix is known. Commonly, chemical treatments are used of both the fiber and the matrix to promote covalent bonding at the interface by a chemical reaction.

In natural fiber composites, the chemical bonding at the interface usually occurs between the hydroxyl groups of cellulose and lignin on the fiber surface and functional groups in the matrix (e.g., maleic anhydride groups in maleic anhydride grafted polypropylene).

5.2.3 MECHANICAL INTERLOCKING

Mechanical interlocking is promoted by the surface roughness of fibers, which may increase or decrease the interfacial adhesion depending on the spreading of matrix on the fiber surface. If the matrix is able to penetrate into the irregularities of the surface, it will not only be locked mechanically to the surface but also help increase the total contact area between the polymer and the fiber. On the other hand, if the matrix is not able to fill the irregularities, the area of contact between the fiber and the matrix will be reduced, producing, in turn, a reduction of adhesion.

5.3 FIBER–MATRIX WETTABILITY AND PHYSICOCHEMICAL INTERACTIONS

5.3.1 WETTING IN NATURAL FIBER COMPOSITES

The wetting of a solid by a liquid is usually described by the physical attraction between the two materials depending on their surface energies. Accordingly, bonding due to wetting involves short-range interactions of electrons on an atomic scale that develop only when the atoms of the constituents approach within a few atomic diameters or are in contact with each other (at equilibrium interatomic distance) [1]. When a good contact between two materials is formed, there will be a certain level of physical adhesion; then, other bonding mechanisms (chemical or mechanical) may also occur and create enhanced interfacial adhesion.

Wetting is usually described by the observable contact angle between the solid substrate and the liquid. Physical adhesion can be quantitatively expressed in terms of the thermodynamic work of adhesion, W_a, of the liquid to the solid using the Dupré equation:

$$W_a = \gamma_S + \gamma_L - \gamma_{SL} \tag{5.1}$$

where γ_S is the solid surface energy, γ_L is the liquid surface energy, and γ_{SL} is the interfacial energy. Here, W_a represents the physical bonding resulting from highly localized intermolecular dispersion forces and polar interactions.

Young was the first to describe the equilibrium contact angle when a sessile drop of liquid is in contact with a solid surface [3]. The relationship between the surface energies of the solid and the liquid through the contact angle θ is expressed as follows:

$$\gamma_S = \gamma_{SL} + \gamma_L \cos\theta \tag{5.2}$$

It can be seen in Figure 5.2 that if the liquid forms contact angles respectively greater than and less than $90°$, then the situation is either nonwetting or favorable for wetting. In case of the presence of a very high surface energy solid (e.g., platinum), the contact angle will usually approach zero, which means a complete spreading of the liquid on the solid [4].

The work of adhesion can be expressed in relation to the equilibrium contact angle by combining Equation 5.1 with Young's equation, resulting in

$$W_a = \gamma_L (1 + \cos\theta) \tag{5.3}$$

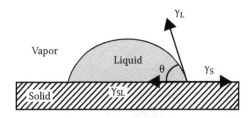

FIGURE 5.2 Schematic of a sessile drop equilibrium contact angle. (Data from Tran L. Q. N. 2012. Polymer composite materials based on coconut fibres, PhD Thesis. KU Leuven, Belgium. With permission [59].)

In composite materials, the wettability and physical interactions between the fiber and the matrix can be investigated when the surface energies of the fiber and the matrix are known. Based on the surface energies, wetting parameters (work of adhesion, interfacial energy, spreading coefficient, and wetting tension) can be calculated to study fiber–matrix wettability and adhesion [5–7]. For this, typically the surface energies of fiber and matrix are split into components, representing either hydrophobic or polar chemical groups on both surfaces. This will be further worked out in the next section.

5.3.2 CONTACT ANGLES AND SURFACE ENERGIES

The surface energy/tension of a liquid and a solid is related to the intermolecular forces (which are physical by nature) between the various molecules present. Then, the equilibrium contact angle of a liquid on a solid is a quantitative measure of the solid–liquid molecular interactions.

Surface energies of fibers and matrices are usually estimated using the contact angles of various fully characterized test liquids on the fibers and matrices. So-called surface energy component (SEC) theories are used to convert contact angles into surface energy (component) values. Regarding the contact angle measurements, several methods can be used to either directly or indirectly measure the contact angle between a solid and a liquid [8–10]. The direct technique is carried out by introducing a drop of liquid on the flat surface of the solid, in which the contact angle is defined at the intersection of the three phases: gas–liquid–solid [3]. The Wilhelmy technique is usually recommended as the indirect method (see Figure 5.3). The principle is to use a microbalance to record the wetting force, which is the capillary force exerted by the liquid on the solid sample. The contact angle is then deduced from the recorded force in relation to the liquid surface tension and the perimeter of the sample as described in Equation 5.4:

$$F_{wet} = p\gamma_{LV}\cos\theta \qquad (5.4)$$

where p is the perimeter, γ_{LV} is the liquid surface tension, and θ is the contact angle.

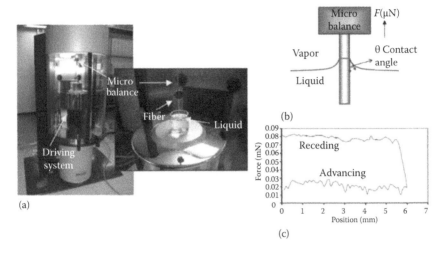

FIGURE 5.3 Dynamic contact angle measurement following Wilhelmy method: (a) ten-siometer, (b) schematic of wetting measurement, and (c) recorded advancing and receding forces. (Data from Tran L. Q. N. 2012. Polymer composite materials based on coconut fibres, PhD Thesis. KU Leuven, Belgium. With permission [59].)

The contact angle measurement of a solid polymer matrix in a liquid is carried out using both the direct and indirect methods, in which either dynamic or static contact angles of a flat polymer plate (with a smooth surface) with various fluids are determined.

In the case of fibers, there are experimental difficulties with the direct method such as improper drop shape and high curvature variation at the interface. The indirect technique, typically using the Wilhelmy method, is commonly preferred for the contact angle measurement. The measurements carried out using the Wilhelmy method usually allow dynamic contact angles to be determined, comprising advancing (θ_{adv}) and receding angles (θ_{rec}), of a test liquid on the fiber. However, to determine surface energy components from contact angles, in principle, equilibrium contact angles should be used, because Young's equation is valid for the thermodynamic equilibrium condition. Figure 5.4 shows how the contact angle evolves as a function of wetting speed for both a receding and an advancing experiment. It can be seen that the equilibrium contact angle is situated somewhere in the middle between the advancing static (zero speed) and receding static angles.

Then, in the case where the surface of the fiber (or the matrix for that matter) is smooth and homogeneously distributed, a model of the arithmetic mean of the corresponding cosines of advancing and receding angles, as shown in Equation 5.5, can be used for the calculation of the equilibrium angles [11–13]:

$$\cos\theta_0 = 0.5\cos\theta_{adv} + 0.5\cos\theta_{rec} \tag{5.5}$$

For most natural fibers, the measured receding angles are usually unstable due to the effect of the fiber surface roughness and sorption (test liquid may remain on the fiber surface after the advancing process). Consequently, only the advancing

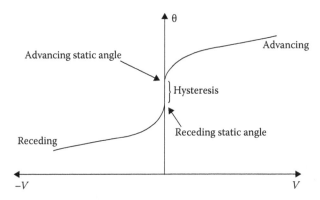

FIGURE 5.4 Schematic presentation of the velocity dependence of the contact angle. (After Dussan, E. B. 1979. On the spreading of liquids on solid surfaces: Static and dynamic contact lines. *Annu. Rev. Fluid Mech.* 11 (1):371. With permission [60].)

contact angles can be determined. In this case, the static advancing angle is the best estimate of the equilibrium angle. To perform a more in-depth study of the wetting behavior, the molecular kinetic theory (MKT) was used to model the dynamic wetting of the fibers following Equation 5.6 [14]. By using experimental data of dynamic advancing angles, the static advancing angle can thus be determined [14]. However, it should immediately be expressed that in reality, low speed advancing angles can also be used with sufficient accuracy, without the need for a full MKT analysis. The methodology described earlier has been applied for characterization of coir and bamboo fibers as reported in [15–16]. The central equation of the MKT analysis is

$$v = 2K_0 \lambda \sin h \left[\frac{\gamma_L (\cos \theta_0 - \cos \theta)}{2nk_B T} \right] \tag{5.6}$$

where v is the wetting speed, K_0 is the equilibrium molecular jump frequency, λ is the distance between two adsorption sites, γ_L is the surface tension of the liquid in contact with vapor, θ is the dynamic contact angle corresponding to the measurement velocity v, n is the number of adsorption sites per unit area, k_B is Boltzmann's constant, and T is the temperature.

Figure 5.5 shows the dynamic advancing contact angles of coir fiber in water and their velocity dependence. The advancing static contact angle can be obtained by fitting the dynamic angles with MKT. Using this method, advancing static contact angles of coir and bamboo fibers in different test liquids were obtained as reported in [15–16]. The angles are presented in Table 5.1. Equilibrium contact angles of flax fibers and several thermoplastic matrices were also determined using the model of the arithmetic mean of the corresponding cosines of advancing and receding angles (Equation 5.5), as shown in Table 5.1.

Contact angles can be used to estimate the surface energy components of both the fiber and the matrix. Several approaches for the surface energy calculation were proposed by Zisman [17], Fowkes [18], Wu [19], and Van Oss et al. [20–21]. Commonly, two methods are used to determine fiber surface energies. The first method based on a geometric mean approach was first provided by Fowkes and later by Owens and Wendt [22], and the second one is the acid–base approach developed by Van Oss et al.

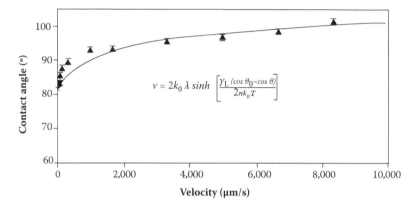

FIGURE 5.5 Fitting dynamic contact angles with molecular kinetic theory. (Adapted from Tran, L. Q. N., C. A. Fuentes, C. Dupont-Gillain, A. W. Van Vuure, and I. Verpoest. 2011. Wetting analysis and surface characterisation of coir fibres used as reinforcement for composites. *Colloids Surf. A* 377 (1):251–260. With permission [15].)

TABLE 5.1
Equilibrium and Static Advancing Contact Angles of Natural Fibers and Polymer Matrices in Water (H₂O), Diiodomethane (DIM), and Ethylene Glycol (EG), Determined by the Two Described Methods, Based on Dynamic Contact Angles

	Test Liquid (°)			Measurement	
Fiber–Matrix	H_2O	DIM	EG	Method	Reference(s)
Coir	77.3	48.2	47.1	MKT fit	[2,15]
Bamboo	60.3	47.9	42.8	MKT fit	[16]
Flax	59.1	48.7	48.5	Arithmetic mean	[56]
Alkali 5% treated coir	70.9	51.5	41.9	MKT fit	[2]
Alkali 3% treated flax	53.9	52.6	48.1	Arithmetic mean	[56]
PP	85.9	57.4	60.5	Arithmetic mean	[2]
MAPP	82.9	59.7	65.4	Arithmetic mean	[2]
PVDF	77.7	55.2	44.3	Arithmetic mean	[2]
PLA	64.0	52.8	53.6	Arithmetic mean	[56]

MAPP, maleic anhydride grafted polypropylene; MKT, molecular kinetic theory; PLA, polylactic acid; PP, polypropylene; PVDF, polyvinylidene fluoride.

In the Owens and Wendt method, the surface energy of the solid is divided into two components, dispersive[*], γ_S^d, and polar, γ_S^p, using a geometric mean approach to combine their contributions:

$$\gamma_S = \gamma_S^d + \gamma_S^p \tag{5.7}$$

The interfacial energy of the two phases can be approximated by

$$\gamma_{SL} = \gamma_S + \gamma_L - 2\left(\sqrt{\gamma_S^d \gamma_L^d} + \sqrt{\gamma_S^p \gamma_L^p}\right) \tag{5.8}$$

By combining with Young's equation, $\gamma_S = \gamma_{SL} + \gamma_L \cos\theta$, one obtains

$$\frac{\gamma_L(1+\cos\theta)}{2\sqrt{\gamma_L^d}} = \sqrt{\gamma_S^p}\left(\frac{\sqrt{\gamma_L^p}}{\sqrt{\gamma_L^d}}\right) + \sqrt{\gamma_S^d} \tag{5.9}$$

If one has obtained contact angle data on the fiber for a series of test liquids with known surface tension components, the two unknowns γ_S^d and γ_S^p are simultaneously solved by linear fitting using Equation 5.9, referred to as the Owens–Wendt equation.

In the approach proposed by Van Oss et al. [20–21], it is suggested that the surface energy consists of two terms (here for the case of the solid): the Lifshitz–van der Waals component[†], γ_S^{LW}, and an acid–base component, γ_S^{AB}, which includes electron acceptor, γ_S^+, and electron donor, γ_S^-, components as shown in Equation 5.10:

$$\gamma_S = \gamma_S^{LW} + \gamma_S^{AB} \text{ and } \gamma_S^{AB} = 2\sqrt{\gamma_S^+ \gamma_S^-} \tag{5.10}$$

When the solid is in contact with a liquid, the relationship between their surface energy components can be described by

$$\gamma_L(1+\cos\theta) = 2\left(\sqrt{\gamma_S^{LW}\gamma_L^{LW}} + \sqrt{\gamma_S^+\gamma_L^-} + \sqrt{\gamma_S^-\gamma_L^+}\right) \tag{5.11}$$

To determine the three unknown surface energy components, contact angle measurements are carried out using various test liquids (at least three liquids) with known surface tension components; thus, a linear set of equations is obtained in the following form:

$$\gamma_{L,i}(1+\cos\theta_i) = 2\left(\sqrt{\gamma_S^{LW}\gamma_{L,i}^{LW}} + \sqrt{\gamma_S^+\gamma_{L,i}^-} + \sqrt{\gamma_S^-\gamma_{L,i}^+}\right), \quad i = 1,\ldots,m \tag{5.12}$$

The set of equations can be written in matrix form:

$$Ax = b \tag{5.13}$$

---

[*] The dispersive component is responsible for London-dispersive interactions, and the polar component represents polar interactions, including hydrogen bonds and acid–base interactions.

[†] The Lifshitz–van der Waals interactions comprise dispersion, dipolar, and induction interactions.

where

$$
A = \begin{bmatrix} \sqrt{\gamma_{L,1}^{LW}} & \sqrt{\gamma_{L,1}^+} & \sqrt{\gamma_{L,1}^-} \\ \sqrt{\gamma_{L,2}^{LW}} & \sqrt{\gamma_{L,2}^+} & \sqrt{\gamma_{L,2}^-} \\ \cdots\cdots\cdots \\ \sqrt{\gamma_{L,m}^{LW}} & \sqrt{\gamma_{L,m}^+} & \sqrt{\gamma_{L,m}^-} \end{bmatrix}, \quad b = \begin{bmatrix} \gamma_{L,1}(1+\cos\theta_1)/2 \\ \gamma_{L,2}(1+\cos\theta_2)/2 \\ \cdots \\ \gamma_{L,m}(1+\cos\theta_m)/2 \end{bmatrix}, \quad x = \begin{bmatrix} \sqrt{\gamma_S^{LW}} \\ \sqrt{\gamma_S^+} \\ \sqrt{\gamma_S^-} \end{bmatrix}
$$

Surface energies of various fibers and matrices estimated by both the Owens–Wendt and van Oss–Good approaches are shown in Table 5.2. According to the Owens–Wendt approach, it can be seen that the coir fibers seem to be hydrophobic with a low polar fraction of the surface energy. Higher polarities are found in bamboo and flax fibers. Similarly, the acid–base components of bamboo and flax fibers are also higher than those of coir fibers following the van Oss–Good approach. All the studied materials seem to have net negative charge on the surface with a higher base component. This indicates that the acid–base components should only be used as relative values, where apparently the flax fibers have a lower acidic aspect but somewhat higher basic aspect than the other two fibers. The alkali treatment on coir and flax fibers shows the improvement of the total surface energy and polar fraction. For the matrices, PP and MAPP have low total surface energy with a small polar fraction. The surface energies of PVDF and PLA are higher than those of PP with a high polar fraction.

TABLE 5.2

Surface Energies of Various Natural Fibers and Matrices Estimated Following the Owens–Wendt and van Oss–Good Approaches

| | Surface Energy (mJ/m²) | | | | | | | | |
| | Disperse–Polar (Owens–Wendt) | | | Acid–Base (van Oss–Good) | | | | | |
Fiber–Matrix	γ_S	γ_S^d	γ_S^p	γ_S	γ_S^{LW}	γ^+	γ^-	γ_S^{AB}	Reference(s)
Coir	40.4	35.1	5.3	37.5	35.5	0.33	3.17	2.0	[2]
Bamboo	44.6	30.9	13.7	38.8	35.4	0.28	10.13	3.4	[24]
Flax	43.7	29.2	14.5	37.0	35.0	0.08	11.7	2.0	[56]
Alkali 5% treated coir	42.2	33.5	8.7	39.6	34.2	0.64	11.27	5.4	[2]
Alkali 3% treated flax	45.2	26.6	18.6	35.9	32.8	0.16	15.2	3.1	[56]
PP	30.7	27.1	3.6	30.9	30.0	0.12	1.87	0.9	[2]
MAPP	28.6	23.6	5.0	28.8	28.3	0.02	3.15	0.5	[2]
PVDF	37.2	30.8	6.4	35.1	31.6	0.88	3.39	3.5	[2]
PLA	40.1	27.4	12.7	34.3	32.7	0.06	10.18	1.6	[56]

There are some remarkable differences between the surface energy components of the different natural fibers and matrices, which will affect the fiber–matrix physical interactions when combining them in a composite system. This will be further explored in the next section.

5.3.3   Wetting Analysis and Physical Adhesion

The wetting analysis is carried out based on wetting parameters comprising spreading coefficient (S), work of adhesion (W_a), interfacial energy (γ_{SL}), and wetting tension (ΔF_i). These parameters can be estimated using the surface energies of the two materials.

The work of adhesion refers to the work required to disjoint a unit area of the solid–liquid interface, thereby creating a unit area of liquid–vacuum and solid–vacuum interfaces [23]. The work of adhesion is defined in terms of surface energies by the Dupré equation:

$$W_a = \gamma_S + \gamma_L - \gamma_{SL} = \gamma_L(1 + \cos\theta) \tag{5.14}$$

The spreading coefficient can be defined as the work required to prevent spreading by exposing a unit area of solid–vacuum interface while destroying corresponding amounts of solid–liquid and liquid–vacuum interfaces [23]. A positive value of S corresponds to instantaneous spreading:

$$S = \gamma_S - (\gamma_{SL} + \gamma_L) \tag{5.15}$$

The wetting tension can be defined as the work needed against wetting a porous network by eliminating a unit area of the solid–liquid interface while exposing a unit area of the solid–vacuum interface [23], which is actually the arithmetic mean of the work of adhesion and the spreading coefficient. A positive wetting tension means that wetting will be favored:

$$\Delta F_i = \gamma_S - \gamma_{SL} \tag{5.16}$$

The interfacial energy is defined as the reversible work necessary to increase the interfacial surface area by unit area. A small interfacial energy will favor wetting. The interfacial energy can be expressed in terms of Lifshitz–van der Waals and acid–base components as follows:

$$\gamma_{SL} = \gamma_S^{LW} + \gamma_L^{LW}$$
$$+ 2\left[\left(\gamma_S^+\gamma_S^-\right)^{1/2} + \left(\gamma_L^+\gamma_L^-\right)^{1/2} - \left(\gamma_S^{LW}\gamma_L^{LW}\right)^{1/2} - \left(\gamma_S^+\gamma_L^-\right)^{1/2} - \left(\gamma_S^-\gamma_L^+\right)^{1/2}\right] \tag{5.17}$$

The four wetting parameters can be considered criteria for optimizing wetting and physical adhesion. For a given substrate, the optimum condition should correspond to the situation in which the work of adhesion is maximum within the region where spontaneous wetting occurs, that is, $S \geq 0$; these two conditions are fulfilled for maximum wetting tension, ΔF.

Fuentes et al. [24] presented a study on wetting of bamboo fibers with various thermoplastic matrices, in which wetting parameters were calculated and plotted as a function of matrix surface energy (Figure 5.6). It showed that the work of adhesion and the wetting tension are increased while increasing the surface energy of the matrix. The spreading coefficient is reduced if the surface energy of the matrix is higher than 10 mJ/m², and becomes negative at around 40 mJ/m², from where it would be difficult for the molten matrix to spread on the surface of the fibers during the impregnation process. This analysis does not take into account the fact that the surface energies of the molten thermoplastic polymers will be somewhat lower than the values used for the solid polymers, which would favor their spreading.

Based on the wetting analysis, the selection of matrix can be done depending on its surface energy components. To have a good wetting and an ease of impregnation, the surface energy of the matrix must be low enough (lower than 37 mJ/m² for the case described earlier) but should not be too low to avoid a low physical adhesion between the fiber and the polymer matrix.

Table 5.3 shows the wetting parameters of several natural fiber composite systems. A low spreading coefficient (e.g., coir/PVDF) indicates difficulties in composite processing in terms of wetting and impregnation. On the other hand, higher surface energy of bamboo fibers may not only increase the wetting with lower surface energy matrices (e.g., bamboo fiber and PP system), but it also results in a low physical adhesion. In short, there is a compromise between the spreading coefficient that represents fiber–matrix wetting during processing and physical adhesion in the consolidated composite measured by the work of adhesion.

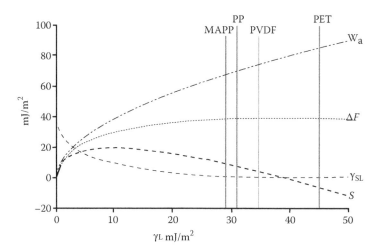

FIGURE 5.6 Wetting parameters as a function of surface energy of matrix; bamboo fiber surface used as substrate; surface energies of studied matrices positioned to predict the wetting with bamboo fiber. (Data from Fuentes, C. A, L. Q. N. Tran, M. Van Hellemont, V. Janssens, C. Dupont-Gillain, A. W. Van Vuure, and I. Verpoest. 2013. Effect of physical adhesion on mechanical behaviour of bamboo fibre reinforced thermoplastic composites. *Colloids Surf. A* 418:7–15. With permission [24].)

Table 5.3
Wetting Parameters Calculated from Surface Energies of Fiber and Matrix of Various Composite Systems

Composite	Wetting Parameters (van Oss–Good Approach)			
	W_a^{ab} (mJ/m²)	γ_{SL} (mJ/m²)	S (mJ/m²)	ΔF (mJ/m²)
Coir/PP	68.1	0.3	6.3	37.2
Coir/MAPP	65.9	0.4	8.3	37.1
Coir/PVDF	72.4	0.2	2.2	37.3
Alkali 5% treated coir/PP	68.6	1.9	6.8	37.7
Alkali 5% treated coir/MAPP	66.0	2.4	8.4	37.2
Alkali 5% treated coir/PVDF	75.0	−0.3	4.8	39.9
Bamboo/PP	68.8	0.9	7.1	37.9
Bamboo/MAPP	66.6	1.3	8.4	37.5
Bamboo/PVDF	74.5	−1.1	5.2	39.9

Sources: Tran, L. Q. N., C. A. Fuentes, C. Dupont-Gillain, A. W. Van Vuure, and I. Verpoest. 2013. Understanding the interfacial compatibility and adhesion of natural coir fibre thermoplastic composites. *Compos. Sci. Technol.* 80:23–30; Fuentes, C. A, L. Q. N. Tran, M. Van Hellemont, V. Janssens, C. Dupont-Gillain, A. W. Van Vuure, and I. Verpoest. 2013. Effect of physical adhesion on mechanical behaviour of bamboo fibre reinforced thermoplastic composites. *Colloids Surf. A* 418:7–15. With permission [2,24].

5.4   FIBER SURFACE CHEMISTRY

The surface chemistry of natural fibers will provide important information on the chemical functionality that determines the interactions with the chemical functionality of the matrix. The chemical groups on the fiber surface also affect the surface energy and level of hydrophilicity of the fibers. For characterization of fiber surface chemistry, X-ray photoelectron spectroscopy (XPS) has proved to be a useful tool and is commonly used to investigate the surface chemistry of several kinds of natural fibers. The characterization is used to determine a quantitative elemental composition and certain functional groups present on the surface [15,25–28].

The surface chemistry of several natural fibers characterized by the XPS technique is presented in Table 5.4, in which the relative atomic percentages of the elements, together with the oxygen-to-carbon (O/C) atomic ratio, are shown.

It can be seen that a high proportion of carbon was found in the three studied natural fibers, especially in coir and flax fibers, which may represent a hydrocarbon-rich waxy layer on the surface. The low O/C ratio also indicates a high proportion of aliphatic and aromatic carbons [29]. There is a decrease in carbon percentage and an increase in O/C ratio for the alkali-treated fibers. It has been reported that the O/C value of alkali-treated coir and flax fibers is close to that of lignin (range of 0.31–0.36) [30–31]. Several studies showed that the surface of flax, bamboo, and coir fibers consists of waxes, fatty substances, and lignin [15–16]. Alkali treatment of the fibers could wash away the waxes and fatty substances, which would lead to exposure of the lignin binding the elementary fibers to the fiber surface.

More details of chemical bonds on the surface of fibers could be investigated by decomposition of the C 1s peak into four subpeaks C1–C4, as presented in Table 5.4 and Figure 5.7. These subpeaks represent the following: carbon solely linked to carbon or hydrogen C–C or C–H (C1), carbon singly bound to oxygen or nitrogen C–O or C–N (C2), carbon doubly bound to oxygen O–C–O or C=O (C3), and carbon involved in ester or carboxylic acid functions O=C–O (C4).

For the studied fibers in Table 5.4, C1 is higher than C2, C3, and C4. The high value of C1 indicates the presence of unoxidized carbon atoms at the surface, which can be attributed to hydrocarbon in waxes and lignin. The alkali-treated coir and flax fibers have lower C1 and higher C2, C3, and C4 than in case of the untreated fibers. This shows that a larger amount of lignin may be present at the surface after removing the waxes, and a partial exposure of cellulose (rich in C2 function) may also take place [15].

TABLE 5.4

Relative Atomic Percentages and Decomposition of C 1s Peaks Obtained by XPS on Untreated and Alkali-Treated Coir and Flax Fibers

					Binding Energy (eV)			
					284.8	286.3	287.5	288.8
					C1 (%)	C2 (%)	C3 (%)	C4 (%)
Fibers	C (%)	O (%)	Si (%)	O/C	(C–C/C–H)	(C–O)	(C=O/O–C–O)	(O–C=O)
Coir	74.9	21.8	0.9	0.29	66.2	23.1	6.2	4.5
Bamboo	74.3	22.9	0.6	0.31	58.0	28.8	7.6	5.6
Flax	88.1	10.7	1.2	0.12	61.7	22.3	4.1	–
Alkali 5% treated coir	72.9	23.2	1.3	0.32	48.0	34.2	11.5	6.4
Alkali 3% treated flax	78.9	18.8	2.3	0.24	58.4	12.9	7.6	–

XPS, X-ray photoelectron spectroscopy.

Sources: Tran, L. Q. N., C. A. Fuentes, C. Dupont-Gillain, A. W. Van Vuure, and I. Verpoest. 2011. Wetting analysis and surface characterisation of coir fibres used as reinforcement for composites. *Colloids Surf. A* 377 (1):251–260; Fuentes, C. A, L. Q. N. Tran, C. Dupont-Gillain, W. Vanderlinden, S. De Feyter, A. W. Van Vuure, and I. Verpoest. 2011. Wetting behaviour and surface properties of technical bamboo fibres. *Colloids Surf. A* 380 (1):89–99; Tran, L. Q. N., X. W. Yuan, D. Bhattacharyya, C. Fuentes Rojas, A. Van Vuure, and I. Verpoest. 2014. The influences of fibre-matrix interfacial adhesion on composite properties in natural fibre composites. In The 11th Asia-Pacific Conference on Materials Processing. Auckland, New Zealand. With permission [15–16,56].

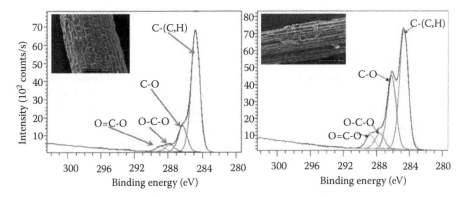

FIGURE 5.7 Typical C 1s spectra that are decomposed into four components C1–C4 for coir fiber (left) and bamboo fiber (right). (Adapted from Tran, L. Q. N., C. A. Fuentes, C. Dupont-Gillain, A. W. Van Vuure, and I. Verpoest. 2011. Wetting analysis and surface characterisation of coir fibres used as reinforcement for composites. *Colloids Surf. A* 377 (1):251–260 [15]; Fuentes, C. A, L. Q. N. Tran, C. Dupont-Gillain, W. Vanderlinden, S. De Feyter, A. W. Van Vuure, and I. Verpoest. 2011. Wetting behaviour and surface properties of technical bamboo fibres. *Colloids Surf. A* 380 (1):89–99 [16]. With permission.)

5.5 PRACTICAL ADHESION

Practical adhesion refers to the sum of a number of physical, mechanical, and chemical bonding forces that overlap and influence one another, acting at the interface of two dissimilar substances in contact. It is usually not easy or even possible to distinguish the individual contributions of these forces from one another and from the total adhesion. Work of physical adhesion can be calculated, but chemical adhesion due to covalent bonding cannot easily be quantified in strength terms. In the case of favorable wetting conditions, mechanical interlocking will boost the strength of both physical and chemical bonds, especially in shear loading.

In Section 5.3, the fiber–matrix interactions at the molecular level were described with surface energies and wetting parameters. At the higher levels (e.g., level of single fiber in matrix and level of composite), the quality of the composite interface also can be characterized by mechanical tests that are performed on either single fiber microcomposites or bulk laminate composites. In the former, a single fiber is embedded in a matrix block of different shapes and sizes. Then, interfacial shear strength (IFSS) is determined in various ways that comprise pullout, fragmentation, and microindentation tests. Regarding bulk laminates, several testing techniques have been developed for unidirectional (UD) fiber composites such as transverse tensile and bending tests, short beam shear tests, and the Iosipescu shear test. In these tests, the interface quality is characterized by either the transverse interfacial tensile strength (mode I) or interlaminar shear strength (ILSS). Every test has shown its advantages and limitations, which are mostly concerned with test sample preparation and properties of the fiber and the matrix, especially of the fiber surface [1].

5.5.1 INTERFACIAL SHEAR STRENGTH OF NATURAL FIBER COMPOSITES MEASURED BY SINGLE-FIBER PULLOUT TESTS

5.5.1.1 Pullout Test Samples and Testing

The single-fiber pullout test is commonly performed on micro-composite samples, in which a single fiber is embedded in the polymer matrix at a defined embedded length. In Figure 5.8, two typical pullout test samples are presented. Figure 5.8a shows coir fiber thermoplastic pullout test samples. In the samples, the fibers are fixed in a polymer plate by the compression molding technique. The embedded fiber length inside the matrix is then controlled by making holes through the fiber and the matrix at a defined distance from the entrance point of the fiber. For thin fibers such as flax or hemp fibers (fiber diameter of 20–50 µm), in which short embedded lengths are required, the pullout test sample can be prepared by forming an axisymmetric polymer droplet on a single fiber, as shown in Figure 5.8b.

The pullout test can be performed on the pullout samples using a mini tensile testing machine. With the sample described in Figure 5.8a, the fiber and the polymer plate are gripped in the tensile testing machine during testing. With the samples in Figure 5.8b, the polymer droplet is placed between two shearing plates that hold the droplet whereas the fiber is pulled up during testing.

5.5.1.2 Evaluation of Interfacial Shear Strength

The pullout test provides pullout load as a function of displacement. Usually, the peak force (F_{max}) is used to calculate the apparent interfacial shear strength (apparent IFSS, τ_{app}) following Equation 5.18:

$$\tau_{app} = F_{max} / \pi dl_e \tag{5.18}$$

where d is the fiber diameter and l is the fiber-embedded length.

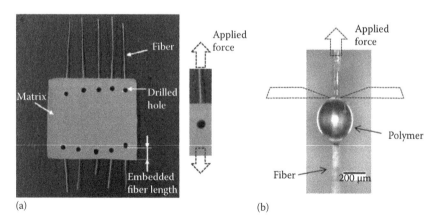

(a)   (b)

FIGURE 5.8 Pullout test samples (a) with fiber embedded length defined by making holes through the microcomposite sample, (b) a single fiber surrounded by a polymer droplet. (Adapted from Tran, L. Q. N., C. A. Fuentes, C. Dupont-Gillain, A. W. Van Vuure, and I. Verpoest. 2013. Understanding the interfacial compatibility and adhesion of natural coir fibre thermoplastic composites. *Compos. Sci. Technol.* 80:23–30. With permission [2].)

The apparent IFSS described earlier would be constant over the whole embedded fiber length in the case of a ductile fracture of the interface, where plastic yielding at the interface takes place under loading. In this case, the interfacial shear stress is uniform and independent of the embedded fiber length. However, in the case of brittle interface fracture, an inhomogeneous stress field appears and the interfacial shear stress is nonuniform. Moreover, partial debonding usually occurs, after which the debonding crack grows along the interface. Therefore, the apparent IFSS calculated from Equation 5.18 can only simply distinguish between "good" and "poor" interfacial bonding [32].

Detailed analysis of the fiber–matrix interface debonding can be carried out by two main approaches: a stress-based model and an energy-based model. In the first model, a local interfacial shear stress is determined, whereas in the second model, a value for the interface fracture toughness or critical energy release rate is derived from the experimental data. In a study on the interface of coir fiber and thermoplastic matrices, Tran et al. [2] presented an investigation of the IFSS by both the average shear stress (apparent IFSS) and the local interfacial shear stress, τ_d. For coir fiber in thermoplastic composites, it is observed that the apparent IFSS is dependent on the embedded length as seen in Figure 5.9. In the same fashion, nonuniform IFSS was also found with bamboo fiber in PP composites, in which the IFSS decreased rapidly on increasing the fiber embedded length (Figure 5.9).

The investigation of the interface quality using local IFSS (τ_d) was reported for various composite systems, in which several models were applied to determine the local IFSS. For coir and bamboo fiber thermoplastic composites, local IFSS was

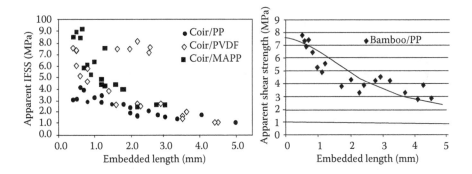

FIGURE 5.9 Apparent interfacial shear strength (IFSS) versus fiber embedded length for coir fiber in thermoplastics (left) and bamboo fiber in PP (right) composites. The curve in the right graph was obtained by fitting the experimental data with the Greszczuk model. (Adapted from Tran, L. Q. N., C. A. Fuentes, C. Dupont-Gillain, A. W. Van Vuure, and I. Verpoest. 2013. Understanding the interfacial compatibility and adhesion of natural coir fibre thermoplastic composites. *Compos. Sci. Technol.* 80:23–30 [2]; Fuentes, C. A, L. Q. N. Tran, M. Van Hellemont, V. Janssens, C. Dupont-Gillain, A. W. Van Vuure, and I. Verpoest. 2013. Effect of physical adhesion on mechanical behaviour of bamboo fibre reinforced thermoplastic composites. *Colloids Surf. A* 418:7–15 [24]. With permission.)

estimated using a model including a fitting procedure proposed by Zhandarov et al. [33], which is a two-parameter (debonding IFSS τ_d and friction stress τ_f) fit of measured and theoretical peak load F_{max} as a function of embedded length l_e using a standard least squares method.

The analysis using the Zhandarov approach is based on a stress-based model that assumes that the process of interfacial crack growth is governed by the local shear stress at the interface. For any crack length, a, the shear stress at the crack tip is assumed to be constant [34]:

$$\tau(a) = \tau_d = \text{const} \tag{5.19}$$

In this consideration, the current load, F, applied to the fiber end can be related to the current crack length as follows:

$$F = f(a, l_e, \tau_d, \Delta T) \tag{5.20}$$

Based on the shear-lag model of stress transfer from the fiber to the matrix, Zhandarov et al. [34–36] have developed several models to describe Equation 5.20. The following expression has proved to be sufficiently accurate for the analysis:

$$F = \frac{\pi d}{\beta} \left\{ \tau_d \tanh\left[\beta(l_e - a)\right] - \tau_T \tanh\left[\beta(l_e - a)\right] \tanh\left[\frac{\left[\beta(l_e - a)\right]}{2}\right] + \beta a \tau_f \right\} \tag{5.21}$$

where a is the crack length, β is the shear-lag parameter as determined by Nayfeh [37], and τ_f is the frictional stress in the already debonded regions.

In pullout tests, the recorded peak force F_{max} indicates the total debonding of the fiber from the matrix, which is attributed to both the adhesion and the friction in the system. A procedure to estimate τ_d and τ_f is proposed by Zhandarov et al. [33], which is a two-parameter (τ_d and τ_f) fit of measured and theoretical peak load F_{max} as a function of embedded length l_e using a standard least squares method. The theoretical peak load is obtained with the help of Equation 5.21, which gives a relationship between the current applied load as a function of the crack length for a sample with a given embedded length. More details of the fitting are described in [2].

Table 5.5 presents the apparent IFSS and debonding IFSS of various natural fiber composite systems. It can be seen that, for the same fiber–matrix system, the determined IFSS values are dependent on the analysis method. Therefore, for better investigation of the interfacial adhesion from pullout tests, it is necessary to understand the fracture behavior of the interface, from which a suitable analysis method will be selected accordingly.

5.5.2 CHARACTERIZATION OF COMPOSITE INTERFACE USING THREE-POINT BENDING TESTS

The interface quality of composites can also be estimated by mechanical testing of UD composites. When UD composites are tested with fibers in the transverse direction, the matrix and interface properties will dominate the final composite

TABLE 5.5

Apparent IFSS and Debonding IFSS (τ_d) from Single-Fiber Pullout Tests of Various Natural Fiber Composites

Composite	Apparent IFSS		Local Debonding IFSS (Zhandarov Model)
	IFSS (MPa)	Method	τ_d (MPa)
Coir/PP	2.4	Average apparent IFSS	9.0
Coir/MAPP	5.6	Average apparent IFSS	15.8
Coir/PVDF	3.3	Average apparent IFSS	15.8
Bamboo/PP	3.2	Apparent IFSS at embedded length of 0.6 mm	3.8
Bamboo/PVDF	8.4	Apparent IFSS at embedded length of 0.6 mm	13.8
Flax/PP	10.9	Apparent IFSS at embedded length of 0.5 mm	—
Alkali 5% treated coir/PP	2.9	Average apparent IFSS	10.4
Alkali 5% treated coir/ MAPP	5.6	Average apparent IFSS	16.8
Alkali 5% treated coir/ PVDF	4.2	Average apparent IFSS	19
Alkali 3% treated flax/PP	11.1	Apparent IFSS at embedded length of 0.5 mm	—

IFSS, interfacial shear strength.

Sources: Tran, L. Q. N., C. A. Fuentes, C. Dupont-Gillain, A. W. Van Vuure, and I. Verpoest. 2013. Understanding the interfacial compatibility and adhesion of natural coir fibre thermoplastic composites. *Compos. Sci. Technol.* 80:23–30; Fuentes, C. A, L. Q. N. Tran, M. Van Hellemont, V. Janssens, C. Dupont-Gillain, A. W. Van Vuure, and I. Verpoest. 2013. Effect of physical adhesion on mechanical behaviour of bamboo fibre reinforced thermoplastic composites. *Colloids Surf. A* 418:7–15; Tran, L. Q. N., X. W. Yuan, D. Bhattacharyya, C. Fuentes Rojas, A. Van Vuure, and I. Verpoest. 2014. The influences of fibre-matrix interfacial adhesion on composite properties in natural fibre composites. In The 11th Asia-Pacific Conference on Materials Processing. Auckland, New Zealand. With permission [2,24,56].

properties. The transverse strength of the composite represents the fiber–matrix interfacial adhesion, or the cohesion of the component materials (fiber or matrix), as seen in Figure 5.10. Therefore, using transverse three-point bending tests (transverse 3PBT) in combination with the investigation of the fracture surface of the tested samples, the interface quality of the composite can be characterized.

The transverse bending strength of various UD natural fiber composites measured by 3PBT is presented in Table 5.6. To estimate the interface strength of the composite using the transverse bending strength, the fracture surface of the tested samples needs to be investigated. Figure 5.11 presents the SEM images of the fracture surface of UD coir fiber thermoplastic composites tested by transverse 3PBT. Clean fiber surfaces can be seen, indicating adhesion failure at the fiber–matrix interface. In this case, the transverse strength can be considered representative for the interfacial tensile strength of the composites. The results show that the higher transverse strength indicates a better interfacial adhesion in the case of coir fibers with PVDF and MAPP as compared with PP.

The flexural properties of UD composites can be assessed by three-point bending in the longitudinal direction. To investigate the influence of interfacial adhesion on the composite strength, the efficiency factor of the longitudinal strength is usually calculated. This factor is the ratio of experimental longitudinal strength over the calculated value following the rule of mixtures. Table 5.6 also presents the longitudinal strength and the efficiency factor of the longitudinal strength of different natural fiber composites. It can be seen that, in most cases, there is a good agreement between the interfacial adhesion and the composite strength.

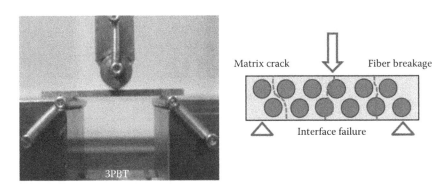

FIGURE 5.10 Transverse three-point bending test on unidirectional (UD) fiber composite (left), and schematic presentation of failure possibilities in the tested sample (right). (Data from Tran L. Q. N. 2012. Polymer composite materials based on coconut fibres, PhD Thesis. KU Leuven, Belgium. With permission [59].)

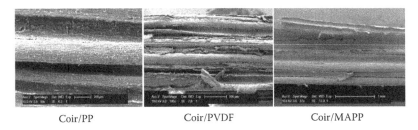

Coir/PP   Coir/PVDF   Coir/MAPP

FIGURE 5.11 Typical scanning electron microscopic images of the fracture surface of coir fiber composites in transverse three-point bending test. (Data from Tran L. Q. N. 2012. Polymer composite materials based on coconut fibres, PhD Thesis. KU Leuven, Belgium. With permission [59].)

TABLE 5.6

Transverse Bending and Longitudinal Tensile Strengths in 3PBT on UD Composites, and Efficiency Factor of the Longitudinal Strength

Composite	Transverse Bending Strength (MPa)	Longitudinal Strength (MPa)	Efficiency Factor of Longitudinal Strength
Coir/PP	3.1	66.4	0.59
Coir/MAPP	21.0	53.8	0.72
Coir/PVDF	16.6	82.8	0.66
Bamboo/PP	14.9	170	0.52
Bamboo/MAPP	18.2	168	0.51
Bamboo/PVDF	24.8	220	0.71
Flax/PLA	19.2	288.6	0.77
Alkali 5% treated coir/PP	4.4	71.5	0.61
Alkali 5% treated coir/ PVDF	21.5	103.4	0.85
Alkali 3% treated flax/ PLA	22.1	308	0.82
Alkali 5% treated coir / MAPP	19.1	49.5	0.71

3PBT, three-point bending test; UD, unidirectional.

Sources: Tran, L. Q. N., C. A. Fuentes, C. Dupont-Gillain, A. W. Van Vuure, and I. Verpoest. 2013. Understanding the interfacial compatibility and adhesion of natural coir fibre thermoplastic composites. *Compos. Sci. Technol.* 80:23–30; Fuentes, C. A, L. Q. N. Tran, M. Van Hellemont, V. Janssens, C. Dupont-Gillain, A. W. Van Vuure, and I. Verpoest. 2013. Effect of physical adhesion on mechanical behaviour of bamboo fibre reinforced thermoplastic composites. *Colloids Surf. A* 418:7–15; Tran, L. Q. N., X. W. Yuan, D. Bhattacharyya, C. Fuentes Rojas, A. Van Vuure, and I. Verpoest. 2014. The influences of fibre-matrix interfacial adhesion on composite properties in natural fibre composites. In The 11th Asia-Pacific Conference on Materials Processing. Auckland, New Zealand. With permission [2,24,56].

5.6   EFFECTS OF FIBER TREATMENTS ON COMPOSITE INTERFACIAL ADHESION

A good understanding of the natural fiber surface is necessary for the development of natural fiber composites. Based on this, matrix systems can be selected or developed to reach the full potential of the composite. The natural fiber surface is complex with heterogeneous substances composed of cellulose, hemicellulose, and lignin. The surface is influenced by bulk morphology, extractive chemicals, and processing conditions. For example, the flax surface is reported to be covered by a waxy layer, whereas lignin is the main component on the surface of bamboo fiber [10,16]. To enhance the fiber–matrix interfacial strength of natural fiber composites, it is necessary to use a physical or chemical treatment to change the surface structure of the fibers as well as the fiber surface energy.

5.6.1 Physical Treatments

Some physical methods, such as stretching [38], thermotreatment [39], and electric discharge [40–42], can be used to change the structure and surface properties of the fibers. Among these treatments, the electric discharge methods such as corona and cold plasma are interesting techniques for surface oxidation activation. A corona treatment process changes the surface energy of cellulose fibers [40] and, in the case of wood surface activation, increases the amount of aldehyde groups [41]. The same effects are reached by cold plasma treatment. Depending on the type and nature of the used gases, a variety of surface modifications have been achieved. Surface cross-links could be introduced, surface energy could be increased or decreased, and reactive-free radicals could be produced [40,41,43].

For example, Gassan and Gutowski [44] studied the corona discharge and ultra-violet (UV) treatments on jute fiber for improving the mechanical properties of jute/epoxy composites. The result showed that the corona treatment increased the polarity of the fiber surface (from 10 to 26 mJ/m^2), whereas the dispersive contribution remained unchanged. The UV treatment of the fiber also resulted in higher polarities of the fiber surface, which led to a better wettability of fibers and a higher composite strength (the composite strength increased 30% after a 10-minute treatment at a distance of 150 mm away from the UV lamp).

5.6.2 Chemical Treatments, Including Alkali Treatment

There are several chemical ways to treat or modify the surface of natural fibers. When the fiber and the matrix are incompatible, it is often possible to bring about compatibility by introducing a third material that operates as a bridge between the other two materials. The treatments should take into account the morphology of the interphase, acid–base reactions at the interface, surface energies, and wetting phenomena [43].

Natural fibers can be impregnated or coated with a polymer that is compatible with the matrix and thus provides a better interfacial adhesion. In this method, polymer solutions or dispersions of low viscosity are used. For instance, natural fibers are impregnated with a butyl benzyl phthalate plastified polyvinylchloride (PVC/BBP) dispersion to create PVC/BBP-coated fibers, which results in a compatible interface between the fibers and polystyrene (PS) [45].

One of the important chemical modification methods is chemical coupling. The fiber surface is treated with a compound that forms a bridge of chemical bonds between the fiber and the matrix. Several methods of chemical coupling can be found in literature, such as graft copolymerization, treatment with isocyanates, triazine coupling agents, and so on. Among these methods, graft copolymerization is an effective method of chemical modification of natural fibers [46–47]. This reaction can be initiated by free radicals on the cellulose molecule. For example, the treatment of cellulose fibers with hot MAPP copolymer provides covalent bonds across the interface [48].

Arbelaiz et al. [49] studied the effect of MAPP treatment on flax fibers in short flax fiber polypropylene composites. The result showed that there was an improvement on the fiber–matrix interface, and the 5% MAPP-treated fibers increased the composite tensile and flexural strength by approximately 35% compared with untreated flax fiber composites.

Brahmakumar et al. [50] investigated the effect of the waxy surface of coir fibers on the fiber–matrix interfacial bonding in low-density polyethylene composites of coir fiber. Removal of the waxy layer resulted in a weaker interfacial bonding and decreased the composite tensile strength by 40% and that of the modulus by 60%. And by grafting a layer of a long alkyl chain molecule to the wax-free fiber, the composite interfacial compatibility and bonding was improved, leading to an improvement of the longitudinal tensile strength and modulus of the coir fiber composite by about 300% and 700%, respectively, by incorporating a 25% fiber volume fraction of 20 mm long fiber.

Alkali treatment is a method that is widely used to change the structure and surface properties of fibers. Depending on the type of alkali (NaOH, KOH, and LiOH) and its concentration, several modifications occur in the fiber, such as swelling, transformation of cellulose inside the fiber (from cellulose-I to cellulose-II), and changes in fiber surface chemistry by removing some substances (e.g., fats and waxes).

The mechanical properties of alkali-treated coir fiber polyester composites were investigated by Rout et al. [51]. It was reported that 2% NaOH-treated fiber polyester composites improved the tensile strength and flexural strength by 26% and 15%, respectively, compared with untreated fiber polyester composites. With a further increase in NaOH concentration from 2% to 5%, the 5% NaOH-treated fiber composite improved the flexural strength by 17% in comparison to the untreated fiber composite. For 10% NaOH-treated fiber composites, both the tensile strength and flexural strength decreased. The enhancement in mechanical properties in alkali-treated (2% and 5%) fiber composites was attributed to the improved wetting of alkali-treated coir with polyester [52]. The decrease in mechanical properties in the case of 10% alkali-treated fiber composites was attributed to cell wall thickening, which leads to poor adhesion with polyester resin.

For other natural fibers, in a study of Borysiak and Garbarczyk [53], flax fibers were treated with different NaOH concentrations. The results showed that there was a degradation of the crystal structure and partial transformation into cellulose-II at a very high alkali concentration. Not only the fatty and waxy layers on the surface of fibers can be removed using strong alkali solution (higher than 5% concentration), but this also results in an increase of water uptake and fiber swelling. A low NaOH concentration can partially remove fats and waxes together with a smoothening of the surface [54–55].

5.6.3   EFFECT OF FIBER TREATMENT ON PHYSICAL INTERACTIONS AND PRACTICAL ADHESION

As mentioned earlier, alkali treatment on natural fibers is a common way to change the fiber surface, in which waxes and fatty substances can be washed away and the fiber surface has higher polarity due to the exposure of lignin and cellulose on the surface. In the study of coir fiber thermoplastic composites, Tran et al. [2] reported the effect of fiber alkali treatment on the coir surface that helped change the surface energy of the fiber. The work of adhesion and the interfacial energy for various untreated and alkali 5%-treated coir fibers in three matrix systems (PP, MAPP, and PVDF) were calculated following the Owens–Wendt and the van Oss–Good approaches. The results showed that the work of adhesion and interfacial energy changed as the fiber surface was modified by alkali treatment. In general, the work of adhesion showed a higher value for coir fiber in PVDF in comparison with that in PP and MAPP (approximately 12%–14% higher), as seen in Figure 5.12, which was mainly due to the higher surface energy and polar component of PVDF. The alkali treatment improved the work of adhesion of all fiber–matrix systems, which was partially attributed to the higher surface energy and polar component of the fibers. For coir fiber in PVDF, the surface energy components of both untreated and treated fibers are quite well matched (equal and high), leading to a high work of adhesion and a low value of interfacial energy; the compatibility is even a little better in the case of the treated fiber.

As discussed in Section 5.3, the work of adhesion directly reflects the significance of the surface energies of the fiber and the matrix, where a higher work of adhesion results in stronger physicochemical interactions. Furthermore, the lower interfacial energy also contributes to a higher work of adhesion, which depends on the matching of surface energy components of the fiber and the matrix. In natural fiber thermoplastic composites, no chemical bonding is involved in the interfacial interactions; physical adhesion is proposed as the main interaction that strongly affects the final practical adhesion of the composites.

The relationship between the physical interactions measured by the work of adhesion and the practical adhesion of coir fiber thermoplastic composites was reported in [2]. The transverse strength of the UD composites was plotted as a function of the work of adhesion, as seen in Figure 5.13. On comparing the interfacial adhesion

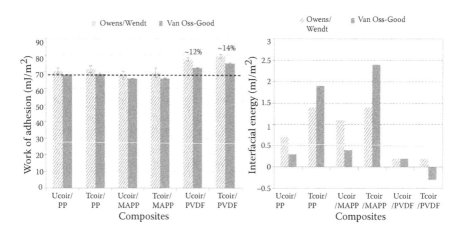

FIGURE 5.12 Work of adhesion (left) and interfacial energy (right) of untreated (Ucoir) and alkali-treated (Tcoir) fibers in thermoplastic composites calculated following the Owens–Wendt and the van Oss–Good approaches. (Data from Tran L. Q. N. 2012. Polymer composite materials based on coconut fibres, PhD Thesis. KU Leuven, Belgium. With permission [59].)

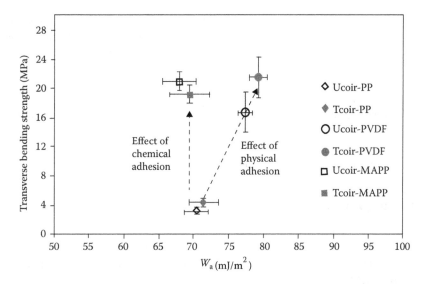

FIGURE 5.13 The transverse bending strength of UD-untreated and alkali-treated coir fiber composites plotted as a function of the work of adhesion. (Data from Tran L. Q. N. 2012. Polymer composite materials based on coconut fibres, PhD Thesis. KU Leuven, Belgium. With permission [59].)

of coir fiber with PP and PVDF, it can be seen that a significantly higher interfacial adhesion for the coir fiber PVDF composites was reached compared with the PP composites, which may be attributed to the higher work of adhesion of coir fiber PVDF. The modification of fiber surface energy by alkali treatment also contributes to a further improvement of adhesion by increasing the fiber surface energy and minimizing the differences between the surface energy components of the fiber and the PVDF matrix.

In a comparison of PP and MAPP systems, the two systems have similar work of adhesion, which suggests the physical adhesion is comparable. However, the interfacial adhesion (measured by interfacial strength in transverse 3PBT) of coir fiber with MAPP is approximately five times higher than that of coir fiber with PP, which may be attributed to covalent bonds between maleic anhydride groups of MAPP and hydroxyl groups of lignin on the fiber surface.

In short, the fiber–matrix adhesion can be improved by increasing the work of adhesion using high surface energy and a compatible matrix, where the physical adhesion plays the main role. Alternatively, creating a chemical interaction across the interface can be used to obtain better interfacial adhesion, which is mainly dominated by covalent bonds.

5.7 CONCLUSIONS

Natural fibers commonly have nonhomogeneous surfaces in terms of physicochemical properties, which influence the wettability and interfacial adhesion when used in composite materials. A better quality of the composite interface can be obtained by tailoring the interface based on an understanding of fiber–matrix interfacial interactions. Several useful tools can be applied for investigating the interfacial interactions, including wetting analysis and measurement of practical adhesion.

Wetting analysis provides the surface energies of the fibers and the matrix, which are used to calculate the fiber–matrix work of adhesion and interfacial energy to predict the physical adhesion and compatibility of the composites. Practical adhesion in single-fiber composites and UD composites can be evaluated using single-fiber pullout tests and transverse three-point bending tests. The combination of different characterization techniques offers a deeper understanding of the interfacial adhesion and compatibility in natural fiber composites.

ACKNOWLEDGMENTS

The authors thank Professor Christine Dupont-Gillain and her research group at the Institute of Condensed Matter and Nanosciences, Université Catholique de Louvain, Louvain-la-Neuve, Belgium, for help in characterizing the surface chemistry of fibers using XPS.

REFERENCES

1. Kim, J. K. 1998. *Engineered Interfaces in Fiber Reinforced Composites.* Oxford, United Kingdom: Elsevier.
2. Tran, L. Q. N., C. A. Fuentes, C. Dupont-Gillain, A. W. Van Vuure, and I. Verpoest. 2013. Understanding the interfacial compatibility and adhesion of natural coir fibre thermoplastic composites. *Compos. Sci. Technol.* 80:23–30.
3. Young, T. 1805. An essay on the cohesion of fluids. *Phil. Trans. R. Soc.* 95:65–87.
4. Erbil, H. Y. 2006. *Surface Chemistry of Solid and Liquid Interfaces.* Oxford, United Kingdom: Blackwell.
5. Bonn, D., J. Eggers, J. Indekeu, J. Meunier, and E. Rolley. 2009. Wetting and spreading. *Rev. Mod. Phys.* 81 (2):739.
6. Mittal, K. L. 1977. The role of the interface in adhesion phenomena. *Polym. Eng. Sci.* 17 (7):467–473.
7. Connor, M., J-E. Bidaux, and J-A. E. Manson. 1997. A criterion for optimum adhesion applied to fibre reinforced composites. *J. Mater. Sci.* 32 (19):5059–5067.
8. Le, C. V. 1996. Measuring the contact angles of liquid droplets on wool fibers and determining surface energy components. *Text. Res. J.* 66 (6):389.
9. Silva, J. L. G. and H. A. Al-Qureshi. 1999. Mechanics of wetting systems of natural fibres with polymeric resin. *J. Mater. Process. Technol.* 93:124–128.
10. Aranberri-Askargorta, I., T. Lampke, and A. Bismarck. 2003. Wetting behavior of flax fibers as reinforcement for polypropylene. *J. Colloid Interf. Sci.* 263 (2):580–589.
11. De Jonghe, V., D. Chatain, I. Rivollet, and N. Eustathopoulos. 1990. Contact angle hysteresis due to roughness in four metal/sapphire systems. *J. Chim. Phys.* 87 (9):1623–1645.
12. Andrieu, C., C. Sykes, and F. Brochard. 1994. Average spreading parameter on heterogeneous surfaces. *Langmuir* 10 (7):2077–2080.
13. Dettre, R. H. 1993. Wetting of low-energy surfaces. *Wettability* 49:1.
14. Blake, T. D. and J. M. Haynes. 1969. Kinetics of liquid-liquid displacement. *J. Colloid Interf. Sci.* 30 (3):421.

15. Tran, L. Q. N., C. A. Fuentes, C. Dupont-Gillain, A. W. Van Vuure, and I. Verpoest. 2011. Wetting analysis and surface characterisation of coir fibres used as reinforcement for composites. *Colloids Surf. A* 377 (1):251–260.
16. Fuentes, C. A, L. Q. N. Tran, C. Dupont-Gillain, W. Vanderlinden, S. De Feyter, A. W. Van Vuure, and I. Verpoest. 2011. Wetting behaviour and surface properties of technical bamboo fibres. *Colloids Surf. A* 380 (1):89–99.
17. Zisman, W. A. 1964. Relation of the equilibrium contact angle to liquid and solid constitution. *Adv. Chem. Ser.* 43 (1):51.
18. Fowkes, F. M. 1963. Additivity of intermolecular forces at interfaces. 1. Determination of contribution to surface and interfacial tensions of dispersion forces in various liquids. *J. Phys. Chem.* 67 (12):2538–2541.
19. Wu, S. 1971. Calculation of interfacial tension in polymer systems. *J. Polym. Sci. Part C: Polym Symposia.* 34, 19.
20. Van Oss, C. J., R. J. Good, and M. K. Chaudhury. 1986. The role of van der Waals forces and hydrogen bonds in "hydrophobic interactions" between biopolymers and low energy surfaces. *J. Colloid Interf. Sci.* 111 (2):378.
21. Van Oss, C. J., R. J. Good, and M. K. Chaudhury. 1988. Interfacial Lifshitz-van der Waals and polar interactions in macroscopic systems. *Chem. Rev.* 88 (6):927.
22. Owens, D. K. and R. C. Wendt. 1969. Estimation of the surface free energy of polymers. *J. Appl. Polym. Sci.* 13 (8):1741.
23. Berg, J. C. 1993. *Wettability.* New York, NY: Marcel Dekker.
24. Fuentes, C. A, L. Q. N. Tran, M. Van Hellemont, V. Janssens, C. Dupont-Gillain, A. W. Van Vuure, and I. Verpoest. 2013. Effect of physical adhesion on mechanical behaviour of bamboo fibre reinforced thermoplastic composites. *Colloids Surf. A* 418:7–15.
25. Belgacem, M. N., G. Czeremuszkin, S. Sapieha, and A. Gandini. 1995. Surface characterization of cellulose fibres by XPS and inverse gas chromatography. *Cellulose* 2 (3):145–157.
26. Johansson, L. S., J. Campbell, K. Koljonen, M. Kleen, and J. Buchert. 2004. On surface distributions in natural cellulosic fibres. *Surf. Interf. Anal.* 36 (8):706–710.
27. Sgriccia, N., M. C. Hawley, and M. Misra. 2008. Characterization of natural fiber surfaces and natural fiber composites. *Compos. Part A Appl. S.* 39 (10):1632–1637.
28. Topalovic, T., V. A. Nierstrasz, L. Bautista, D. Jocic, A. Navarro, and M. M. C. G. Warmoeskerken. 2007. XPS and contact angle study of cotton surface oxidation by catalytic bleaching. *Colloids Surf. A* 296 (1–3):76–85.
29. Panthapulakkal, S. and M. Sain. 2007. Agro-residue reinforced high-density polyethylene composites: fiber characterization and analysis of composite properties. *Compos. Part A Appl. S.* 38 (6):1445–1454.
30. Dorris, G. M. and D. G. Grey. 1978. The surface analysis of paper and wood fibers by ESCA. I. Application to cellulosics and lignin. *Cellulose Chem. Technol.* 12:9–23.
31. Dorris, G. M. and D. G. Grey. 1978. The surface analysis of paper and wood fibers by ESCA. II. Surface composition of mechanical pulp. *Cellulose Chem. Technol.* 12:721–734.
32. Hampe, A. and C. Marotzke. 1992. Adhesion of polymers to reinforcing fibres. *Polym. Int.* 28 (4):313–318.
33. Zhandarov, S. F., E. Mader, and O. R. Yurkevich. 2002. Indirect estimation of fiber/polymer bond strength and interfacial friction from maximum load values recorded in the microbond and pull-out tests. Part 1: local bond strength. *J. Adhes. Sci. Technol.* 16 (9):1171–1200.
34. Zhandarov, S., E. Pisanova, and E. Mader. 2000. Is there any contradiction between the stress and energy failure criteria in micromechanical tests? Part II. Crack propagation: Effect of friction on force-displacement curves. *Compos. Interf.* 7 (3):149–175.

35. Zhandarov, S. F. and E. V. Pisanova. 1997. The local bond strength and its determination by fragmentation and pull-out tests. *Compos. Sci. Technol.* 57 (8):957–964.
36. Zhandarov, S., E. Pisanova, and K. Schneider. 2000. Fiber-stretching test: a new technique for characterizing the fiber-matrix interface using direct observation of crack initiation and propagation. *J. Adhes. Sci. Technol.* 14 (3):381–398.
37. Nayfeh, A. H. 1977. Thermo-mechanically induced interfacial stresses in fibrous composites. *Fibre Sci. Technol.* 10 (3):195–209.
38. Zeronian, S. H., H. Kawabata, and K. W. Alger. 1990. Factors affecting the tensile properties of nonmercerized and mercerized cotton fibers. *Text. Res. J.* 60 (3):179–183.
39. Semsarzadeh, M. A. 1986. Fiber matrix interactions in jute reinforced polyester resin. *Polym. Compos.* 7 (1):23–25.
40. Belgacem, M. N., P. Bataille, and S. Sapieha. 1994. Effect of corona modification on the mechanical properties of polypropylene/cellulose composites. *J. Appl. Polym. Sci.* 53 (4):379–385.
41. Sakata, I., M. Morita, N. Tsuruta, and K. Morita. 1993. Activation of wood surface by corona treatment to improve adhesive bonding. *J. Appl. Polym. Sci.* 49 (7):1251–1258.
42. Wang, Q., S. Kaliaguine, and A. Ait-Kadi. 1993. Catalytic grafting: A new technique for polymer-fiber composites. III. Polyethylene-plasma-treated KevlarTM fibers composites: Analysis of the fiber surface. *J. Appl. Polym. Sci.* 48 (1):121–136.
43. Bledzki, A. K. and J. Gassan. 1999. Composites reinforced with cellulose based fibres. *Prog. Polym. Sci.* 24 (2):221–274.
44. Gassan, J. and V. S. Gutowski. 2000. Effects of corona discharge and UV treatment on the properties of jute-fibre epoxy composites. *Compos. Sci. Technol.* 60 (15):2857–2863.
45. Gatenholm, P., H. Bertilsson, and A. Mathiasson. 1993. The effect of chemical composition of interphase on dispersion of cellulose fibers in polymers. I. PVC-coated cellulose in polystyrene. *J. Appl. Polym. Sci.* 49 (2):197–208.
46. Ugboule, S. C. O. 1990. Structure/property relationships in textile fibres. *Text. Prog.* 20 (4):1–43.
47. Kroschwitz, J. I. 1990. *Polymers: Fibers and Textiles: A Compendium.* Wiley.
48. Felix, J. M. and P. Gatenholm. 1991. The nature of adhesion in composites of modified cellulose fibers and polypropylene. *J. Appl. Polym. Sci.* 42 (3):609–620.
49. Arbelaiz, A., B. Fernandez, G. Cantero, R. Llano-Ponte, A. Valea, and I. Mondragon. 2005. Mechanical properties of flax fibre/polypropylene composites. Influence of fibre/matrix modification and glass fibre hybridization. *Compos. Part A Appl. S.* 36 (12):1637–1644.
50. Brahmakumar, M., C. Pavithran, and R. M. Pillai. 2005. Coconut fibre reinforced polyethylene composites: Effect of natural waxy surface layer of the fibre on fibre/matrix interfacial bonding and strength of composites. *Compos. Sci. Technol.* 65 (3–4):563.
51. Rout, J., M. Misra, S. S. Tripathy, S. K. Nayak, and A. K. Mohanty. 2001. The influence of fibre treatment on the performance of coir-polyester composites. *Compos. Sci. Technol.* 61 (9):1303–1310.
52. Prasad, S. V., C. Pavithran, and P. K. Rohatgi. 1983. Alkali treatment of coir fibres for coir-polyester composites. *J. Mater. Sci.* 18 (5):1443–1454.
53. Borysiak, S. and J. Garbarczyk. 2003. Applying the WAXS method to estimate the supermolecular structure of cellulose fibres after mercerisation. *Fibres Text. East. Eur.* 11 (5):104–106.
54. Bismarck, A., A. K. Mohanty, I. Aranberri-Askargorta, S. Czapla, M. Misra, G. Hinrichsen, and J. Springer. 2001. Surface characterization of natural fibers: Surface properties and the water up-take behavior of modified sisal and coir fibers. *Green Chem.* 3 (2):100–107.

55. Bismarck, A., I. Aranberri Askargorta, J. Springer, T. Lampke, B. Wielage, A. Stamboulis, I. Shenderovich, and H. H. Limbach. 2002. Surface characterization of flax, hemp and cellulose fibers: Surface properties and the water uptake behavior. *Polym. Compos.* 23 (5):872–894.

56. Tran, L. Q. N., X. W. Yuan, D. Bhattacharyya, C. Fuentes Rojas, A. Van Vuure, and I. Verpoest. 2014. The influences of fibre-matrix interfacial adhesion on composite properties in natural fibre composites. In The 11th Asia-Pacific Conference on Materials Processing. Auckland, New Zealand.

57. Charlet, K., C. Baley, C. Morvan, J. P. Jernot, M. Gomina, and J. Bréard. 2007. Characteristics of Hermès flax fibres as a function of their location in the stem and properties of the derived unidirectional composites. *Compos. Part A Appl. S.* 38 (8):1912–1921.

58. Osorio, L., E. Trujillo, A. W. Van Vuure, and I. Verpoest. 2011. Morphological aspects and mechanical properties of single bamboo fibers and flexural characterization of bamboo/epoxy composites. *J. Reinf. Plast. Compos.* 30 (5):396–408.

59. Tran L. Q. N. 2012. Polymer composite materials based on coconut fibres, PhD Thesis. KU Leuven, Belgium.

60. Dussan, E. B. 1979. On the spreading of liquids on solid surfaces: Static and dynamic contact lines. *Annu. Rev. Fluid Mech.* 11 (1):371.

61. Greszczuk, L. B. 1969. Theoretical studies of the mechanics of the fiber-matrix interface in composites. *Interf. Compos.*, ASTM STP 452:42–58.

6 Processing of Natural Fiber Composites

S. Shibata

CONTENTS

6.1 INTRODUCTION

Natural fiber composites have been focused on as lightweight environmentally friendly materials possessing high stiffness [1]. However, the production of these materials requires special processing techniques. There are some differences between natural and chemical synthetic fibers. Then, some special points in processing have to be considered to produce high-quality materials. For instance, there is a limit of around 200°C in processing temperature, and when exceeded, fiber carbonization by thermal decomposition, mechanical properties degradation, coloring of resin, and unpleasant odor occur. In addition, when the natural fiber pellets are not well dried, gas is generated at the time of injection molding, causing pinholes, blowholes, and unpleasant odors. This chapter will discourse the technology of mixing natural fibers such as bagasse, bamboo, jute [2], kenaf [3], flax, hemp [4], and pineapple [5] with conventional thermoplastic resins such as polypropylene (PP), polyethylene (PE), polyethylene terephthalate (PET), and injection molding and press forming of the composites.

6.2 FACTORS AFFECTING THE PROCESSING OF NATURAL FIBER COMPOSITES

Thermal denaturation of natural fibers occurs when they are heated. At temperatures above 180°C, hemicellulose, lignin, protein, and sugar content make reactions such as condensation, oxidation, and reduction, changing the color of the fiber from white to brown and degrading the strength of the final composites. Further, at temperatures above 250°C, the fibers degrade at the point of their carbonization, turning the color of the composites into black. For these reasons, the temperature control at the time of mixing and molding is important. The same reactions described earlier affect the natural fiber at the time of its preparation, that is, just after they are harvested, natural fibers contain water, sugar, oil, and protein; it is better for some of them to be removed before processing, otherwise their presence may possibly decrease the processing temperature. Alkali treatment is also effective in improving the adhesion between the fiber and the matrix [6–7].

Natural fibers have stiffness, thus their kneading with polymers may be difficult due to the high viscosity of the mixture, which decreases the processing formability. To solve this problem, the fiber length must be shortened by cutting. However, fiber lengths below their critical length will decrease the mechanical properties. Generally, 50 wt% fibers in injection molding and 80 wt% fibers in press forming constitute the upper limit in natural fiber composites [8]. Above these values, the viscosity becomes so high that the matrix resin cannot cover the entire surface of the composite, which results in a composite with lower mechanical properties than those expected and calculated. Moreover, homogeneous mixing is not achieved. Natural fiber does not function as a reinforcing material when its length is less than the critical length; thus, it is recommended to prepare the composites with longer fibers than the critical length. However, if the fibers are used only as fillers instead of reinforcing materials, the use of short length fibers, such as powder, in some cases achieves excellent formability.

It is also necessary to consider the fiber orientation. The orientation factor largely affects the mechanical properties of the natural fiber composites. For instance, in the case of unidirectional fiber orientation, in accordance with the rule of mixtures, the elastic modulus and strength of the composites should be dependent on orientation coefficient × fiber strength × fiber content + matrix strength × matrix content, and then the fiber orientation coefficient may be 1. In the case of two- and three-dimensional alignment, the orientation coefficient is 0.275 and 0.18, respectively, and strengthening effects through the natural fibers are reduced. Therefore, the orientation of the fibers can be considered a very important factor in the evaluation of the mechanical properties of the natural fiber composite. The orientation coefficient is determined by fiber orientation as described in the Section 6.6.

The size of polymer pellets before mixing with natural fibers is not a negligible factor either. Generally, natural fiber is hydrophilic and polymer is hydrophobic, hence compatibility between them is so low that natural fibers and polymers do not mix homogeneously [9]. If pellet size is much larger than the natural fiber diameter, the fibers themselves will be adhesive to each other. Then, the expected modulus and strength in composites cannot be obtained. In this case, maleic acid–modified polymer or silane coupling agent can be helpful in increasing the adhesion bond between the natural fiber and the polymer matrix [10].

6.3 PROCESSING METHOD FOR NATURAL FIBER COMPOSITES

The first industrial processing method for composites is injection molding, in which, to provide high productivity and designability, it is necessary to reduce the fibers' length to a given value, otherwise longer fibers can block the exit of the extruder and also decrease formability due to the high viscosity. After making the fiber length uniform, natural fibers are mixed with the polymer by the extruder to produce natural fiber composite pellets, which are used as raw material for injection molding.

Another processing method is press forming, in which the preparation of pellets is not always necessary. Pellets and natural fibers can be layered directly in the mold and then hot pressed. However, in this method, the shapes of the products are limited. The latter is the hot-pressing method, where a nonwoven mat, which is a mixture of natural fibers and thermoplastic polymer fibers, is hot pressed [11]. It is possible to produce a lightweight rigid board with a hollow internal structure using this method. However, both natural and polymer fibers should be long enough to get entangled together.

6.4 PREPARATION OF NATURAL FIBERS

When producing a natural fiber composite material by injection molding, the length of natural fibers must be made uniform before being mixed with polymers, as shown in Figure 6.1. When natural fibers, such as bagasse and palm fibers, are discharged in factories, they exist in various lengths from powder to long fibers. When the fibers are very long, for instance more than tens of millimeters, there is the possibility of clogging the nozzle of the extruder or molding machine. Thus, it is necessary to mill the fibers before mixing with polymers. After milling, the sieving process is preferable [12].

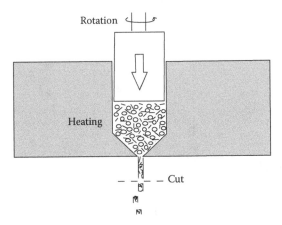

FIGURE 6.1   A processing method that makes the pellets of bagasse/PP using a single extruder.

Figure 6.2 shows the relationship between the melt flow rate (MFR) of 40 wt% sieved bagasse/PP composites and temperature at various fiber lengths. The MFR is close to zero at the average fiber length of 10.5 mm, and liquidity could not be identified. Thus, the average fiber length should be below 10 mm when the fiber is mixed with polymers, to avoid low liquidity, which is needed for mixing.

Figure 6.3 shows the microphotographs of bagasse/PP composites made by the hot-pressing method. Item (a) shows the original pellet specimens, and (b) through (d) are the hot-pressed specimens after an injection at 185°C–260°C. It was found that the fiber lengths in specimens (b) through (d) were apparently shorter than those in the original pellets (a). Moreover, at a higher temperature, the fiber length decreased. Thus, it is considered that the bagasse fibers were shortened by fiber attrition and mechanical shearing in the cylinder [13–14].

Furthermore, the matrix color was more darkened, and smoky odors became evident with increasing temperature, as observed in [15–16]. Figure 6.4 shows the relationship between average fiber length and injection temperature [17]. The average fiber length, which had been 3.3 mm in the original pellets, was shortened to 2.1 mm at 165°C and to 1.5 mm at 260°C. Generally, there was a strong correlation between fiber length and flexural modulus in composites. Figure 6.5 shows the calculated result for flexural modulus using the Cox model as a function of fiber length at 40 wt% bagasse /PP composites [18]. According to the obtained result, the flexural modulus decreased from 3000 MPa at 2.1 mm

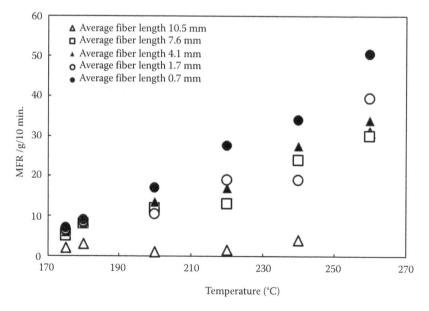

FIGURE 6.2 Relationship between MFR and temperature in 40 wt% bagasse/PP composites.

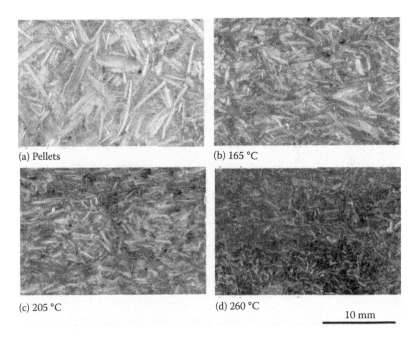

(a) Pellets

(b) 165 °C

(c) 205 °C

(d) 260 °C

10 mm

FIGURE 6.3 Mircrophotographs of bagasse/PP composites after hot pressing.

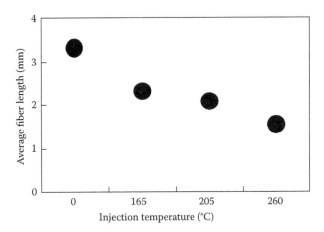

FIGURE 6.4 Relationship between average fiber length in composites and injection temperature.

to 2600 MPa at 1.5 mm. Therefore, the mechanical properties decrease if the aspect ratio of the fiber is below 10.

Figure 6.6 shows the calculated result for flexural modulus using the Cox model as a function of fiber length at 40 wt% bagasse/PP composites [17]. Theoretically, as the fiber length approaches zero, the flexural modulus of the molded article approaches the flexural modulus of the matrix (PP).

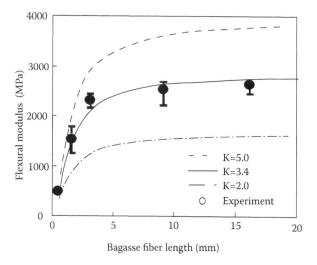

FIGURE 6.5 Relationship between flexural modulus and bagasse fiber length.

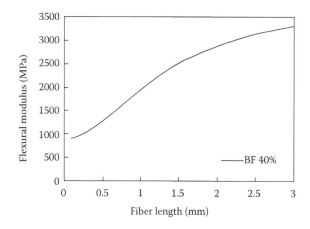

FIGURE 6.6 Relationship between flexural modulus and fiber length calculated using the Cox model.

6.5 EFFECTS OF PROCESSING TEMPERATURE ON MECHANICAL PROPERTIES OF NATURAL FIBER COMPOSITES

Natural fibers undergo thermal pyrolysis when heated at temperatures above 160°C–180°C. Generally, thermal pyrolysis occurs in the order of hemicellulose, lignin, and finally cellulose. Figure 6.7 shows the glass transition temperature (T_g) profiles of bagasse and palm fiber. As shown in the figure, the thermal pyrolysis of bagasse occurs at temperatures above 200°C, and at 150°C for palm fibers [19].

Figure 6.8 shows hot-press molded products of 40 wt% bagasse/PP composites at various temperatures. The various hot-pressed (pressure 10 kg/cm², 5 minutes)

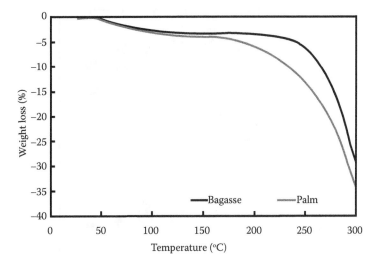

FIGURE 6.7 T_g profiles for bagasse and palm fibers.

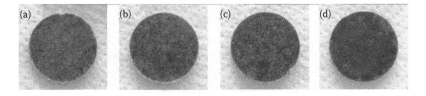

FIGURE 6.8   Appearances of bagasse PP composites. (a) 200°C, (b) 220°C, (c) 240°C, and (d) 260°C.

temperatures apparently had an effect on the appearances of the composites. Item (d) was especially brittle and the results indicated that the limit of processing temperature could be around 260°C. The color of the composites turned from thin brown to close to black. The presence of carbohydrates, such as glucose, fructose, and sucrose, proteins, and fatty acids is evident in natural fibers. Hence, these chemicals react with one another and can result in reactions such as caramelization, the Maillard reaction, or melanization. All these reactions change the white and thin brown color into brown and black color, whose latter color depends on the quantity of the reaction. Furthermore, the condensation reaction between cellulose, hemicellulose, and lignin results in the carbonization of natural fibers.

Figure 6.9 shows the relationship between the mechanical properties of 40 wt% bagasse fiber/PP composites and the injection temperature. The injection temperatures ranged from 165°C to 260°C, which corresponded to the range recommended by the polymer suppliers. All the mechanical properties decreased with the increasing of injection temperature. The intervals in the figure were the injection interval time from shot to shot. During the interval, the melted composite was kept in the injection cylinder.

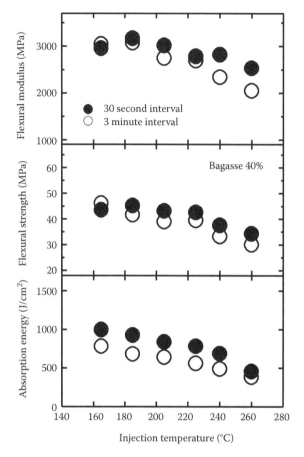

FIGURE 6.9 Relationship between mechanical properties and injection temperature.

Figure 6.10 shows scanning electron microscopy (SEM) micrographs of the fracture surfaces just after the impact test in 40 wt% bagasse specimens at (a) 165°C and (b) 260°C. At 165°C, some bagasse fibers were pulled out, leaving holes behind, as indicated by the white arrows in the figure. On the other hand, at 260°C, no pulled-out fibers were found. The fibers were cut at those roots due to their brittleness, which was a result of thermal degradation [20]. On the fractured surface, small cavities were generated by the gas during fiber thermal degradation. Hence, at high injection temperature, the thermal degradation of fibers affects not only the fiber length and brittleness but also the porosity. This is the reason that the flexural strength, modulus, and impact strength decreased.

Figure 6.11 shows the relationship between the specific gravity in 40 wt% bagasse specimens and the injection temperature. The specific gravity clearly decreased with increasing injection temperature. The percentage decrease at 260°C was 15% in comparison to that at 165°C. Thus, the fiber degradation can be considered a cause that generates the porosity and decreases the specific gravity in the composites.

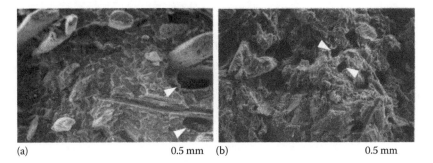

(a)   0.5 mm (b)   0.5 mm

FIGURE 6.10 Fracture surfaces of bagasse/PP composites after impact test. (a) 165°C and (b) 260°C.

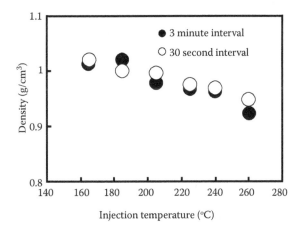

FIGURE 6.11 Relationship between density of bagasse/PP composites and injection temperature.

6.6 EFFECT OF FIBER ORIENTATION FACTOR ON THE MECHANICAL PROPERTIES IN NATURAL FIBER COMPOSITES

Kenaf and bamboo fibers were used to investigate the effect of the fiber orientation factor on the mechanical properties. The kenaf fiber diameter and length were 0.1 and 70 mm, respectively. The fiber was extracted by retting and dried at room temperature. Bamboo, which was 0.27 mm in diameter and 300 to 500 mm long, had been extracted by an alkali treatment and also dried. The biodegradable resin, CP-300, was a cornstarch based resin that was a blend of starch and poly caprolactone (PCL), supplied from Miyoshi Oil Fat, Japan (T_g: 60°C, softening temperature 55°C–62°C). The resin was chopped into uniform pellets of 1 to 2 mm diameter. The mechanical properties of kenaf, bamboo, and resin are shown in Table 6.1. The composite specimen was fabricated by a cylindrical steel mold, and the press forming was performed at 160°C and 10 MPa for 10 minutes. The fiber was put into the mold without any previous mixing. In the case of fiber-oriented composites, the fabrication steps were as follows. (1) Half of the fibers were put into the mold through the clearances of the slit jig, as shown in Figure 6.12.

TABLE 6.1

Mechanical Properties of Kenaf Fiber and Biodegradable Resin

	Young's Modulus (MPa)	Tensile Strength (MPa)	Specific Gravity (kg/m³)	Diameter (mm)
Bamboo	18,500	450	1,310	0.270
Kenaf	22,000	335	970	0.106
CP-300	494	9.4	1,160	—

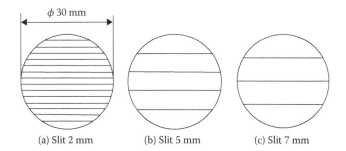

(a) Slit 2 mm   (b) Slit 5 mm   (c) Slit 7 mm

FIGURE 6.12   Slit jigs for aligned fiber composites: (a) 2mm, (b) 5 mm, and (c) 7 mm..

At this time, the bottom of the cylindrical slit jig was placed in contact with the mold bottom to maintain fiber orientation, (2) the slit jig was pulled out carefully and all resin, CP-300, was put into the mold, and (3) the slit jig was put on the resin again, and the rest of the fibers were put on the resin through the clearance of the slit jig. The clearances between the slits were 2, 5, and 7 mm, respectively. A wider slit jig would be expected to produce composites with wider fiber orientation distribution. The obtained specimens were disc shaped (30 mm diameter and 1.8–1.9 mm thickness), and the fiber orientation distribution was two-dimensional. Flexural tests were conducted in accordance to ISO 178 specifications on five flexural specimens. The flexural specimens (18 mm × 15 mm × 1.8–1.9 mm) were cut out of the original specimen. Figure 6.13a and b shows the random and the fiber-oriented composites in the flexural test, respectively.

6.6.1 CALCULATION

It is well known [21] that the Young's modulus of a short fiber-reinforced composite is determined by

$$E_{comp} = \eta_\vartheta \eta_f V_f \cdot E_f + (1 - V_f) \cdot E_m \qquad (6.1)$$

where E_f, E_m, and V_f denote the Young's modulus of the fiber, the matrix, and the volume fraction of the fiber in the composite, respectively. η_f and η_θ denote efficient factors of fiber length and orientation. η_f is given by [22–23].

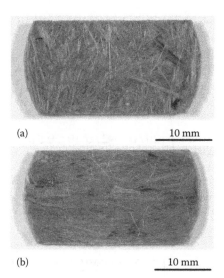

(a)   10 mm

(b)   10 mm

FIGURE 6.13   Appearance of flexural specimens: (a) kenaf, random (b) kenaf, fiber oriented with a slit size of 2 mm.

$$\eta_f = 1 - (\tanh \frac{1}{2}\beta L) / \frac{1}{2}\beta L \tag{6.2}$$

$$\beta = \left(\frac{2G_m}{E_f r_f^2 \ln(R / r_f)} \right)^{\frac{1}{2}} \tag{6.3}$$

Equation 6.2 means the Young's modulus of the composite decreases with decrease in the fiber length l, where r_f and R denote the radius of the fiber and the interval among fibers. If the distribution of the fibers is homogeneous in an ideal packing square composite, R is given by

$$R = \frac{r_f}{2} \sqrt{\frac{\pi}{V_f}} \tag{6.4}$$

The shear modulus G_m, assuming that the composite is isotropic, is given by

$$G_m = \frac{E_m}{2(1 + \nu_m)} \tag{6.5}$$

where r_f and ν_m denote the radius of the fiber and Poisson's ratio, respectively. The value ν_m can be assumed as 0.3.

On the other hand, the efficiency factor η_θ has been analyzed [24] using a probabilistic theory and assumed orientation distribution functions. The distribution functions are categorized by rectangular, sinusoidal, and triangular distribution, respectively. Judging from the experimental result as shown later in Figure 6.14, we assumed the triangular distribution as an orientation distribution function in the specimen shown in Figure 6.13b. This function is given by

$$g(\theta) = \begin{cases} -2\theta / \alpha^2 + 2 / \alpha & (0 \leq \theta \leq \alpha) \\ 0 & (\alpha < \theta) \end{cases} \tag{6.6}$$

where α denotes the limit angle of fiber orientation. If we experimentally determine the angle α, the orientation efficiency factor is given by

$$\eta_\vartheta = 4 \cdot \frac{1 - \cos\alpha}{\alpha^2} \left(\frac{3 - \nu}{4} \cdot \frac{1 - \cos\alpha}{\alpha^2} + \frac{1 + \nu}{4} \cdot \frac{1 - \cos 3\alpha}{9\alpha^2} \right) \tag{6.7}$$

This distribution function assumes that the fiber distribution is two-dimensional. Figure 6.14 shows the relationship between η_θ and limit angle α. The efficiency factor η_θ is 1.0 when α is equal to zero degrees (unidirectional fiber distribution), whereas η_θ is 0.47 at 90 degrees.

The compression ratio is defined by the ratio of the original fiber volume to the fiber volume in the composite. Hence, the compression ratio, K, can be calculated by

$$K = \frac{V_f'}{\left[V - \left(\dfrac{W - W_f}{\rho_m} \right) \right]} \tag{6.8}$$

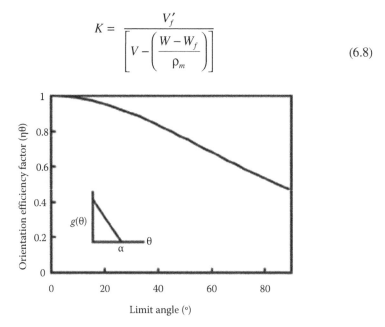

FIGURE 6.14 Relationship between orientation efficiency factor and limit angle.

where V_f', V, W, W_f, and ρ_m denote original fiber volume, volume of the composite, weight of the composite, weight of the fiber, and density of the matrix, respectively. Therefore, the final equation that predicts the flexural modulus of the composite is

$$E_{comp} = K\eta_\vartheta \eta_f V_f \cdot E_f + (1 - V_f) \cdot E_m \qquad (6.9)$$

Table 6.2 shows these parameters determined by the experiments mentioned earlier.

6.6.2 EFFECT OF FIBER ORIENTATION DISTRIBUTION ON FLEXURAL MODULUS

Figure 6.15 shows the fiber orientation distributions in kenaf composites by slit jigs. The orientation angle was defined as the deviation angle between the horizontal direction of the flexural specimen in Figure 6.13b and the longitudinal direction of the fiber on the composite surface. Thus, if the deviation angles of all fibers are

TABLE 6.2

Parameters Used for the Flexural Modulus Prediction in the Fiber-Oriented Composites

	Random	Slit 7 mm	Slit 5 mm	Slit 2 mm
α	-	25	29	45
η_θ	0.27	0.61	0.75	0.82
η_l	0.91	0.91	0.91	0.91
K_{kf}	1.35	1.35	1.35	1.35
K_{bf}	1.05	1.05	1.05	1.05

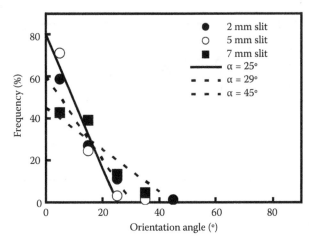

FIGURE 6.15 Fiber orientation distribution of kenaf fiber in fiber-oriented composites with various slit clearances.

zero degrees, the specimen would be unidirectional. In each composite, 600 fibers randomly chosen were measured. Frequency on the *y*-axis in Figure 6.15 means existing fiber possibility in every 10-degree range. The fiber orientation distributions were widened by increasing the slit clearance from 2 to 7 mm. In all composites, the fiber frequencies were the highest at 0–10 degrees, and they decreased linearly with increasing fiber orientation angle. Hence, it was found that the fiber orientation distribution function in the slit jig composite was triangular, as shown in Figure 6.14.

Figure 6.16 shows the relationship between flexural modulus in kenaf composites and slit clearances. The fiber volume fraction was 60%. The black and white plots represent experimental results in kenaf and bamboo, whereas the solid and the broken lines are the calculated values. The flexural modulus increased considerably with decreasing slit clearance and reached 14,900 MPa for kenaf and 14,400 MPa for bamboo. These values were 3.1 and 2.7 times higher than those in kenaf and bamboo random composites. It was clear that the calculated flexural modulus in bamboo is apparently below to the experimental, whereas both the experimental and calculated flexural modulus in kenaf is close. This same reason is described in the previous paragraph while referring to random bamboo fiber composites.

Figure 6.17 shows the relationship between the flexural modulus in kenaf composites and slit clearance in the case of the cross-ply composites. This cross-ply means that the fiber distribution in the lower half composites was orthogonal to that in the upper half layer. In the cross-ply composites, the resin was put between the upper and lower fiber layers. The flexural modulus in the experimental cross-ply was apparently lower than that in the calculated and even random composite. Figure 6.18a and b are the microphotographs of the fiber-oriented kenaf composite with a 2 mm slit and cross-ply kenaf composite. The top and the bottom in the figures are surfaces, respectively. It was found that the resin was clearly segregated in the middle, as indicated by the arrow. In the case of cross-ply composites, the fibers did not mix with

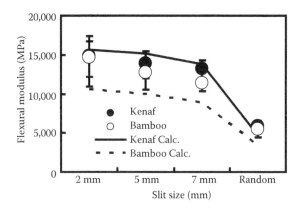

FIGURE 6.16 Relationship between flexural modulus in bamboo and kenaf fiber-oriented composites.

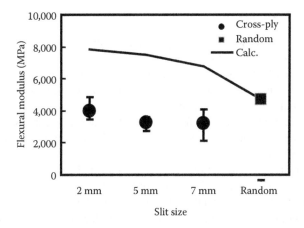

FIGURE 6.17   Relationship between flexural modulus in kenaf cross-ply composites.

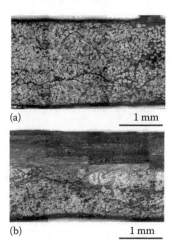

FIGURE 6.18   Optical microphotographs of (a) kenaf fiber-oriented composites and (b) kenaf cross-ply composites in the cross-section.

each other, because the upper and lower layers were cross-plied. On the other hand, the fiber in the fiber-oriented composites mixed well and no segregation was found. This may be the reason that the flexural modulus was considerably lower than that as calculated. From these results, we can conclude that the Cox model that incorporates the effect of fiber compression can be used for effectively predicting both random and aligned fiber composites under a good fiber–matrix wetting condition.

6.7   CONCLUSIONS

1. Natural fibers ensure thermal pyrolysis is above 180°C in processing. However, this degradation temperature can be increased up to 260°C by removing carbohydrate, and impurities of the natural fiber before processing. The thermal degradation causes not only decreased mechanical properties but also uncomfortable odor.
2. In the processing of natural fiber composites, thermal degradation of natural fibers causes pinholes and blowholes and the decrease of mechanical properties in composites.
3. Fiber orientation distribution in composites strongly affects the mechanical properties of composites. The effect of fiber orientation distribution should be paid attention to especially while testing mechanical properties of natural fiber composites. However, the ideal strength and elastic modulus can be obtained only with sufficient wetting among fibers.
4. The natural fiber strength decreases with increasing processing temperature, especially above 180°C. This causes a decrease of fiber length in composites. The average fiber length below the critical length dramatically decreases mechanical properties in composites due to the decrease of stress transfer among fibers.

REFERENCES

1. Mishra, S., Mohanty, A.K., Drzal, L.T., Misra, M., and Hinrichsen, G. 2004. A review on pineapple lead fibers, sisal fibers and their biocomposites. *Macromolecular Materials and Engineering* 289:955–74.
2. Gassan, J. 2002. A study of fiber and interface parameters affecting the fatigue behavior of natural fiber composites. *Composites Part A* 33:369–74.
3. Nishino, T., Hirao, K., Kotera, M., Nakamae, K., and Inagaki, H. 2003. Kenaf reinforced biodegradable composite. *Composites Science and Technology* 63:1281–6.
4. Shanks, R.A., Hodzic, A., and Wong, S. 2004. Thermoplastic biopolyester natural fiber composites. *Journal of Applied Polymer Science* 91:2114–21.
5. Mukherjee, P.S. and Satyannarayana, K.G. 1986. Structure and properties of some vegetable fibers. *Journal of Materials Science* 21:51–6.
6. Ray, D., Sarkar, B.K.A., Rana, K., and Bose, N.R. 2001. The mechanical properties of vinylester resin matrix composites reinforced with alkali-treated jute fibers. *Composites Part A* 32:119–27.
7. Herrera-Franco, P.J., and Valdez Gonzalez, A. 2004. Mechanical properties of continuous natural fiber-reinforced polymer composites. *Composites Part A* 35:339–45.
8. Keller, A. 2003. Compounding and mechanical properties of biodegradable hemp fiber composites. *Composites Science and Technology* 63:1307–16.
9. Aranberri-Askargorta, I., Lampke, T., and Bismarck, A. 2003. Wetting behavior of flax fibers as reinforcement for polypropylene. *Journal of Colloid and Interface Science* 263:580–9.
10. Karnani, R., Krishnan, M., and Narayan, R. 1997. Biofiber-reinforced polypropylene composites. *Polymer Engineering Science* 37:476–83.
11. Gowda, T.M., Naibu, A.C.B., and Chhaya, R. 1999. Some mechanical properties of untreated jute fabric-reinforced polyester composites. *Composites Part A* 30:277–84.

12. De Sousa, M.V., Monteiro, S.N., and d'Almeida, J.R.M. 2003. Evaluation of pre-treatment, size and moulding pressure on flexural mechanical behavior of chopped bagasse-polyester composites. *Polymer Testing* 23:253–8.
13. Karmaker, A.C., and Youngquist, J.A. 1996. Injection moulding of polypropylene reinforced with short jute fibers. *Journal of Applied Polymer Science* 62:1147–51.
14. Wollerdorfer, M., and Bader, H. 1998. Influence of natural fibers on the mechanical properties of biodegradable polymers. *Industrial Crops and Products* 8:105–12.
15. Espert, A., De las Heras. L.A., and Karlsson, S. 2005. Emmision of possible odourous low molecular by head space SPME-GC-MS. *Polymer Degradation and Stability* 90:555–62.
16. Tsai, W.T., Lee, M.K., and Chang, Y.M. 2006. First pyrolysis of rice straw, sugarcane bagasse and coconut shell in an induction-heating reactor. *Journal of Analytical and Applied Pyrolysis* 76:230–7.
17. Shibata, S., Bozlur, R.M., Fukumoto, I., and Kanda, Y. 2011. Effects of injection temperature on mechanical properties of bagasse/polypropylene injection moulding composites. *Bioresources* 5:2097–111.
18. Shibata, S., Cao, Y., and Fukumoto, I. 2005. Effect of bagasse fiber on the flexural properties of biodegradable composites. *Polymer composites* 26:689–94.
19. Shibata, S. 2014. Effect of time-dependent moisture absorption on surface roughness of bagasse and oil palm fibers/polypropylene composites. *Bioresources* 9:4430–40.
20. Nabi Saheb, D. and Jog, J.P. 1999. Natural fiber polymer composites, A review. *Advances in Polymer Technology* 18:351–63.
21. Cox, H.L. 1952. The elasticity and strength of paper and other fibrous materials. *British Journal Applied Physics* 55:72–9.
22. Hull, D. 1982. *Introduction to Composite Materials*, Cambridge University Press, UK.
23. Sanomura, Y., and Kawaura, M. 2003. Fiber orientation control of short-fiber reinforced thermoplastics by ram extrusion. *Polymer Composites* 24:587–96.
24. Fukuda, H., and Chou, T.W. 1982. A probabilistic theory for the strength of short-fiber with variable fiber length and orientation. *Journal of Materials Science* 17:1003–11.

7 Testing and Characterization of Natural Fiber-Reinforced Composites

H.N. Dhakal, J. MacMullen, and Z.Y. Zhang

CONTENTS

7.1 INTRODUCTION

The use of reinforcements as discrete components play an important role in improving thermo-mechanical properties such as strength, modulus, elongation, creep, and fatigue resistance of composites. Moreover, the structure and the properties of natural fiber composites largely depend on the fiber characteristics, such as their strength-to-weight ratio, chemical compositions, and morphology. The matrix retains the fibers within the composite and transmits applied load to the fibers while offering a protective coating. Therefore, the mechanical properties of the fibers determine the stiffness and tensile strength of the composites in a decisive manner. Moreover, the optimum fiber volume fraction is important to obtain better mechanical performance (tensile strength, failure strain, and modulus) as shown in Figure 7.1.

Natural fibers are widely used in composites as reinforcements, owing to their advantageous properties such as light weight, low energy-intensive manufacturing, and environmentally friendly attributes. However, natural fibers and their initial properties are inadequate for structural application unless they are suitably modified for integration into the matrix. Suitable fiber treatment and some additives are applied to enhance their fundamental properties, which are discussed in Sections 7.4.1 and 7.4.2.

Natural fibers are classified according to their source, which includes plants, animals, and minerals (Figure 7.2). Animal and mineral fibers have extensive processing and material limitations constraining their effectiveness and use in a much similar manner to plant-based sources. Substantial research is being conducted on how to integrate bird feathers, for example, into biomatrices, which seem to have problematic processing issues due to varied fiber profiles that limit their overall usefulness. Furthermore, being brittle, mineral fibers have fundamental processing limitations. In general, plant fibers are the most attractive and abundant natural reinforcement in polymer composites. Many varieties of plant fibers exist, and they can be divided into five main categories depending on the part of the plant from which they are extracted [1–2], as shown in Figure 7.2.

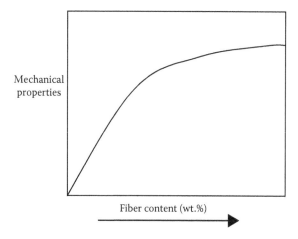

FIGURE 7.1 The relationship between the fiber volume fraction and mechanical performance.

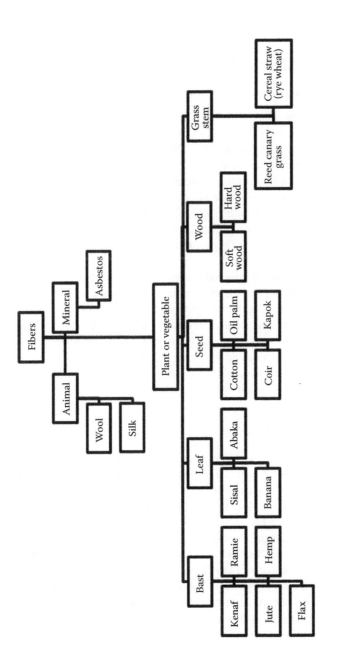

FIGURE 7.2 Classification of natural fibers.

Testing and characterization of fibers and resulting composites using appropriate methods is important to establish credible data for desired structural applications. In this chapter, issues related to the testing of important properties and characterization of various natural fibers and their composites are discussed.

7.2 NATURAL PLANT FIBERS AS REINFORCEMENTS

7.2.1 NATURAL PLANT FIBERS: COMPOSITION AND STRUCTURE

All plant species are built up of cells and when a cell is very long in relation to its width it is called a "fiber." The fiber is similar to a microscopic tube with walls surrounding a central void referred to as the "lumen." The cell wall is made up mainly (85% or more) of cellulose, hemicellulose, and lignin (referred to as lignocellulosic fibers). Key plant fiber sources for reinforcement include straws, flax, hemp, jute, and sisal. Knowledge about fiber length to width (aspect ratio) is important for characterizing and comparing different types of natural fibers. A high aspect ratio is crucial in natural fiber composites, as it provides an indication of possible mechanical properties. Hemp single fiber, for example, has an aspect ratio of 1000, which is considered good for mechanical properties.

The performance of natural fiber-reinforced composites largely depends on the properties of the reinforcement and the matrix and their interfacial compatibility. Cellulose is the main structural element of bast fibers such as flax and hemp and it is strongly polar due to hydroxyl groups, acetal, and ether linkages (C–O–C) [3]. This renders cellulose more compatible with polar, acidic, or basic groups, compared with nonpolar polymers such as thermoplastic and thermosets. Natural fibers are hydrophilic and have low moisture resistance, whereas many matrices are hydrophobic, which leads to poor interfacial compatibility between the two phases.

7.2.2 ADVANTAGES OF NATURAL PLANT FIBERS

Natural plant fibers have been used since the beginning of composite materials. Some early composites included sun-dried bricks and pottery containing organic matter such as straw to reduce cracking. These came into existence at approximately 5000 BC and it is believed that they were created due to the fact that the reinforcement stopped the bricks from drying out too quickly in the sun, which minimized the cracking. Later, Mesopotamia river boats at approximately 3000 BC were constructed from tied bundles of papyrus reeds and embedded with a bitumen matrix; they may (if a little embellished) be considered the precursor to modern-day fiberglass boats [4].

As material technology advanced, synthetic reinforced composites became evermore popular due to imparting better mechanical properties as a result of less variation in their structure and composition. However, recent environmental implications have now allowed natural fiber composites to regain their status as a viable material for the future. They have the potential to be effectively used in different industries, because they offer several advantages compared with synthetic fibers. The key features of plant fibers include high specific strength because of their low density and relative low cost as well as their sustainability. Other highly desirable qualities of natural fibers include their being nontoxic, ease of processing, low initial outlay, and manufacturing costs. This makes these materials advantageous for structural components that now have to encompass sustainability as an underlying design criterion. Other advantages include reduced tool wear in machining operations, enhanced energy recovery, and reduced dermal and respiratory irritation during processing [5–7].

7.2.3 DISADVANTAGES OF PLANT FIBERS

Disadvantages of natural fiber include high moisture uptake and moisture regain potential, susceptibility to fungal and insect attack, poor interfacial adhesion between fiber and matrix, and lower strength properties compared with their synthetic counterparts. Unpredictable influences such as weather and growing conditions make natural plant fibers highly variable, leading to nonuniform properties in the final composite. This creates batch-to-batch variation that is problematic for structural applications [8–10].

7.3 NATURAL FIBERS AND THEIR STRUCTURE

To characterize natural fiber composites, it is vital that the fiber structures are characterized as the properties of the fibers play a significant role in the overall properties of the composite. However, the microstructure of natural fibers is extremely complex, especially through their cross-section, due to cell morphology. To calculate the properties of natural fiber-reinforced composites, it is important to accurately determine the physical, mechanical, and chemical properties of the fibers. Figure 7.3 shows fibrous structures of oil palm and flax fiber exhibiting a high degree of morphological variation on their outer surface and cross-section, respectively.

The cell walls of natural fibers are predominantly made up of a number of layers, including a primary wall (the first layer deposited during cell development) and a secondary wall (S), which comprises three sublayers (S1, S2, and S3). As in

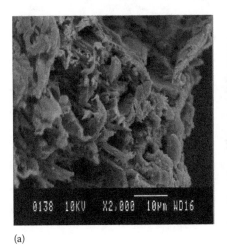
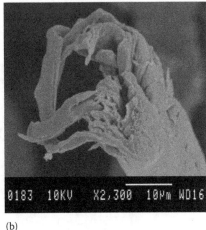

(a)   (b)

FIGURE 7.3 Scanning electron microscopic (SEM) images: (a) oil palm fiber; (b) flax fiber at high magnification showing a fibrous hollow structure at the end.

all lignocellulosic fibers, these layers mainly contain cellulose, hemicellulose, and lignin to varying concentrations. The individual fiber is bonded together by lignin-rich regions within the middle lamella. High concentrations of cellulose are found in the S2 layer (about 50%), and lignin is almost entirely concentrated within the middle lamella (about 90%). The S2 layer is, by far, the thickest layer (32–150 laminas), and it dominates the properties of the fiber. This fiber structure is illustrated in Figure 7.4.

The natural plant fibers are generally characterized by the same parameters and properties as the synthetic fibers. However, due to their natural origin, they show much higher variability with regard to chemical composition, crystallinity, surface roughness and profile properties, length, strength, and stiffness. Property variation is attributed to growing conditions (climate), harvesting methods, and processing techniques [11]. This natural variation poses difficulty in characterizing the fibers and the problems encountered in traditional synthetic composite theories [12].

7.3.1 STRUCTURAL CHARACTERIZATION OF NATURAL FIBERS

A suitable structural characterization of natural fibers is important to use natural fibers as reinforcements in structural composites. It is vital that specialized equipment and techniques are carefully selected and used to assess fiber attributes. In the following sections, established techniques that characterize natural fibers and their corresponding composites are discussed.

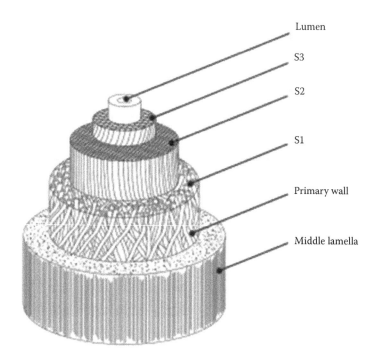

FIGURE 7.4   Illustration of bast fiber cross-section.

7.3.1.1   Scanning Electron Microscopy

A scanning electron microscope (SEM) works by emitting electrons produced by a high-tension filament under vacuum and capturing returned electrons from the surface by means of a receiver. This enables a qualitative image of the surface to be produced at higher resolutions than a conventional light-based microscope could produce effectively. Furthermore, the electron signature received from the surface allows energy-dispersive X-ray diffraction spectroscopy (EDX) to be used. This spectroscopy output is interpreted by the user to characterize the elements present at the surface of the sample. This is achieved through assessing possible elemental orbital configurations that match the electron diffraction pattern, and it is dependent on prior knowledge of the sample materials. From this information, a specimen's surface topography, composition, and other properties such as damage patterns can be evaluated. Different types of SEM are available with various capacities ranging from secondary electron imaging (SEI) to specialized field-emission SEM (FESEM). An SEM image of oil palm fiber is illustrated in Figure 7.5. From this figure, variation in cross-section diameter can be clearly seen and is useful to relate composite tested properties to morphological characteristics and composition.

7.3.1.2   Fourier Transform Infrared Spectroscopy

Fourier transform infrared spectroscopy (FTIR) is used to typically assess bonding characteristics between fiber and resin phases. This analytical process involves using an infrared source to produce transmission patterns from the sample surface at different wavelengths to characterize molecular composition. As different element–element bond patterns produce unique infrared patterns, if these bond wavelengths are known, then molecular characteristics may be determined.

The FTIR is sensitive to different conformations and local bond characteristics, enabling composite bonding to be evaluated through first assessing phases individually and then assessing them in relation to the composite. As it can be used to assess

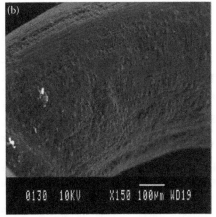

FIGURE 7.5 SEM images of oil palm fiber obtained from a JEOL 6100 SEM (a) at lower magnification and (b) at higher magnification.

what type of bond is formed, it is highly suitable for the determination of cellulose allomorphs and hydrogen-bonding patterns on fiber substrates [13]. Figure 7.6 shows an example of FTIR being used to characterize nanoparticles in ethanol before incorporation into a matrix system.

7.3.1.3 Wide-Angle X-Ray Diffraction

X-ray diffraction (XRD) and wide-angle X-ray diffraction (WAXD) are techniques that are commonly used to evaluate the crystalline index (CI) of natural fibers because of their availability and ease of use. The XRD method enables an understanding of gallery distance between cellulose polymers through use of Bragg's law = $\sin \theta = n\lambda/2d$, where n is an integer, d is the interlayer d spacing, and λ is the wavelength. Table 7.1 shows the definition.

The XRD patterns of untreated and treated hemp fiber are illustrated in Figure 7.7. The major peaks observed for all fiber samples are at 2θ diffraction angles of $15.1°$, $16.6°$, and $22.8°$, indicating the presence of cellulose. Table 7.2 presents the

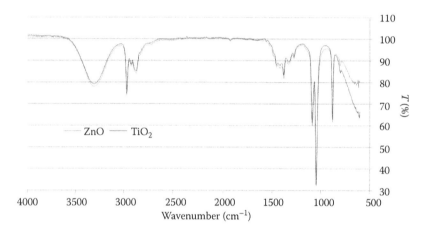

FIGURE 7.6 Fourier transform infrared spectroscopy used for the identification of titanium and zinc oxide nanoparticles in ethanol.

TABLE 7.1
Bragg's Law Description

Symbols	Description
Λ	Corresponds to the wavelength of the X-ray radiation used in the diffraction experiment
D	The spacing between diffraction lattice planes
Θ	The measured diffraction angle or glancing angle

Source: Dhakal, H.N., Zhang, Z.Y., Bennett, N., and Reis, P.N.B. 2012. Low-velocity impact response of non-woven hemp fiber reinforced unsaturated polyester composites: Influence of impactor geometry and impact velocity. *Composite Structures* 94:2756–63. With permission [14].

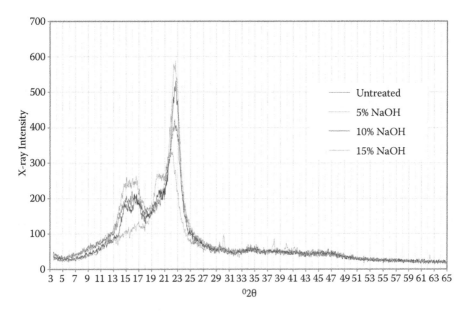

FIGURE 7.7   X-ray diffraction pattern of hemp fiber.

TABLE 7.2
Crystallinity Index of Treated and Untreated Hemp Fiber

Treatment Solution Concentration (%)	Crystallinity Index (%)
0	84
5	87
10	85
15	76

crystallinity index calculated according to the Segal empirical method [15]. The crystallinity index CI_{XRD} can be calculated from the resolved peak areas:

$$CI_{XRD}(\%) = \frac{I_{cp} - I_{am}}{I_{am}} \times 100 \qquad (7.1)$$

where I_{cp} is the intensity of the crystalline peak, and I_{am} is the height of the minimum (I_{am}) between the minimum and maximum peaks.

For example, hemp fiber was treated with different concentrations of sodium hydroxide to assess the influence of surface modification in relation to the crystallinity index. The crystallinity index was found to increase for 5% NaOH-treated samples compared with unmodified samples. This increase could be related to the removal of noncellulosic compounds from the surface of the fiber, which would then allow better packing factors for cellulose chains.

7.3.1.4   Inverse Gas Chromatography

Inverse gas chromatography (IGC) is a variation of conventional gas chromatography but, unlike analytical chromatography, the stationary phase is the sample under study whereas a known substance in the mobile phase acts as a probe molecule [16]. So, it is a versatile and powerful technique that can be used in characterizing polymers and fibers that provide information about various adsorption properties, including surface energy (dispersive and polar), free energy of adsorption, acid–base characteristics, monolayer capacity, surface area, adsorption isothermal analysis, surface heterogeneity, and permeability [17].

7.3.2   Surface Energy Characterization

In the assessment of surface characteristics of natural fibers, the contact angle is frequently used to characterize the wettability of the fiber by the matrix in an effort to characterize the efficacy of impregnation. Contact angle is an invaluable metric for understanding material surface properties, including adhesion, wettability, and solid surface-free energy. Contact angle is used to measure cleanliness, the effects of surface treatments, adhesiveness, and water repellence.

After treatment with 5% sodium hydroxide, the contact angle of the hemp fibers is reduced from 80.0° to 70.0°, whereas 10% and 15% treatment produces a greater effect on the advancing contact angle. After treatment with 10% sodium hydroxide (NaOH), the advancing contact angle of the hemp fibers is reduced to 65.0°. All the results are presented in Tables 7.3 and 7.4.

From these results, we can conclude that the 5% NaOH treatment is the most likely to have a better interface with the unsaturated polyester (UP) resin than the untreated hemp fiber. Moreover, it is the treatment that yields better impermeability between the three concentrations.

7.3.3   Thermal Stability Characterization

The processing and service temperatures of natural fiber composites are of fundamental importance for their use. It is important to assess the thermal attributes of composite materials to predict their thermal characteristics and they may be evaluated by thermogravimetric analysis (TGA) and/or differential scanning calorimetry (DSC) techniques. In TGA testing, thermal decomposition with regard to mass loss is measured through heating the sample in an inert atmosphere. Analysis is typically carried out by calculating mass percentage loss against temperature.

The DSC can produce information regarding phase changes at different temperatures and heat flow characteristics to achieve this. The higher dispersion of fibers and crystallinity between phases modifies the final DSC heat flow characteristics and is useful in general thermal performance assessment. Results from DSC may identify the glass transition temperature (T_g) of materials along with other phase change characteristics such as melting and crystallization points, when they are relevant.

TABLE 7.3

Contact Angle Results for Different Treated and Untreated Hemp and Unsaturated Polyester (UP) Composites

Sample	Contact Angle	Water Drop Shape
4L Hemp + UP composite–no treatment	≈80.00°	
UP resin	≈107.00°	
4L Hemp–no treatment, no resin	≈70.00°	
4L Hemp + UP composite–5% NaOH treatment	≈70.00°	
4L Hemp + UP composite–10% NaOH treatment	≈65.00°	
4L Hemp + UP composite–15% NaOH treatment	≈60.00°	

TABLE 7.4

Contact Angle Results for Treated and Untreated Hemp Fiber

Sample	Contact Angle	Drop Shape
Hemp fiber–no treatment	≈105.00°	
Hemp fiber–5% NaOH treatment	≈90.00°	
Hemp fiber–10% NaOH treatment	≈100.00°	
Hemp fiber–15% NaOH treatment	≈113.50°	

7.3.4 Chemical Composition of Natural Fibers

The chemical composition and some structural attributes of commonly used natural plant fibers are shown in Table 7.5. The comparison of physical and mechanical properties of different biofibers (bundles) and glass fibers is presented in Table 7.6 [20–21].

Tables 7.5 and 7.6 show the chemical, physical, and mechanical properties of various natural fibers. It is interesting to observe that seed hair cotton and flax fibers have the highest cellulose content and also show the highest tensile strength. However, this does not correspond to the modulus values. As far as modulus values are concerned, ramie, flax, and hemp show superior values than other fibers. There is no doubt that chemical composition plays an important role in the mechanical and thermal properties of natural fiber composites. But it is difficult to correlate the direct effect of chemical composition on mechanical properties. Nevertheless, from Table 7.6, it can be seen that the specific modulus of hemp and flax fibers is higher than that of E-glass fibers. This shows that these two bast fibers are potentially competitive with glass fiber [22].

TABLE 7.5

Chemical Composition and Some Structural Parameters of Some Important Biofibers

Fiber Types	Cellulose (wt.%)	Hemicellulose (wt.%)	Lignin (wt.%)	Pectin (wt.%)	Wax (wt.%)	Microfibril Angle (°)	Moisture Content (%)
Ramie	68.6–76.2	13.1–16.7	0.6–0.7	1.9	0.3	7.5	8.0
Flax	81	14	3	2	—	10	10.0
Hemp	74	18	4	1	—	8	10.8
Jute	70–75	12–15	10–15	1	—	6.2	12.6
Seed hair cotton	89–99	3–6	—	—	—	—	
Coir	38–45	24–39	22–28	—	—	0.5	8
Kenaf	39–45	30–33	26–34	—	—	0.5	
Sisal	73	13	11	1	1	20	

Sources: Satyanarayana, K.G., Arizaga, G.G.C., and Wypych, F. 2009. Biogradable composites based on lignocellulosic fibers—An overview. *Progress in Polymer Science* 34:982–1021; Nayak, S.K., Tripahy, S.S., Rout, J., and Mohanty, A.K. 2000. Coir–polyester composites: Effect on fiber surface treatment on mechanical properties of composites. *International Plastics Engineering Technology* 4:79–86. With permission [18–19].

TABLE 7.6

Physical and Mechanical Properties of Some Commonly Used Plant Fibers

Fiber Type	Density (g/cm³)	Diameter (μm)	Elongation at Break (%)	Tensile Strength (MPa)	Young's Modulus (GPa)	Specific Modulus $\left(\dfrac{E}{\rho}\right)$ (GPa)
Jute	1.3–1.45	25–200	1.5–1.8	393–773	10–30	7–21
Flax	1.50	—	2.7–3.2	345–1035	60–80	26–46
Hemp	1.14		1.6	690	30–70	21–50
Cotton	1.5–1.6	287–800	7.0–8.0	287–800	5.5–12.6	
Ramie	1.50	—	1.2–3.8	400–938	61.4–128	
E-glass	2.50		2.5	2000–3500	70.0	28

7.4 TESTING OF NATURAL FIBER COMPOSITES

Natural fiber-reinforced composites based on polymer matrices containing different weight percentages of reinforcement usually show significant improvement in various properties compared with those of unreinforced neat matrices. Property improvements include increased strength, modulus, hardness, and thermal stability.

7.4.1 MECHANICAL PROPERTIES

7.4.1.1 The Reinforcing Effects on the Strength and Modulus

The properties of natural fiber composites depend on the properties of the individual components (i.e., fibers and matrix) as for any composite system. However, it is recognized that the inherent heterogeneous nature of composite materials means that the interface region between the fiber and matrix also plays an important role in defining the composite properties [23]. The renewed interest in the use of natural fiber composites in primary structural applications has also resulted in attention being focused on the role of defects, and in particular voids, on the reduction of the performance of these composite materials. With most synthetic composite materials, such as glass/polyester and carbon/epoxy, considerable knowledge has been built over the years to control and optimize the fabrication process and the void content, which is usually low (<2%) [24]. As far as void content in natural fiber composites is concerned, fabrication techniques are in their infancy and the natural origin of the fiber component exacerbates the problem. Both these factors contribute to the creation of voids, which make a noteworthy contribution to the overall composite volume and compromise overall composite performance [25].

Fibers act to reinforce and oppose deformation within the matrix due to their high strength and moduli. The matrix transfers applied loads to reinforcement provided that the adhesion between the two phases is sufficient. From this phenomenon, it can be assumed that the higher the aspect ratio of the reinforcement in contact with the polymer, the larger the reinforcing effect should be.

The structural reliability of a composite depends on the interface between the matrix and the fibers. One of the most significant factors to be considered is its load-bearing capacity and mechanical properties. It is obvious that the reinforcement must be the strongest component if it is to give good mechanical properties. So, the reinforcement must have a higher elastic modulus, and the bond between the matrix and the reinforcement is critical, since good interfacial adhesion between the matrix and the fibers transfers the stress from the matrix to the fibers, improving the mechanical properties of the composites [26].

Overall, the properties of the composite are determined by the following features:

- The properties of the fiber
- Void content
- The properties of the resin
- The ratio of fiber/resin in the composite (fiber volume fraction [FVF])
- The geometry and orientation of the fibers in the composite
- Manufacturing processes

The restriction on the ratio of the fiber to resin used is largely derived from the manufacturing process. As the volume of natural fiber increases in relation to the composite volume, the wettability of the resin decreases. This reduced wetting contributes to an increased void content, and as a result the practical maximum obtainable fiber volume fraction in the composite will be notably decreased. This is one of the problems associated with natural fiber composite fabrication. However, it is also influenced by the type of resin system used, and the form in which the fibers are incorporated. In general, since the mechanical properties of the fibers are much higher than those of the resin, a higher fiber volume fraction provides greater mechanical properties of the resultant composite. In practice, there are limits to this, since the fibers need to be fully coated in the resin to be effective, and there will be an optimum packing of generally circular cross-sectional fibers. In addition, the manufacturing process used to combine fibers with the resin may lead to varying amounts of air inclusions that are attributed to defects.

The geometry of the fibers in a composite is also important, since fibers have their highest mechanical properties along their length, rather than across their widths. This leads to the highly anisotropic properties of composites, where unlike metals the mechanical properties of the composite are likely to be very different when tested in different directions. This means that it is very important when considering the use of composites to understand at the design stage both the magnitude and the direction of the applied loads. When correctly accounted for, these anisotropic properties can be very advantageous, since it is only necessary to put material where loads will be applied, and thus redundant material is mitigated.

The tensile and flexural testing is an important part in the process of developing natural fiber composites. From an engineering point of view, the three most important tensile properties are tensile modulus (stiffness), ultimate stress, and strain to failure. The evidence of the influence of the reinforcements on the tensile and flexural properties of hemp fiber composites has been previously studied [27]. An experimental analytical approach was used to reveal how the effect of fiber volume fraction influences the mechanical properties. The tensile strength and modulus of hemp composites were analyzed by using the standard practice presented within BS EN 2747: 1998, ASTM D 2256, ISO 2062.

Their results suggested that the tensile strength and modulus is gradually increased with increasing fiber content, most notably in the range of 0.15 and 0.26 fiber volume fractions. The increasing trend in the tensile strength and tangent modulus shows a direct correlation between fiber weight fraction and tensile strength and stiffness. For low loading fiber content, especially for a 0.1 volume fraction of hemp, the increase in the tensile strength and modulus is very small compared with higher fiber loading. This effect could be due to the poor distribution of the fibers in the matrix with a tendency to form resin-rich areas in low fiber loading; hence, the tensile loads are not effectively transmitted between the fibers.

It is believed that at low loading levels, imperfections in the laminate such as misalignment, fiber crimp, and void inclusion introduce discontinuities in the reinforcement and would account for the decrease in strength and stiffness. This disproportionality of reinforced fiber at low loading levels leads to lower quality specimens with high degrees of stress concentrations. Such defects act as a stress raiser, which leads to lowering the stiffness. It is also possible that they contain a greater proportion of porosity than for higher fiber loadings. It is also worth noting that the addition of the fiber (above a threshold value) seems to reduce the wettability of fiber, hence reducing fiber matrix interfacial adhesion. For stresses to be transferred more effectively from the polymer matrix to the reinforcements, good adhesion between the fiber and matrix is, therefore, important.

7.4.2   FACTORS AFFECTING THE STRENGTH AND MODULUS

There are many factors that may affect the structure and properties of natural fiber composites. The first factor that plays an important role in the properties of natural fiber composites could be the fiber volume fraction. It is generally acceptable that the reinforcement has an optimal level that should be identified for the tensile strength and modulus. It was suggested that the fiber loading affects the resulting structure. Furthermore, the stress–strain behavior of the fiber and matrix as well as fiber geometry and orientation also play important roles. Tensile strength is sensitive to matrix fiber interfacial characteristics, whereas the modulus is more dependent on fiber properties and fiber volume fraction.

Fiber defects can play a major role in the tensile and flexural properties of natural fiber composites. Depending on the degree of resin impregnation, the flexural strength can decrease with the increased fiber content; however, the flexural modulus can increase with the increased fiber content. So, it can be emphasized that strength is more dependent on the fiber matrix adhesion whereas the modulus is predominantly related to the volume of reinforcement.

7.4.3   EFFECTS OF SURFACE TREATMENT ON COMPOSITE ATTRIBUTES

As with synthetic fiber composites, integration of reinforcement through proper wetting of the fiber is vital for enhanced performance of the composites [28]. Much work has been done pertaining to chemically modifying active sites of reinforcement using acids or bases to enhance bonding potential and reduce hydrophilic tendency. Work has also been conducted using natural white rot fungi and other biological organisms in bagging methods to strip active sites [29].

However, in most cases, this surface modification process is reversible, making it undesirable for extended service life during prolonged water exposure [30]. Conventional sizing agents work through either achieving a chemical bond between the matrix and reinforcement or alternatively bonding to the reinforcement and altering the surface energy to aid resin wetting. Different resins have different surface energy characteristics as well as bonding mechanisms, and, therefore, different sizing agents are required specifically for each composite system. Processing of such materials during sizing and matrix integration then becomes a major challenge as temperature, pressure, and humidity constrain the type of sizing agent used. Natural reinforcement materials provide an even larger challenge, as most are porous and highly polar in nature. This makes complete wetting by the resin phase problematic. However, if achieved, this would enhance overall performance characteristics extensively [31].

Work has been conducted on the use of natural-based surface modifiers such as methylolated lignin, which seems to show improved fiber matrix compatibility in phenolic resin and reduced water ingress. Further work in this area is of key interest in advancing sustainable composites further for structural applications. Currently, work has shown that after processing the methylolated lignin, water absorption was reduced further. However, it is apparent that as such materials had substantially lower impact strength (25%) while still possessing some hydroxyl sites on the sizing agent, it may be concluded that synthetic sizing agents still have an important role to play in the foreseeable future with regard to structural applications [32–33].

The majority of sizing agents currently used are silanes, which are small molecules based on silicon–oxygen backbone chemistry. Due to larger bond angles and covalent bond potential similar to carbon–carbon polymer chains, highly diverse molecules with various and multiple functional groups may be produced. Many silanes (alkyl) containing ethoxy or methoxy groups work by reacting with water to produce ethanol/methanol as a by-product with water, leaving reactive oxygen sites on the molecule for hydrogen bonding. At this stage, the modified molecules (called silanol) then start reacting with each other to produce networks (assuming multiple functional sites) and with the substrate, producing a thin film. Monolayers can be produced depending on the molecule shape and number of functional groups. Typically, this is a two-stage hydrolysis and condensation reaction. These alkyl functionalities are exceptionally useful for bonding silanes to active hydroxyl sites on natural fibers, although others are ideally needed for chemical bonding later with the matrix resin. Silanes have excellent thermal stability and low viscosity favoring complete wetting and impregnation of porous structures [32]. Silane functionalities vary considerably as shown in Table 7.7. Their diverse compatibility potential in conventional matrix materials highlights their integration potential in natural composite systems.

This makes them ideally suited as tailored sizing agents for a variety of processing conditions and service life requirements [34]. For this reason, silanes are also used in many diverse applications, including water-repellent poreliner treatments for masonry and timber, a key component in caulking silicone sealant cross-linkers, internal florescent light tube coatings for enhanced energy transmittance, and antifog mirror coatings. Silanes have also shown to be of interest for superhydrophobic thin films using nanoparticulates that aid hydrophobic attributes. Applications for nanoparticulate-enhanced treatments may include chemical-resistant textiles and self-cleaning surfaces [35]. Copper nanoparticulate-enhanced silane solutions have been used in timber treatments to reduce biofouling and biodegradation [36]. Due to the nature of the treatments, a low viscosity allowed for excellent infiltration characteristics while also reducing the amount of biocide material required to protect the timber. This makes silanes and nanoparticulate incorporation into sizing agents a more attractive technique in improving the properties of natural fiber-reinforced composites.

Nanoparticulates could not only aid interfacial energy adhesion but also depending on the particulates used be tailored for the composite-specific task for low quantities of modifying agent. For instance, nanoclays could be used to impart a better fire resistance allowing for a structure to perhaps still be mechanically stable for longer during a fire [37]. This could be taken further with self-extinguishing materials also added to help make greener materials comparable to the fire-resistant properties that phenolic resins provide [38].

Figure 7.8 shows how nanoparticulate modification of such treatments has great potential for practical use and development in the field. The aim of the investigation given below was to assess the use of two sizes of silica nanoparticulates (7 nm, 14 nm) at different concentrations for use in silane/siloxane treatments to modify water sorptivity potential of natural composite reinforcements. Water absorption in natural fiber composites is a major challenge. Three main processes have been

TABLE 7.7

Silane Functionalities Applied Successfully to Some Known Matrices

Functionality	Known Matrices Where Functionality Has Been Successfully Applied
Amino	Epoxy
	Polyethylene
	Butyl rubber
	Polyacrylate
	Polyvinyl chloride
Vinyl	Polypropylene
	Polyethylene
	Polyacrylate
Methyacryl	Polyethylene
	Polyester
Mercapto	Natural rubber
	Poly vinyl chloride
Glycidoxy	Epoxy
	Butyl rubber
	Polysulfide
Chlorine	Polyvinyl chloride
	Polyethylene
Azide	Polyethylene
	Polypropylene
	Polystyrene
Alkyl	Natural rubber
	Polyethylene

Source: George, J., Sreekala, M.S., and Thomas, S. 2001. A review on interface modification and characterization of natural fiber reinforced plastic composites. *Polymer Engineering and Science* 41:1471–85. With permission [32].

observed: (1) diffusion of water molecules in micropores between polymer chains, (2) capillary transport between interfacial gaps attributed to incomplete wetting, and (3) matrix microcrack water transportation, which is negligible and may be discounted in most scenarios. In general, a significant reduction in mechanical properties and degradation occurs due to swelling and debonding [39]. Such treatments, although they do not solve the issue of chemical adhesion to the respective phases of the composite, highlight how influential nanoparticulates may be in such sizing treatments when properly functionalized. Four treatments were formulated and based on the following materials: $1H,1H,2H,2H$-perfluorooctyltriethoxysilane (FAS), n-iso-octyltriethoxysilane (ITES), trimethoxyhexadecylsilane (THS), and OH-terminated polydimethylsiloxane (PDMS) 3500 cSt. The FAS was selected for its low surface energy characteristics on smooth surfaces, although its expense makes it currently not suitable for any practical applications. The ITES and THS were selected for their water-repellent characteristics and cost-effective nature. The PDMS was selected

FIGURE 7.8   Superhydrophobic nanoparticulate-enhanced polydimethylsiloxane thin film on glass substrate.

as a cost-effective control. The mentioned materials were mixed with white spirit at 8%wt concentration and then decanted and mixed to respective concentrations and ratios of silica nanoparticles (0%wt, 1%wt, 1.5%wt, 2%wt, 2.5%wt, and 3%wt). These treatments were then sprayed on glass microscope slides and left to dry for 1 week before contact angle assessment.

For a flat surface such as glass, the theoretical water contact angle that may be achieved is 120°. For a rough surface, the theoretical water contact angle is 180°. Results showed that 25%wt 7 nm and 75%wt 14 nm nanoparticulates exhibited the best surface modification when hybridized at different ratios. The best results produced were achieved at 2.5%wt nanoparticulate concentration overall. Table 7.8 shows the contact angles achieved by each treatment at 2.5%wt concentration.

Table 7.8 also shows that PDMS did very well at imparting hydrophobicity of neat and modified solutions, although its viscosity would limit its use for impregnation of natural reinforcements. Due to its trifunctional ethoxy groups, free ITES hydroxyl groups may partially be present at the interface; this is due to its cured structure providing a multilayer interpenetrating network. Despite its suitability for water-repellent solutions, it may not be the best material for a monolayer sizing agent that is ideally required for better composite performance. FAS worked effectively, whereas THS showed comparable performance attributes when used in combination with the nanoparticles. This means that the commercially viable THS has performance characteristics that are comparable to FAS in this scenario, showing exciting possibilities for other functional silanes for phase grafting and nanoparticulate enhancement.

TABLE 7.8
Silane Treatment Results

Treatment Type	Mean Water Contact Angle (°)
THS no nanoparticles	93.47
FAS no nanoparticles	104.10
ITES no nanoparticles	81.00
PDMS no nanoparticles	109.14
THS with 2.5%wt silica (25% 7 nm:75% 14 nm)	150.76
FAS with 2.5%wt silica (25% 7 nm:75% 14 nm)	149.00
ITES with 2.5%wt silica (25% 7 nm:75% 14 nm)	64.00
PDMS with 2.5%wt silica (25% 7 nm:75% 14 nm)	174.00

THS, trimethoxyhexadecylsilane; FAS, $1H,1H,2H,2H$-perfluorooctyltriethoxysilane; ITES, n-isooctyltriethoxysilane; PDMS, polydimethylsiloxane.

Testing and evaluation of such treatments in future studies would require surface energy assessment of the matrix and reinforcement and then a silane selected with suitable functional groups for both phases. Aspect ratios and morphological characteristics of the silane would also need to be considered to get suitable coverage and bonding to active sites at the interface. Assuming an emulsion delivery system is used, surface tension of the silane would have to be investigated to help ensure that suitable wetting of the reinforcement could be achieved after emulsion deposition. Interestingly, nanoparticulate-stabilized emulsions are currently of interest in a variety of fields and could be useful for reducing excessive stabilizers that are being integrated for such systems [40].

7.5   CONCLUSIONS AND FUTURE OUTLOOK

Due to environmental legislation, consumer awareness, and economic considerations, natural fiber composites are becoming viable alternatives within several industrial sectors as they make changes to their design criteria and material selection to achieve sustainable material and product development. Natural fiber composites show great potential in structural and semi-structural applications provided their inherent property variation and hydrophilic tendency are addressed through diligent processing, suitable surface modification, testing, and evaluation throughout production.

Mechanical properties through efficient transfer of load from matrix to fiber reinforcement are highly desired critical attributes in all fiber-reinforced composites. The structure and the chemical composition of fibers play an integral role in the final properties of the corresponding composites. Therefore, it is important to use suitable testing and characterization methods to evaluate the properties of natural fiber-reinforced composites. Specifically for natural composites, high susceptibility to water ingress currently limits their use in load-bearing applications. It is important that this challenge is met to deploy these materials effectively in structural applications.

Surface modification is critical in addressing natural composite hydrophilic behavior in structural applications. To achieve this, correct treatments must be selected not only for water-repellent attributes but also for enhanced matrix-reinforcement integration. Such modification ideally should be a monolayer in nature to enhance mechanical and thermal load distribution in such composites.

Surface modification may be further enhanced through the integration of relevant nanoparticles to further protect and enhance structural and service life properties. Such attributes may include water sorptivity reduction within the fiber through high aspect ratio of nanoplatelets, and enhancement of the biodeterioration resistance through copper or silver nanoparticulate incorporation. It is assumed that such integration would lead to synergistic traits for the overall composite. From this chapter, it can be seen that exciting advances may be developed within this field in the near future.

ACKNOWLEDGMENT

The authors express their gratitude to Nicolas Kerlidou for his assistance and hard work on the development of the nanoparticulate-enhanced silane treatments considered within this chapter.

REFERENCES

1. Richardson, M.O.W., Santana, T.J., and Hague, J. 1998. Natural fiber composites—The potential for the Asian markets. *Progress in Rubber and Plastics Technology* 14:174–88.
2. Kalia, S., Kaith, B.S., and Kaur, I. 2009. Pre-treatment of natural fibers and their applications as reinforcing material in polymer composites: A review. *Polymer Engineering and Science* 49:1253–72.
3. Bismarck, A., Askargorta, I.A., Springer, J., Lampke, T., Wielage, B., Stamboulis, A., Shenderovich, I., and Limbach, H.H. 2002. Surface characterization of flax, hemp and cellulose fibers: Surface properties and the water uptake behaviour. *Polymer Composites* 23:872–94.
4. Richardson, M.O.W. 1977. *Polymer Engineering Composites.* Essex, England: Applied Science Publishers Ltd.
5. Bledzki, A.K. and Gassan, J. 1997. Natural Fibre Reinforced Plastics. In: Cheremisinoff, N.P., ed., *Handbook of Engineering Polymeric Materials.* New York: Marcel Dekker, Inc.
6. Joshi, S.V., Drzal, L.T., Mohanty, A.K., and Arora, S. 2004. Are natural fibers composites environmentally superior to glass fibers reinforced composites? *Composites: Part A* 35:371–6.
7. Eichhorn, S.J. and Young, R.J. 2004. Composite micromechanics of hemp fibers and epoxy resin microdroplets. *Composites Science and Technology* 64:767–72.
8. Eichhorn, S.J., Baillie, C.A., Zafeiropoulos, N., Maikambo, L.Y., Ansell, M.P., Dufresne, A., Entwistle, K.M., Herrera-Franco, P.J., Escamilla, G.C., Groom, L., Hughes, M., Hill, C., Rials, T.G., and Wild, P.M. 2001. Review. Current international research into cellulosic fibers and composites. *Journal of Materials Science* 36:2107–31.

9. Summerscales, J., Dissanayake, N. P. J., Virk, A. S., and Hall, W. 2010. A review of bast fibers and their composites. Part 1—Fibres as reinforcements. *Composites: Part A* 41:1329–35.

10. Dhakal, H.N., Zhang, Z.Y., and Bennett, N. 2012. Influence of fiber treatment and glass fiber hybridisation on thermal degradation and surface energy characteristics of hemp/unsaturated polyester composites. *Composites: Part B* 43:2757–61.

11. Kalia, S., Kaith, B.S., and Kaur, I. 2009. Pre-treatment of natural fibers and their applications as reinforcing material in polymer composites: A review. *Polymer Engineering and Science* 49:1253–72.

12. Yan, L., Chouw, N., and Jayaraman, K. (2014). Flax and its composites: A review. *Composites Part B: Engineering* 56:296–317.

13. Olsson, A.M. and Salmen, L. 2004. The association of water to cellulose and hemicellulose in paper examined by FTIR spectroscopy. *Carbohydrate Research* 339:813–8.

14. Dhakal, H.N., Zhang, Z.Y., Bennett, N., and Reis, P.N.B. 2012. Low-velocity impact response of non-woven hemp fiber reinforced unsaturated polyester composites: Influence of impactor geometry and impact velocity. *Composite Structures* 94:2756–63.

15. Reddy, N. and Yang, Y. 2005. Structure and properties of high quality natural cellulose fibers from cornstalks. *Polymer* 46:5494–500.

16. Jam, M. S. and Waters, K.E. 2014. Inverse gas chromatography applications: A review. *Advances in Colloid and Interface Science* 212:21–44.

17. Cordeiro, N., Gouveia, C., Moraes, A.G.O., and Amico, S.C. 2011. Natural fiber characterisation by inverse gas chromatopraphy. *Carbohydrate Polymers* 84:110–7.

18. Satyanarayana, K.G., Arizaga, G.G.C., and Wypych, F. 2009. Biogradable composites based on lignocellulosic fibers—An overview. *Progress in Polymer Science* 34:982–1021.

19. Nayak, S.K., Tripahy, S.S., Rout, J., and Mohanty, A.K. 2000. Coir–polyester composites: Effect on fiber surface treatment on mechanical properties of composites. *International Plastics Engineering Technology* 4:79–86.

20. Faruk, O., Bledzki, A. K., Fink, H. P., and Sain, M. 2012. Biocomposites reinforced with natural fibers: 2000–2010. *Progress in Polymer Science* 37:1552–96.

21. Wambua, P.W., Ivens, J., and Verpoest, I. 2003. Natural fibers: Can they replace glass in fiber reinforced plastics? *Composites Science and Technology* 63:1259–64.

22. Mohanty, A. K., Misra, M., and Hinrichsen, G. 2000. Biofibers, biodegradable polymers and biocomposites: An overview. *Macromolecular Materials and Engineering* 276/277:1–24.

23. Thomason, J.L. 1995. The interface region in glass fiber-reinforced epoxy resin composites: Water absorption, voids and the interface. *Composites* 26:477–85.

24. Lystrup A. 1998. Hybrid yarns for thermoplastic fiber composites. Riso-R-1034 (EN). Roskilde, Denmark: Riso National Laboratory.

25. Madsen, B. and Lilholt, H. 2003. Physical and mechanical properties of unidirectional plant fiber composites: An evaluation of the influence of porosity. *Composites Science and Technology* 63:1265–72.

26. Bisinda, E.T.N. and Ansell, M.P. 1991. The effect of silane treatment on the mechanical and physical properties of sisal-epoxy composites. *Composites Science and Technology* 41:165–78.

27. Dhakal, H.N., Zhang, Z.Y., and Richardson, M.O.W. 2007. Effect of water absorption on the mechanical properties of hemp fiber reinforced unsaturated polyester composites. *Composites Science and Technology* 67:1674–83.

28. Guo, L., Chen, F., Zhou, Y., Liu, X., and Xu, W. 2015. The influence of interface and thermal conductivity of filler on the nonisothermal crystallization kinetics of polypropylene/natural protein fiber composites. *Composites: Part B* 68:300–9.

29. Li, Y, Pickering, K.L., and Farrell, R.L. 2009. Analysis of green hemp fiber reinforced composites using bag retting and white rot fungal treatments. *Industrial Crops and Products* 29:420–6.
30. Thakur, V.K. and Thakur, M.K. 2014. Processing and characterization of natural cellulose fibers/thermoset polymer composites. *Carbohydrate Polymers* 109:102–17.
31. Xie, Y., Hill, C.A.S., Xiao, Z., Militz, H., and Mai, C. 2010. Silane coupling agents used for natural fiber/polymer composites: A review. *Composites: Part A* 41:806–19.
32. George, J., Sreekala, M.S., and Thomas, S. 2001. A review on interface modification and characterization of natural fiber reinforced plastic composites. *Polymer Engineering and Science* 41:1471–85.
33. Mwaikambo, L. Y. and Ansell, M. P. 2004. Chemical modification of hemp. Sisal, jute and kopok fibers by alkalisation. *Journal of Applied Polymer Science* 84: 2222–34.
34. Luo, H., Xiong, G., Ma, C., Chang, P., Yao, F., Zhu, Y., Zhang, C., and Wan, Y. 2014. Mechanical and thermo-mechanical behaviors of sizing-treated corn fiber/polylactide composites. *Polymer Testing* 39:45–52.
35. MacMullen, J., Zhang, Z., Dhakal, H.N., Radulovic, J., Karabela, A., Tozzi, G., Hannant, S., Alshehri, M.A., Buhe, V., Herodotou, C., Totomis, M., and Bennett, N. 2014. Silver nanoparticulate enhanced aqueous silane/siloxane exterior facade emulsions and their efficacy against algae and cyanobacteria biofouling. *International Biodeterioration & Biodegradation* 93:54–62.
36. Mantanis, G., Terzi, E., Kartal, S.N., and Papadopoulos, A.N. 2014. Evaluation of mould, decay and termite resistance of pine wood treated with zinc- and copper-based nanocompounds. *International Biodeterioration & Biodegradation* 90:140–4.
37. Ayana, B., Suin, S., and Khatua, B.B. 2014. Highly exfoliated eco-friendly thermoplastic starch (TPS)/poly (lactic acid) (PLA)/clay nanocomposites using unmodified nanoclay. *Carbohydrate Polymers* 110:430–9.
38. Megiatto Jr., J.D., Silva, C.G., Rosa, D.S., and Frollini, E. 2008. Sisal chemically modified with lignins: Correlation between fibers and phenolic composites properties. *Polymer Degradation and Stability* 93:1109–21.
39. Pan, Y., and Zhong, Z. 2014. A nonlinear constitutive model of unidirectional natural fiber reinforced composites considering moisture absorption. *Journal of the Mechanics and Physics of Solids* 69:132–42.
40. MacMullen, J., Radulovic, J., Zhang, Z., Dhakal, H.N., Daniels, L., Elford, J., Leost, M.A., and Bennett, N. 2013. Masonry remediation and protection by aqueous silane/ siloxane macroemulsions incorporating colloidal titanium dioxide and zinc oxide nanoparticulates: Mechanisms, performance and benefits. *Construction and Building Materials* 49:93–100.

8 Environment-Related Issues

Samrat Mukhopadhyay

CONTENTS

8.1 INTRODUCTION

The industrial-scale production of synthetic polymers started in the 1940s. Since then, the production, consumption, and waste generation rate of plastic solid waste (PSW) has increased considerably. Thus, waste recycling has been a center of attention of many researchers in the past few decades. Such research is often driven by changes in regulatory and environmental issues. Plastics are used in various applications—from coating and wiring, to packaging, films, covers, bags, automobiles, containers, and aircrafts. Thus, there is a huge amount of waste generated, which for environmental reasons needs to be recycled and reused. The increasing cost of landfill disposal and interest in support of recycling has meant that such recycling of materials should increase. Natural fibers have gained importance as reinforcement in composite materials. The majority of the research has been conducted on natural fibers used as reinforcement with plastics. Plastics are made from crude oil. Recycling of plastics would also mean saving of this crucial natural resource. Several issues have gained relevance—recycling of such materials, use of natural fibers in recycled polymers to improve mechanical properties, their water absorption and effect on properties, issues such as incineration, and life cycle analysis (LCA) for this class of composites. In this chapter, these issues and the latest research in this field are discussed.

Any strategy of waste management is principled on three basic guidelines [1–2]:

1. Avoidance—avoiding production of waste at source
2. Reclamation—the recovery of materials from the waste stream for recycling
3. Elimination—disposal of nonrecyclable materials, for example, in landfills

8.2   RECYCLING

There are three popular kinds of recycling: primary, secondary, and tertiary. Primary recycling involves use of the same component again through an extrusion process and is often called "re-extrusion." Low-density polyethylene (LDPE) products that may not have met the specifications are often pelletized and reused through injection molding. The properties are slightly degraded, though in primary recycling the end products are of the same type as the material from which the polymer is extracted for recycling.

The second in the category is secondary recycling. It can be performed on single-polymer plastic, for example, polyethylene (PE), polypropylene (PP), and polystyrene (PS). Mechanical recycling refers to operations that aim at recovering plastic waste via mechanical processes (grinding, washing, separating, drying, re-granulating, and compounding), thus producing recyclates that can be converted into new plastic products and often substituting for virgin plastics. For mechanical recycling, only thermoplastic materials are of interest, for example, polymeric materials that may be remelted and reprocessed into products via techniques such as injection molding or extrusion [3].

Thus, the difference between the two processes lies in their end use. Primary recycling, also known as closed-loop recycling, ensures that products are recycled into products of the same type, for example, aluminum cans into aluminum cans. In secondary recycling (also known as open-loop recycling), products are converted into different products, for example, tires are converted into other rubber products [4].

There is another class called tertiary recycling and it involves a chemical route. The term implies a set of advanced processes that convert plastic materials into smaller fragments, usually liquids or gases, which are suitable for use as a feedstock for the production of new petrochemicals and plastics. An alteration is bound to occur to the chemical structure of the polymer. Products of chemical recycling have proved to be useful as fuel. The technology behind its success is the depolymerization processes that can result in a very lucrative and sustainable industrial scheme, providing a high product yield and minimum waste. Under the category of chemical recycling, advanced process appear, for example, pyrolysis, gasification, liquid–gas hydrogenation, viscosity breaking, steam or catalytic cracking, and the use of PSW as a reducing agent in blast furnaces [5–6]. The main advantage of chemical recycling is the possibility of treating heterogeneous and contaminated polymers with limited use of pre-treatment. If a recycler is considering a recycling scheme with a 40% target or more, one should deal with materials that are very expensive to separate and treat.

As discussed by Zia et al. [7], increasing cost and decreasing space of landfills are forcing considerations of alternative options for PSW disposal. Howard [8] suggested economically and environmentally viable methods in this regard, though his study was based on polyurethane (PU). The plastic industry has successfully identified workable technologies for recovering, treating, and recycling of waste from discarded products. In 2002, 388,000 tonnes of PE were used to produce various parts of textiles, of which 378,000 tonnes were made from PE discarded articles. The plastic industry is committed to meeting current needs without compromising future needs. In the United Kingdom, 95% of PSW arising from process scrap ($\approx$250,000 tonnes) was recycled in 2007, which was a significant achievement. The recycling is done for a number of end-products, including automobile parts, appliances, textiles, mulches, greenhouses, and films [5]. As discussed by Kang et al. [9], sorting is the most crucial step in the recycling loop. Even with natural fiber composites, which have been used outside of the automobile industry, the serious challenge that a recycler would be facing is removal of paint on the finished material. Grinding and solvent stripping are some of the popular ways in which the surface is made free of such exterior coatings.

8.3   USING NATURAL FIBERS IN RECYCLED POLYMERS TO IMPROVE MECHANICAL PROPERTIES

The use of natural fiber as an alternative reinforcement material has increased and received a lot of attention in the past decade. The interest comes from a combination of very good properties such as high-specific strength and stiffness, natural availability, ease of processing, and environmental friendliness. A very interesting use of natural fibers can be as reinforcement of recycled plastics. Recycled plastics, which are close to the end of their lifetime, can be reused with natural fibers. This can be clearly considered an environmentally friendly option because of advantages with respect to end-of-time disposal by incineration or thermal recycling. A huge amount of polyolefins used in packaging can be potentially recovered for recycling. However, contamination of these materials with paper such as from labels on containers, sales receipts left in plastic bags, and so on is common. The tensile strength of the recycled plastics is usually good, whereas the materials generally have fairly low stiffness and creep resistance, limiting their use in a number of applications. If these recycled plastics are combined with wood or other natural fibers in a composite structure, the stiffness and creep resistance can be substantially improved.

Much work has been done on virgin thermoplastic and natural fiber composites, which have successfully proved their applicability to various fields of technical applications, especially for load-bearing application. Thermoplastics such as PE, PP, PS, and poly(lactic acid) (PLA) have been compounded with natural fibers such as sisal, jute, kenaf, flax, hemp, cotton, Kraft pulp, coconut husk, areca fruit, pineapple leaf, oil palm, henequen leaf, ovine leather, banana, abaca, and straw to manufacture composites. Work on recycled plastic/natural fiber systems is still limited. The subsequent section discusses the work done with recycled plastic used as a matrix or recycled natural fiber/material used as reinforcement.

In a work by Singleton et al. [10], composite laminate based on natural flax fiber and recycled high-density polyethylene (HDPE) was manufactured by a hand layup and compression molding technique. The researchers investigated the mechanical properties of composites under tensile and impact loading. Changes in the stress–strain characteristics, of yield stress, tensile strength, and tensile modulus, of ductility and toughness, all as a function of fiber content, were determined experimentally. A significant enhancement of toughness of the composite was observed. This can be qualitatively explained in terms of the principal deformation and failure mechanisms identified by optical microscopy and scanning electron microscopy (SEM). These mechanisms were dominated by delamination cracking, crack-bridging processes, and extensive plastic flow of polymer-rich layers and matrix deformation around fibers. Improvements in strength and stiffness combined with high toughness can be achieved by varying the fiber volume fraction and controlling the bonding between layers of the composite.

In another work by AlMaadeed et al. [11], recycled polypropylene (RPP)-based hybrid composites of date palm wood flour/glass fiber were prepared by different weight ratios of the two reinforcements. The mixing process was carried out in an extruder, and samples were prepared by an injection molding machine. The RPP properties were improved by reinforcing it by wood flour. The tensile strength and Young's modulus of wood flour-reinforced RPP were increased further by adding glass fiber. Glass fiber-reinforced composites showed higher hardness than other composites. Morphological studies indicated that glass fiber has good adhesion with RPP, supporting the improvement of the mechanical properties of hybrid composites with glass fiber addition. Addition of as little as 5 wt.% glass fiber to wood flour-reinforced RPP increases the tensile strength by about 18% relative to the wood flour reinforcement alone. An increase in wood particle content in the PP resulted in a decrease in the degree of crystallinity of the polymer. The tensile strength of the composites increased with an increase in the percentage of crystallinity when adding the glass fiber. The improvement in the mechanical properties with the increase in crystallinity percentage (and with the decrease of the lamellar thicknesses) can be attributed to the constrained region between the lamellae, because the agglomeration was absent in this case.

Najafi et al. [12] studied water absorption of wood-plastic composites (WPC) prepared from sawdust and virgin and/or recycled plastics such as HDPE and PP. Wood flour was prepared from sawdust and mixed with different virgin or recycled plastics at 50% by weight fiber loading. The mixed materials were compression molded into panels. Long-term water absorptions of manufactured WPCs were evaluated by immersing them in water at room temperature for several weeks (1750 hours). Water diffusion coefficients were also calculated by evaluating the water absorption isotherms. Results indicated that maximum water absorption and diffusion coefficients increased by increasing the proportion of recycled HDPE and/or PP. Time to reach saturation also decreased at higher recycled plastic contents. Water absorption of the studied composites was proved to follow the kinetics of a Fickian diffusion process.

In another article by Khalil et al. [13], T-compounding of RPP and wood sawdust (WSD) were carried out using five different filler loadings (0, 10, 20, 30, 40, and 50%) with three WSD filler sizes (100, 212, and 300 mm). The composites were mixed and extruded using a Haake Rheodrive 500 twin screw extruder. The mechanical and water absorption properties of composites were studied. The results show that composites with a smaller particle size (100 mm) had remarkably higher properties compared with others (212 and 300 mm). Composites filled up to 30% WSD exhibit improved mechanical properties, but the value dramatically decreased above 30% filler loading. The evidence of the fiber–matrix interphase was analyzed using an SEM.

The use of surface modifiers has also been studied by researchers. In research by Cui et al. [14], wood/recycled plastic composite (WRPC) material was fabricated with postconsumer HDPE and wood fiber using a single-screw extruder. To improve the interfacial adhesion between the wood fiber and the HDPE, a coupling agent, maleic anhydride-modified polypropylene (MAPP), together with three surface treatments (an alkaline method, a silane method, and an alkaline followed by silane method) were used to treat the wood fibers. The surface chemistry of the treated fibers was evaluated using Fourier transform infrared spectroscopy (FTIR) techniques. The effects of wood fiber length, weight fraction, and surface treatment on the mechanical properties of WRPC materials were investigated. Fracture surfaces of tested WRPC specimens were examined, and the fracture mechanism of WRPC materials was also discussed in this paper. Test results indicate that WRPC material with wood fiber treated by the alkaline followed by silane treatment method together with the MAPP coupling agent possesses good mechanical properties. The content of wood fiber affects the flexural strength, flexural modulus, and impact strength of these WRPC materials.

Wang et al. [15] used wood chips that were from recycled wood wastes from different wood species. The chips were subsequently divided into coarse chips with dimensions of a 5–8 mesh and fine chips of an 8–20 mesh. These chips were immersed in water-soluble phenol formaldehyde (PF) resin solution at concentrations of 4.5, 6.5, and 10%. After 5 minutes, they were removed from the PF solution and dried in an oven until in a half-hardened condition. Three-layer mats with target densities of 0.70 and 0.80 g/cm^3 were formed by using fine chips for the face layer (25%) and back layer (25%) and coarse chips for the core layer (50%). A conventional hot press was used for fabrication of the particleboard, and the temperature and pressure were 453 K, 2.9 MPa. The researchers observed that PF resin absorption content of chips increased linearly with an increase in the concentration of PF solution. Their relationship could be represented by the linear regression formula. The bending strength, internal bonding strength, and thickness swelling (percentage) of the PF-impregnated particleboards exhibited excellent performance and the retention rate (percentage) of the modulus of rupture (MOR) and modulus of elasticity (MOE) of PF-impregnated particleboard after treatment in hot water (343 K) for 2 hours were maintained at 48.7–84.5% and 49.2–82.7%, respectively.

Najafi et al. [16] studied mechanical properties of WPC manufactured from sawdust and virgin and/or recycled plastics, namely HDPE and PP. Sawdust was prepared from beech industrial sawdust by screening to the desired particle size and was mixed with different virgin or recycled plastics at 50% by weight fiber loading. The mixed materials were then compression molded into panels. Flexural and tensile properties and impact strength of the manufactured WPCs were determined according to the relevant standard specifications. The researchers found that although composites containing PP (virgin and recycled) exhibited higher stiffness and strength than those made from HDPE (virgin and recycled), they had lower unnotched impact strengths. Mechanical properties of specimens containing recycled plastics (HDPE and PP) were statistically similar and comparable to those of composites made from virgin plastics. This was considered a possibility to expand the use of recycled plastics in the manufacture of WPC.

Composite materials derived from waste wood sawdust and recycled plastic have significant potential for use in the manufacture of pallets. In this study, wood filler (WF)–RPP composite was used to fabricate a one-fifth scale model of a pallet. For mechanical property testing, several composite boards were prepared with filler of particle sizes of 100 µm and 300 µm and filler content varying from 10 to 40% (by weight). The mechanical properties were used in the finite element analysis (FEA) to predict the deformation of the pallet under static load.

In a study by Gosselin et al. [17], fiber-reinforced microcellular foams were produced via injection molding and studied as a function of mold temperature as well as wood, blowing agent, and coupling agent concentrations. Birch wood fibers were added to a post-consumer recycled HDPE/PP matrix (85:15 ratio) in proportions ranging from 0 to 40 wt.% and then foamed. The MAPP was also used as a coupling agent in proportions ranging from 0 to 10 wt.% of wood content. In this first part, the morphological analysis of composite foams was presented, including cell roundness, cell size, skin thickness, void fraction, and fiber aspect ratio.

Composites based on recycled high-density polyethylene (RHDPE) and natural fibers were studied by Lei et al. [18]. The composites were made through melt blending and compression molding. The effects of the fibers (wood and bagasse) and coupling agent type/concentration on the composite properties were studied. The use of maleated polyethylene (MAPE), carboxylated polyethylene (CAPE), and titanium-derived mixture (TDM) improved the compatibility between the bagasse fiber and RHDPE, and mechanical properties of the resultant composites could be compared well with those of virgin HDPE composites. The modulus and impact strength of the composites had an increase in maxima with MAPE content. The composites had lower crystallization peak temperatures and a wider crystalline temperature range than neat RHDPE, and their thermal stability was lower than RHDPE.

8.4 WATER ABSORPTION PROPERTIES OF NATURAL FIBER COMPOSITES

Since natural fibers are highly hydrophilic in nature, the natural fiber composites have a tendency to absorb moisture over a long period of time. There have been several studies on the water absorption behavior of natural fibers. In a study by Dhakal et al. [19], hemp fiber-reinforced unsaturated polyester composites (HFRUPE) were subjected to water immersion tests to study the effects of water absorption on the mechanical properties. HFRUPE composite specimens containing 0, 0.10, 0.15, 0.21, and 0.26 fiber volume fractions were prepared. Water absorption tests were conducted by immersing specimens in a deionized water bath at 25°C and 100°C for different time durations. The tensile and flexural properties of water-immersed specimens subjected to both aging conditions were evaluated and compared alongside dry composite specimens. The percentage of moisture uptake increased as the fiber volume fraction increased due to the high-cellulose content. The tensile and flexural properties of HFRUPE specimens were found to decrease with an increase in percentage moisture uptake. Moisture-induced degradation of composite samples was significant at elevated temperature. The water absorption pattern of these composites at room temperature was found to follow Fickian behavior, whereas at elevated temperature it exhibited non-Fickian behavior. The nature of fiber matrix bonding was investigated through SEM micrographs (Figure 8.1) [19].

The behavior of jute fiber reinforced with unsaturated polyester (UP) composites was studied by Akil et al. [20]. These composites were subjected to water immersion tests to study the effects of water absorption on its mechanical properties. Water absorption tests were conducted by immersing composite specimens into three different environmental conditions, including distilled water, sea water, and acidic solutions at room temperature for a period of 3 weeks. The study obtained water absorption curves and determined the characteristic parameter "D" diffusion coefficient and maximum moisture content. The water absorption of jute fiber-reinforced UP composites was found to follow so-called pseudo-Fickian behavior. The effects of the immersion treatment on the flexural and compression characteristics were investigated. The flexural and compression properties were found to decrease with the increase in percentage water uptake as illustrated in Figure 8.2 [20]. These flexural and compression behaviors were explained by the plasticization of the matrix–fiber interface and swelling of the jute fibers.

Wang et al. [21] did work on fundamental modeling of the diffusion process in composites. Generally, the traditional diffusion theory was used to understand the mechanism of moisture absorption, but it cannot address the relationship between the microscopic structure-infinite 3D network and the moisture absorption. The researchers used the percolation theory to conduct some preliminary work. First, two new concepts, accessible fiber ratio and diffusion-permeability coefficient, were defined; second, a percolation model was developed to estimate the critical accessible fiber ratio; and finally, the moisture absorption and electrical conduction behavior of composites with different fiber loadings were investigated. At high fiber loading when accessible fiber ratio was high, the diffusion process was the dominant mechanism; whereas at low fiber loading close to and below the percolation threshold,

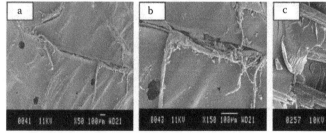

FIGURE 8.1 Failure showing (a) matrix cracking, (b) fracture running along the interface, and (c) fiber–matrix debonding due to attack by water molecules. (From Dhakal, H.N., Zhang, Z.Y., and Richardson, M.O.W. 2007. Effect of water absorption on the mechanical properties of hemp fibre reinforced unsaturated polyester composites. *Composites Science and Technology* 67:1674–83. With permission [19].)

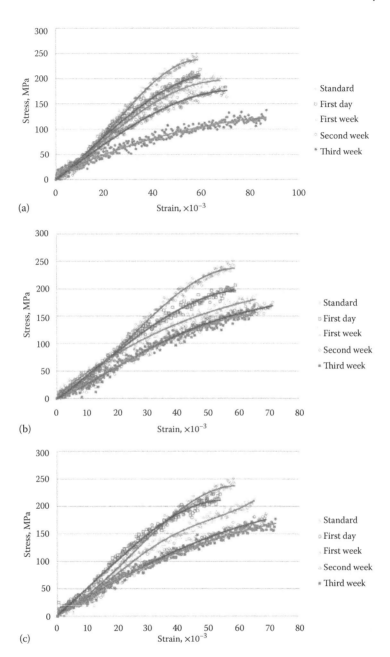

FIGURE 8.2 Flexural stress–strain curves of pultruded jute fiber-reinforced unsaturated polyester composite after exposure to (a) distilled water, (b) sea water, and (c) acidic solution. (From Akil, H.M., Cheng, L.W., Ishak, Z.A.M., Bakar, A.A., and Rahman M.A.A. 2009. Water absorption study on pultruded jute fibre reinforced unsaturated polyester composites. *Composites Science and Technology* 69:1942–8. With permission [20].)

percolation was the dominant mechanism. The overestimation of accessible fiber ratio led to discrepancies between the observed and model estimates.

8.5 THERMAL DEGRADATION PROPERTIES AND ACCELERATED WEATHERING STUDIES OF NATURAL FIBER COMPOSITES

The thermal degradation and fire resistance of different natural fiber composites were studied by Manfredi et al. [22]. UP and modified acrylic resins (Modar) were used as matrix composites. The smoke emission of the materials was analyzed, and the performance against fire of the biocomposites and glass-reinforced composites was compared. Thermal degradation indicated that the Modar matrix composites were more resistant to temperature than the composites with the UP matrix. Flax fibers, due to their low lignin content, exhibit the best thermal resistance among the natural fibers studied. From the results obtained about the thermal and fire resistance of the composites, it was possible to conclude that the flax fiber seems to be the most adequate, due to the long time to ignition and the long period before reaching the flashover. On the other hand, the jute fiber composites showed a short duration but a quick growing fire with the lowest smoke emission. The low smoke was an important advantage, as it reduces one of the main hazards of fire.

In another study by Azwa et al. [23], characteristics of kenaf fiber/epoxy composites, both treated and untreated using the alkalization process, exposed to high temperature were studied. Thermogravimetric analysis (TGA) was used to study the thermal decomposition behavior of treated and untreated kenaf/epoxy composites as well as their components, kenaf fiber and neat epoxy, from room temperature to 600°C. The weight loss and physical changes of these samples were observed through furnace pyrolysis. Surface morphology of the composites after degradation was observed using SEM. The results from the TGA showed that the addition of kenaf fibers to the epoxy slightly improves both the charring and thermal stability of the samples. However, it was observed that alkalization causes a reduction in these behaviors for the kenaf/epoxy composite. Generally, increased exposure time causes higher weight loss of the composites only to 150°C. At higher temperature, duration of exposure has little influence on the weight loss. Fiber–matrix debondings were observed in degraded samples, implying that mechanical degradation of the composites had occurred.

In a fundamental study on the thermal stability of the natural fibers, Yao et al. [24] performed dynamic TGA under nitrogen, which was used to investigate the thermal decomposition processes of 10 types of natural fibers commonly used in the polymer composite industry. These fibers included wood, bamboo, agricultural residue, and bast fibers. Various degradation models, including the Kissinger, Friedman, Flynn–Wall–Ozawa, and modified Coats–Redfern methods, were used to determine the apparent activation energy of these fibers. For most natural fibers, approximately 60% of the thermal decomposition occurred within a temperature range between 215°C and 310°C. The result also showed that apparent activation energy of 160–170 kJ/mol was obtained for most of the selected fibers throughout the polymer processing temperature range. These activation energy values allow development of a simplified approach to understand the thermal decomposition behavior of natural fibers as a function of polymer composite processing.

In a study by Sharkh et al. [25], date palm leaves were compounded with PP and ultraviolet (UV) stabilizers to form composite materials. The stability of the composites in the natural weathering conditions of Saudi Arabia and in accelerated weathering conditions was investigated. The composites were found to be much more stable than PP under the severe natural weathering conditions of Saudi Arabia and in accelerated weathering trials. Compatibilized samples were generally less stable than uncompatibilized ones as a result of the lower stability of the maleated polypropylene. In addition to enhanced stability imparted by the presence of the fibers in the composites, enhanced interfacial adhesion resulting from oxidation of the polymer matrix can be the source of retention of mechanical strength.

8.6   INCINERATION

Incineration with energy recovery is at first sight an ecologically acceptable way of using carbon-based polymer wastes due to their high calorific value. However, there is a widespread distrust of incineration by the general public due to the possibility of toxic emissions from some polymers, particularly poly(vinyl chloride) (PVC) [26], which may produce dioxins during combustion. On the other hand, waste plastics are increasingly regarded as resources to be reused. Fiber-reinforced composites have a low caloric value due to their high fiber content. Therefore, in many cases, incineration is not suitable for energy recovery of composites. An ideal way of maximizing the energy recovery of the recycling stage is not downcycling but closed-loop recycling, in which recycled fibers can be used in the production of other polymeric composites such as short fiber-reinforced composites and sheet molding compound (SMC) without losing their performance characteristics. Not much research is conducted on the effect of incineration on composites and plastics. In fact, many of the commercial packaging plastics still used are not biodegradable, all of them because their molecular weights are too high and their structures are too rigid for assimilation by organisms, and most of them also because they have substituents that prevent biodegradation via the enzymatic fatty acid oxidation mechanism. Linear PE is the only commercial packaging plastic with potential for biodegradation when its molecular weight has been reduced drastically by photodegradation. Degradable plastics are not a satisfactory solution to the problems of municipal solid waste. For those problems, multiple approaches need to be used, including especially recycling and incineration [27].

In one interesting study on polymeric nanocomposites, the researchers [28] commented that there is no evidence that all nanoobjects are safely removed from the off-gas when incinerating nanocomposites. The researchers found that nanoobject emission levels will increase if bulk quantities of nanocomposites end up in municipal solid waste. Many primary and secondary nanoobjects arise from the incineration of nanocomposites and removal seems insufficient for objects that are smaller than 100 nm. For the nanoobjects studied in this paper, risks occur for aluminum oxide, calcium carbonate, magnesium hydroxide, polyhedral oligomeric silsesquioxane (POSS), silica, titanium oxide, zinc oxide, zirconia, mica, montmorillonite, talc, cobalt, gold, silver, carbon black, and fullerenes. Since this conclusion was based on a desktop study without accompanying experiments, further research is required to reveal which nanoobjects will actually be emitted to the environment and to determine their toxicity to human health.

8.7 LIFE CYCLE ANALYSES OF NATURAL FIBER COMPOSITES

The use of natural fibers in a composite material does not by design make it a sustainable material. It is necessary to evaluate the entire life cycle of a material, such as the stages of extraction, use, and disposal, to make a judgment about the environmental characteristics of the material. It determines the environmental impacts of products, processes, or services, through production, usage, and disposal. It is a technique that is used for assessing the potential environmental aspects and potential aspects associated with a product (or service) by:

1. Compiling an inventory of relevant inputs and outputs
2. Evaluating the potential environmental impacts associated with those inputs and outputs
3. Interpreting the results of the inventory and impact phases in relation to the objectives of the study [29]

As concerns the end-of-life (EOL) disposal of green composites, it is worth pointing out that a bio-based material is not, by definition, biodegradable, that is, biobased and biodegradable are independent properties. The bio-based concept refers to the environmentally friendly origin of the material, whereas biodegradability is a final property defined by a number of international norms such as EN 13432, which indicate the requisites that plastic products must possess to be biodegradable and compostable. Hence, although in green composites the natural fiber component is intrinsically biodegradable, the bio-based polymer matrix can be either biodegradable or not, as schematically illustrated in Figure 8.3 [30].

In a study by Nicolier et al. [31], environmental performance of China reed fiber used as a substitute for glass fiber was investigated as reinforcement in plastics. An LCA was performed on these two materials for an application to plastic transport pallets. Transport pallets reinforced with China reed fiber prove to be ecologically advantageous if they have a minimal lifetime of 3 years compared with the 5-year lifetime of the conventional pallet. The energy consumption and other environmental impacts are strongly reduced by the use of raw renewable fibers, due to three important factors: (a) the substitution of glass fiber production by natural fiber production; (b) the indirect reduction in the use of PP linked to the higher proportion of China reed fiber used; and (c) the reduced pallet weight, which reduces fuel consumption during transport. Considering the whole life cycle, the PP production process and the transport cause the strongest environmental impacts during the use phase of the life cycle. Since thermoplastic composites are hardly biodegradable, incineration has to be preferred to discharge in landfills at the end of the useful life cycle. The potential advantages of the renewable fibers will be effective only if a purer fiber extraction were obtained to ensure an optimal material stiffness, a topic

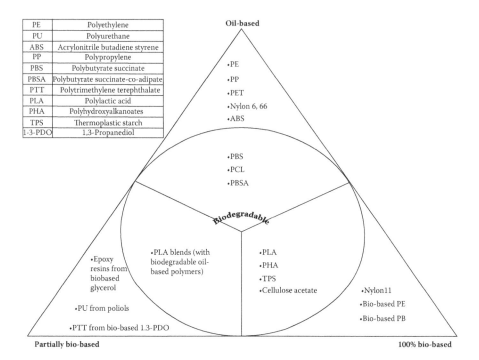

FIGURE 8.3 Biodegradability of oil-based, bio-based, and partially bio-based polymers. (From Zini, E. and Scandola, M. 2011. Green composites: An overview. *Polymer Composites* 32:1905–15. With permission [30].)

for further research. China reed bio fibers are finally compared with other usages of biomass and biomaterials and, in general, can enable a 3 to 10 times more efficient valorization of biomass than mere heat production or biofuels for transport.

Xu et al. [32] carried out LCA for wood-fiber-reinforced PP composite preforms produced by compression molding in comparison with those of PP. Three levels of fiber contents, 10, 30, and 50% by mass, were used. The level of environmental impact caused by transportation was also studied. This study introduces a new term called "material service density," which was defined as the volume of material satisfying a specific strength requirement (tensile strength in this study). The rationale behind this was that specific volumes of different materials are required to withstand a given mechanical load (tensile load in this case). Comparisons of the material service density for two materials, wood-fiber-reinforced composites and PP, were conducted. The results showed that when material service density was used as the functional unit, wood-fiber-reinforced composites demonstrated superior environmental friendliness compared with PP.

In a very interesting study by Pervaiz et al. [33], environmental performance of hemp-based natural fiber mat thermoplastic (NMT) was evaluated by quantifying carbon storage potential and CO_2 emissions and comparing the results with commercially available glass fiber composites. Nonwoven mats of hemp fiber and PP matrix were used to make NMT samples by the film-stacking method without using any binder aid. The results showed that hemp-based NMT has compatible or even better strength properties as compared with conventional flax-based thermoplastics. A value of 63 MPa for flexural strength was achieved at a 64% fiber content by weight. Similarly, impact energy values (84–154 J/m) were also promising. The carbon sequestration and storage by hemp crop through photosynthesis was estimated by quantifying dry biomass of fibers based on one metric ton of NMT. A value of 325 kg carbon per metric ton of hemp-based composite was estimated, which can be stored by the product during its useful life. An extra 22% of carbon storage can be achieved by increasing the compression ratio by 13% while maintaining the same flexural strength. Further, net carbon sequestration by an industrial hemp crop was estimated as 0.67 ton/h/year, which was compatible with all urban trees in the United States and very close to naturally regenerated forests. A comparative LCA focused on nonrenewable energy consumption of natural and glass fiber composites shows that a net saving of 50,000 MJ (~3 ton CO_2 emissions) per ton of thermoplastic can be achieved by replacing 30% glass fiber reinforcement with 65% hemp fiber. The authors further estimated that 3.07 million ton CO_2 emissions and 1.19 million m^3 crude oil can be saved by substituting 50% fiber glass plastics with natural fiber composites in North American auto applications. However, to compete with glass fiber effectively, further research is needed to improve natural fiber processing and interfacial bonding, and control moisture sensitivity in the longer run.

Dow chemicals have been keenly working on their key polymer PLA. One of the reported works by Vink et al. [34] overviews applications of LCA to PLA production and provides insights into how they were used. The first application reviews the contributions to the gross fossil energy requirement for PLA (54 MJ/kg). In the second one, PLA was compared with petrochemical-based polymers using fossil energy use, global warming, and water use as the three impact indicators. The last application gives more details about the potential reductions in energy use and greenhouse gases (GHGs). Cargill Dow's 5- to 8-year objective was to decrease the fossil energy use from 54 MJ/kg PLA to about 7 MJ/kg PLA. The objective for GHGs was a reduction from +1.8 to −1.7 kg CO_2 equivalents/kg PLA.

In a study by Duigou et al. [35], environmental impacts of the productions of hackled flax fibers destined for reinforcement of composite materials were compared with those for the production of glass fibers. The ecological advantages of flax fiber production compared with glass fibers were shown. Most of the environmental indicators used (climate change, acidification, and nonrenewable energy consumption) were favorable to flax fibers. However, the eutrophication indicator remains high, mainly due to the use and production of fertilizers. Possible optimization was studied by means of reduction of mechanical extraction steps and chemical product use. A sensitivity analysis highlights that the calculation method used (CML2000 modified) was robust. The hypothesis of carbon dioxide sequestration influences the climate change indicator. The choice of allocation method, by weight or by economic value, also has a strong influence on the environmental impacts. Globally, the production of flax fibers appears to be an environmentally attractive alternative to glass fibers.

In a study by Alves et al. [36], LCA analysis was used to investigate replacement of glass fibers by jute fibers as reinforcement of composite materials to produce automotive structural components. With regard to the composite materials, an off-road vehicle case study demonstrated that jute fiber composites present the best solution in enhancing the environmental performance of the off-road vehicle's enclosures, hence improving the environmental performance of the whole vehicle. Despite the fuel consumption becoming lower using jute fibers, due to the weight reduction of the vehicle, LCA pointed out some unknown impacts in the production and disposal phases of the bonnets, specifically related to the logistics of the transport of the jute fibers and the recycling scenario of the bonnets. It gave the design team an overview scenario of the problem, besides traditional inputs generally used in the design of products, providing results that help the design team make decisions, still working in partnership with suppliers to improve the logistics of the jute fibers, and focusing on the more pollutant phases to prevent potential environmental effects.

The life cycle impact of kenaf board has been studied by Ardente et al. [37] The research group compared the performances of various replaceable products, such as polyurethane, glass wool, flax rolls, stone wool, mineral wool, and paper wool. Such a comparison showed that the highest impacts are related to synthetic materials, whereas the better performances are due to mineral wools. Interestingly, kenaf-fiber-based products become the least impactful by a large margin if different disposal scenarios are adopted. In particular, incineration with energy recovery and electricity production could decrease the global energy requirements to 17 MJ. A further reduction can be obtained with introduction of recycled materials into the manufacturing process or if the kenaf plants are locally grown. This established the large energy and environmental advantages related to the employment of insulating materials. This study shows that the overall energy impact of the building could be more simply evaluated with an LCA approach. Embodied energy data and LCA should be included in energy certification schemes to effectively lead the building sector toward sustainability. The life cycle thinking approach seems to be more "reliable" with regard to the achieved results, encompassing all the life phases of the building. The researchers commented that LCA should be considered in the most advanced building energy legislation.

In an LCA case study by Pegoretti et al. [38], the use of recycled cotton fibers in automotive acoustic components in replacement of components made mainly of PU was investigated. The so-called DL-cotton panel option, which combines two layers of recycled fibers of different densities and for which natural fibers represent a majority of its raw materials, was identified as the most environmentally friendly option, on average with 30% less impact than the other two solutions. This component combines good acoustic performance with lower weight, economy of fossil resources, heat and energy savings during production; it also avoids textile disposal in landfills. The paper therefore confirms the advantages of the cotton fiber and the performance of natural fibers, in general, used in vehicle applications, in particular thanks to lower weight, as reported in other publications. Moreover, this work proved the limited performance of current and future possible EOL scenarios and highlights the necessity to work on future scenarios based on recycling. The researchers stressed the need for leading multiobjective (coupling technical and environmental performance) optimization for ecodesign of these acoustic components coupling environmental and acoustic models.

In a study by Luz et al. [39], sugarcane bagasse (SCB)-reinforced PP substitutes for talc-filled PP were studied for their automotive components. A comparative LCA was performed for the two alternatives, from raw extractions to the EOL phase of SCB-PP and talc-PP composite, where data collected in different industries in Brazil were included in the LCA GaBi software. The research analysis showed that in addition to similar mechanical performance, natural fiber composites showed superior environmental performance throughout the entire life cycle. This superior performance, according to the authors, was because (1) in the cultivation phase, sugarcane absorbs carbon through the photosynthesis process while growing, thus reducing the global warming impact of the materials used; (2) the production process was cleaner; (3) SCB-reinforced composites were found to be lighter for equivalent performance, which reduced the amount of PP used; and (4) the economic reuse proposed for the EOL SCB-PP composite was found to be the best alternative to minimize environmental impacts.

In a work by Simões et al. [40], an attributional LCA study was performed to identify the best practicable environmental option of three lighting columns made in alternative materials. The materials used were glass fiber-reinforced polymer composites, along with two traditional materials, steel and aluminium. Analysis of an alternative composite lighting column with the same matrix, but reinforced with natural fibers (jute) was also performed. The results showed that the lighting column made of steel has the worst overall environmental profile, although steel production is less energy intensive. The composite lighting column is environmentally preferable due to its extended lifespan and lack of maintenance. The production of the raw material is responsible for most of the impacts in all three cases. The glass fiber production is the main contributor for the overall environmental burden of the composite column. In this case, the results also show that there are environmental advantages in replacing glass fiber for jute fiber.

Paweljik et al. [41] did a critical assessment of the LCA process per se. The authors argues that internationally agreed LCA standards provide generic recommendations on how to evaluate the environmental impacts of products and services but generally do not address details that are specifically relevant for the life cycles of bio-based materials. They further argued that treatment of biogenic carbon storage is critical for quantifying the GHGs emissions of bio-based materials in comparison with petrochemical materials. Although the authors acknowledged that biogenic carbon storage remains controversial, they still recommended accounting for it, depending on product-specific life cycles and the likely duration of carbon storage. If carbon storage is considered, coproduct allocation is nontrivial and should be chosen with care to (i) ensure that carbon storage is assigned to the main product and the coproduct(s) in the intended manner and (ii) avoid double counting of stored carbon in the main product and once more in the coproduct(s). Land-use change, soil degradation, water use, and impacts on soil carbon stocks and biodiversity are important aspects that have recently received attention. They explained various approaches to account for these and conclude that substantial methodological progress is necessary, which is, however, hampered by the complex and often case- and site-specific nature of impacts. With the exception of soil degradation, they recommend preliminary approaches for including these impacts in the LCA of bio-based materials.

The authors further comment that use of attributional versus consequential LCA approaches is particularly relevant in the context of bio-based materials. They concluded that it is more challenging to prepare accurate consequential LCA studies, especially because these should account for future developments and secondary impacts around bio-based materials that are often difficult to anticipate and quantify. Although hampered by complexity and limited data availability, the application of the proposed approaches to the extent possible would allow for obtaining a more comprehensive insight into the environmental impacts of production, use, and disposal of bio-based materials.

In a work by Rosa et al. [42], an application of LCA methodology was used to explore the possibility of improving the eco-efficiency of glass fiber composite materials by replacing part of the glass fibers with hemp mats. The LCA results showed that hemp mats in glass fiber-reinforced thermosets were more eco-efficient than conventional glass fiber alternatives. The advantage of the hybrid material was mostly found in the reduction of glass fiber and resin content compared with standard glass fiber composites. A significant environmental advantage was found during the production phase, and it was mainly attributable to the "green" origin of the hemp mat. The limits of renewable raw materials were because of the fact that they score better than petrochemical polymers with regard to fossil energy use and GHG emissions whereas they score worse with regard to land use, ecotoxicity, and eutrophication.

Qiang et al. [43] studied PLA-based WPC. Based on the input–output model, the LCA was carried out to evaluate the energy demand and environmental impacts, as well as the water requirement of the PLA-based WPC during the cradle-to-gate stages. An attribute hierarchy model (AHM) was used to determine the weighting factors of the different environmental impact categories on the environmental impact load (EIL). The results showed that the energy demand for 1000 kg of the unmodified PLA-based WPC (Sample A, 20 wt.% WF + 80 wt.% PLA) and their polyhydroxyalkanoates (PHA) toughened counterpart (Sample B, 20 wt.% WF + 55 wt.% PLA + 25 wt.% PHA) was 3.08×10^8 kJ and 3.05×10^8 kJ, respectively. The water requirement for Sample A was more than that of Sample B. The non-dimensional indicator EIL was 1.9 and 1.6 for Sample A and Sample B, respectively. Photochemical oxidation potential constituted the largest part among the six environmental impact categories (global warming potential, acidification potential, photochemical oxidation potential, eutrophication potential, smog potential, and eco-toxicity potential), whereas eutrophication potential was the least for both Samples A and B. This cradle-to-gate LCA will contribute to optimizing the eco-design, thus reducing the energy consumption and pollutant emissions during the eco-profiles of the PLA-based WPC. As far as the energy demand, EIL, water requirement consumption, and solid waste are concerned, the PLA-based WPC toughened with PHAs have environmental advantages compared with their untoughened counterparts. The authors commented that the LCA results can only be one of the references for the manufacturers and decision makers, since the weighting factors for different environmental impact categories depend on subjective expert opinion. On the other hand, the cradle-to-gate assessment of the PLA-based WPC does not include the product use and EOL phases, which limits the ability to identify the burden shifting.

Using recycled plastics can significantly reduce the environmental impacts by avoiding exploration, mining, and transportation of natural gas and oil. Adding fibrous reinforcement is a potential way to increase the recycling rate and also to locate high-value applications to recycled plastics. In a work by Rajendran et al. [44], environmental impact assessment was carried out using the LCA approach. The glass fibers and flax fibers were chosen as fibrous reinforcement. The resource depletion potential and global warming potential were evaluated for glass fiber and flax fiber filled recycled plastics and compared with their respective virgin alternatives. The properties of the recycled and virgin composites were estimated from the literature and semiempirical models. The resource depletion and global warming were compared at equivalent stiffness. The results indicate that reinforcing recycled plastics can significantly reduce the resource consumption and global warming in civil and infrastructural applications. However, the composites made from virgin plastics remain a greener and sustainable alternative in the automotive applications.

In the case of recycled plastics, increase of fiber volume fraction (V_f) increases the elastic stiffness along with the increase in abiotic depletion for glass fiber reinforcement, but an inverse relationship exists for flax fiber reinforcement. This is due to composites based on recycled plastics consuming less energy than glass fibers, but not less than flax fibers. The contribution analysis indicated that flax fiber composites had huge energy credit from energy recovery, which reduced the impact created by raw materials. The raw materials in virgin plastic composites also play a dominant role in maximizing the impacts. In summary, with glass fibers and flax fibers, recycled plastics are found to consume significantly less fossil fuels and materials when compared with virgin plastics at an equivalent stiffness. This indicates that replacing virgin plastics composites with recycled plastics composites that meet the functional requirements can significantly save on renewable resources and reduce emission levels. In the cradle-to-gate analysis, the potential contribution to global warming is mainly from fibers, matrix, and transport involved in shipping fibers and matrix to the manufacturing site.

Outstanding mechanical, thermal, and environmental properties of kenaf, as well as successful applications of kenaf-based boards in structural insulated panels (SIP), inspired Batouli et al. [45] to investigate the incorporation of kenaf core in the PU insulation core of SIP to create environmentally friendly building material. Three composites made of rigid PU reinforced with 5, 10, and 15% kenaf core were prepared and analyzed. The three composites and pure rigid PU were then used as insulation cores of SIP with the same kenaf-based structural boards. An LCA was conducted to determine the environmental profiles of the four SIP at 10°C and 50°C. It is shown that although kenaf has much less environmental impact than PU, increasing the amount of kenaf core in PU composites does not necessarily result in less environmental impact. In fact with the current practice of making the composites, kenaf core does not replace the PU; instead, it mostly fills the void space, which is initially filled with air and, hence, the kenaf core decreases the porosity of PU composites and increases the density without improving thermal resistance.

It is interesting to note that inclusion of a small amount of kenaf in PU composites makes it more porous and less dense. The resulting material has good thermal resistance, such as the 5% case in this study. However, the authors found that such composites performed poorly with regard to flexural strength. The authors commented that since the insulation core of SIP has to withstand shear forces, in addition to providing thermal insulation, composites with low-kenaf core mix should be either used carefully in places where it is structurally adequate or be avoided completely. On the other hand, increasing the amount of kenaf core in PU composites will increase the bending strength of the composites; however, the environmental impact of composites also increases. Thermal conductance of kenaf-based SIP increases when the temperature rises. Therefore, to provide the same thermal resistance at higher temperatures, thicker insulation is needed, which causes greater environmental impacts.

In a work by Pietrini et al. [46], a cradle-to-grave environmental LCA of a few poly(3-hydroxybutyrate) (PHB)-based composites was performed and compared with commodity petrochemical polymers. The end-products studied are a cathode ray tube (CRT) monitor housing (conventionally produced from high-impact polystyrene, HIPS) and the internal panels of an average car (conventionally produced from glass-fiber-filled polypropylene, PP–GF). The environmental impact is evaluated on the basis of nonrenewable energy use (NREU) and global warming potential over a 100-year time horizon (GWP100). The SCB and nanoscaled organophilic montmorillonite (OMMT) are used as PHB fillers. The results obtained show that, despite the unsatisfying mechanical properties of PHB composites, depending on the type of filler and on the product, it is possible to reach lower environmental impacts than by use of conventional petrochemical polymers. These savings are mainly related to the PHB production process, whereas no improvements are related to composite preparation.

Shien et al. [47] reviewed LCA studies to gain insights of the environmental profiles of polysaccharide products (e.g., viscose or natural fiber polymer composites) in comparison with their conventional counterparts (e.g., cotton or petrochemical polymers). The application areas covered are textiles, engineering materials, and packing. It is found that for each stage of the life cycle (production, use phase, and waste management) polysaccharide-based end products show better environmental profiles than their conventional counterparts in terms of NREU and GHG emissions. Cotton is an exception, with high-environmental impacts that are related to the use of fertilizers, herbicides, pesticides, and high-water consumption. The available literature for man-made cellulose fibers shows that they result in a reduced amount of NREU and GHG emissions in the fiber production phase.

Vidal et al. [48] performed cradle-to-grave life cycle inventory studies for 1 kg of each of the three new composites: PP and cotton linters, PP and rice husks, and HDPE and cotton linters. Inventory data for the recycling of thermoplastics and cotton were obtained from a number of recycling firms in Spain, whereas environmental data regarding rice husks were obtained mainly from one rice-processing company located in Spain. Life cycle inventory data for virgin thermoplastics were acquired from Plastics Europe. Two different scenarios—incineration and landfilling—were considered for the assessment of the disposal phase. A quantitative impact assessment was performed for four impact categories: global warming over a hundred years, nonrenewable energy depletion, acidification, and eutrophication.

The composites subject to analysis exhibited a significantly reduced environmental impact during the material acquisition and processing phases compared with conventional virgin thermoplastics in all of the impact categories considered. The use of fertilizers for rice cultivation, however, impaired the results of the rice husk composite in the eutrophication category, where it, nevertheless, outperformed its conventional counterparts.

8.8   CONCLUSIONS

This chapter critically discusses the environment-related issues pertaining to natural fiber-reinforced composite materials. With recycling being one of the options, however, there is more scope to see the viability of recycling such materials. It is evident, as critically discussed, that natural fibers significantly improve the usability of recycled polymers. However, research on the use of recycled natural fibers as reinforcement has been scant. Absorption of moisture has always been a concern with this class of composites, and researchers have reported significant change in mechanical properties with such moisture uptake. Several studies on LCA, especially with respect to glass fibers, show the superiority of natural fibers in terms of their environmental friendly quotient. However, as is evident from the chapter, there needs to be further careful enquiries into the environmental aspect of these natural fiber-reinforced plastics.

REFERENCES

1. Goodship, V. 2007. *Introduction to Plastic Recycling*, 2nd ed. Smithers Rapra Publishers, United Kingdom.
2. Brandrup, J. 1996. *Recycling and Recovery of Plastics*. Hanser Publishers, Germany.
3. http://www.plasticsrecyclers.eu/mechanical-recycling [accessed August 9, 2014].
4. http://www.epa.gov/climatechange/wycd/waste/downloads/recycling-chapter10-28-10.pdf [accessed August 9, 2014].
5. Al-Salem, S.M., Lettieri, P., and Baeyens, J. 2009. Recycling and recovery routes of plastic solid waste (PSW): A review. *Waste Management* 29:2625–43.
6. Gullon, I.M., Esperanza, M., and Font, R. 2001. Kinetic model for the pyrolysis and combustion of poly-(ethylene terephthalate) (PET). *Journal of Analytical and Applied Pyrolysis* 58–59:635–50.
7. Zia, K.M., Bhatti, H.N., and Bhatti, I.A. 2007. Methods for polyurethane and polyurethane composites, recycling and recovery: A review. *Reactive & Functional Polymers* 67:675–92.
8. Howard, G.T. 2002. Biodegredation of polyurethane: A review. *International Biodeterioration and Biodegradation* 49:245–52.
9. Kang, H. and Schoenung, J.M. 2005. Electronic waste recycling: A review of U.S. infrastructure and technology options. *Conservation and Recycling* 45:368–400.
10. Singleton, A.C.N., Baillie, C.A., Beaumont, P.W.R., and Peijs, T. 2003. On the mechanical properties, deformation and fracture of a natural fibre/recycled polymer composite. *Composites Part B: Engineering* 34:519–26.
11. AlMaadeed, M.A., Kahraman, R., Khanam, P.N., and Madi, N. 2012. Date palm wood flour/glass fibre reinforced hybrid composites of recycled polypropylene: Mechanical and thermal properties. *Materials & Design* 42:289–94.

12. Najafi, S.K., Kiaefar, A., Hamidina, E., and Tajvidi, M. 2007. Water absorption behavior of composites from sawdust and recycled plastics. *Journal of Reinforced Plastics and Composites* 26:341–8.

13. Khalil, H.P.S.A., Shahnaz, S.B.S., Ratnam, M.M., Ahmad, F., and Fuaad, N.A.N. 2006. Recycle polypropylene (RPP) wood saw dust (WSD) composites—Part 1: The effect of different filler size and filler loading on mechanical and water absorption properties. *Journal of Reinforced Plastics and Composites* 25:1291–303.

14. Cui, Y., Lee, S., Noruziaan, B., Cheung, M., and Tao, J. 2008. Fabrication and interfacial modification of wood/recycled plastic composite materials. *Composites Part A: Applied Science and Manufacturing* 39:655–61.

15. Yang, T.H., Lin, C.J., Wang, S.Y., and Tsai, M.J. 2007. Characteristics of particleboard made from recycled woodwaste chips impregnated with phenol formaldehyde resin. *Building and Environment* 42:189–95.

16. Najafi, S.K., Hamidinia, E., and Tajvidi, M. 2006. Mechanical properties of composites from sawdust and recycled plastics. *Journal of Applied Polymer Science* 100:3641–5.

17. Gosselin, R., Rodrigue, D., and Riedl, B. 2006. Injection molding of postconsumer wood-plastic composites I: Morphology. *Journal of Thermoplastic Composite Materials* 19:639–57.

18. Lei, Y., Wu, Q., Yao, F., and Xu, Y. 2007. Preparation and properties of recycled HDPE/natural fiber composites. *Composites Part A: Applied Science and Manufacturing* 38:1664–74.

19. Dhakal, H.N., Zhang, Z.Y., and Richardson, M.O.W. 2007. Effect of water absorption on the mechanical properties of hemp fibre reinforced unsaturated polyester composites. *Composites Science and Technology* 67:1674–83.

20. Akil, H.M., Cheng, L.W., Ishak, Z.A.M., Bakar, A.A., and Rahman M.A.A. 2009. Water absorption study on pultruded jute fibre reinforced unsaturated polyester composites. *Composites Science and Technology* 69:1942–8.

21. Wang, W., Sain, M., and Cooper, P.A. 2006. Study of moisture absorption in natural fiber plastic composites. *Composites Science and Technology* 66:379–86.

22. Manfredi, L.B., Rodríguez, E.S., Przybylak, M.W., and Vázquez, A. 2006. Thermal degradation and fire resistance of unsaturated polyester, modified acrylic resins and their composites with natural fibres. *Polymer Degradation and Stability* 91:255–61.

23. Azwa, Z.N. and Yousif, B.F. 2013. Characteristics of kenaf fibre/epoxy composites subjected to thermal degradation. *Polymer Degradation and Stability* 98:2752–9.

24. Yao, F., Wu, Q., Lei, Y., Guo, W., and Xu, Y. 2008. Thermal decomposition kinetics of natural fibers: Activation energy with dynamic thermogravimetric analysis. *Polymer Degradation and Stability* 93:90–8.

25. Sharkh, B.F.A. and Hamid, H. 2004. Degradation study of date palm fibre/polypropylene composites in natural and artificial weathering: mechanical and thermal analysis. *Polymer Degradation and Stability* 85:967–73.

26. Scott, G. 2000. 'Green' polymers. *Polymer Degradation and Stability* 68:1–7.

27. Klemchuk, P.P. 1990. Degradable plastics: A critical review. *Polymer Degradation and Stability* 27:183–202.

28. Roes, L., Patel, M.K., Worrell, E., and Ludwig, C. 2012. Preliminary evaluation of risks related to waste incineration of polymer nanocomposites. *Science of the Total Environment* 417–418:76–86.

29. http://www.gdrc.org/uem/lca/lca-define.html [assessed August 11, 2014].

30. Zini, E. and Scandola, M. 2011. Green composites: An overview. *Polymer Composites* 32:1905–15.

31. Nicollier, T.C., Laban, B.G., Lundquist, L., Leterrier, Y., Månson, J.A.E., and Jolliet O. 2001. Life cycle assessment of biofibres replacing glass fibres as reinforcement in plastics. *Resources, Conservation and Recycling* 33:267–87.

32. Xu, X., Jayaraman, K., Morin, C., and Pecqueux, N. 2008. Life cycle assessment of wood-fibre reinforced polypropylene composites. *Journal of Materials Processing Technology* 198:168–77.

33. Pervaiz, M., and Sain, M.M. 2003. Carbon storage potential in natural fiber composites. *Resources, Conservation and Recycling* 39:325–40.

34. Vink, E.T.H., Rábago, K.R., Glassner, D.A., and Gruber, P.R. 2003. Applications of life cycle assessment to NatureWorks™ polylactide (PLA) production. *Polymer Degradation and Stability* 80:403–19.

35. Duigou, A.L., Davies, P., and Baley, C. 2011. Environmental impact analysis of the production of flax fibres to be used as composite material reinforcement. *Journal of Biobased Materials and Bioenergy* 5:153–65.

36. Alves, C., Ferrão, P.M.C., Silva, A.J., Reis, L.G., Freitas, M., Rodrigues, L.B., and Alves, D.E. 2010. Ecodesign of automotive components making use of natural jute fiber composites. *Journal of Cleaner Production* 18:313–27.

37. Ardente, F., Beccali, M., Cellura, M., and Mistretta, M. 2008. Building energy performance: A LCA case study of kenaf-fibres insulation board. *Energy and Buildings* 40:1–10.

38. Pegoretti, T.S., Mathieux, F., Evrard, D., Brissaud, D., and Arruda, J.R.F. 2014. Use of recycled natural fibres in industrial products: A comparative LCA case study on acoustic components in the Brazilian automotive sector. *Resources, Conservation and Recycling* 84:1–14.

39. Luz, S.M., Pires, A.C., and Ferrão, P.M.C. 2010. Environmental benefits of substituting talc by sugarcane bagasse fibers as reinforcement in polypropylene composites: Ecodesign and LCA as strategy for automotive components. *Resources, Conservation and Recycling* 54:1135–44.

40. Simões, C.L., Pinto, L.M.C., and Bernardo, C.A. 2012. Modelling the environmental performance of composite products: Benchmark with traditional materials. *Materials & Design* 39:121–30.

41. Pawelzik, P., Carus, M., Hotchkiss, J., Narayan, R., Selke, S., Wellisch, M., Weiss, M., Wicke, B., and Patel, M.K. 2013. Critical aspects in the life cycle assessment (LCA) of bio-based materials—Reviewing methodologies and deriving recommendations. *Resources, Conservation and Recycling* 73:211–28.

42. Rosa, A.D., Cozzo, G., Latteri, A., Recca, A., Björklund, A., Parrinello, E., and Cicala, G. 2013. Life cycle assessment of a novel hybrid glass-hemp/thermoset composite. *Journal of Cleaner Production* 44:69–76.

43. Qiang, T., Yu, D., Zhang, A., Gao, H., Li, Z., Liu, Z., Chen, W., and Han, Z. 2014. Life cycle assessment on polylactide-based wood plastic composites toughened with polyhydroxyalkanoates. *Journal of Cleaner Production* 66:139–45.

44. Rajendran, S., Scelsi, L., Hodzic, A., Soutis, C., and Al-Maadeed, M.A. 2012. Environmental impact assessment of composites containing recycled plastics. *Resources, Conservation and Recycling* 60:131–9.

45. Batouli, S.M., Zhu, Y., Nar, M., and D'Souza, N.A. 2014. Environmental performance of kenaf-fiber reinforced polyurethane: A life cycle assessment approach. *Journal of Cleaner Production* 66:164–73.

46. Pietrini, M., Roes L., Patel, M.K., and Chiellini, E. 2007. Comparative life cycle studies on Poly(3-hydroxybutyrate)-based composites as potential replacement for conventional petrochemical plastics. *Biomacromolecules* 8:2210–8.

47. Shen, L. and Patel, M.K. 2008. Life cycle assessment of polysaccharide materials: A review. *Journal of Polymers and the Environment* 16:154–67.

48. Vidal, R., Martínez, P., and Garraín, D. 2009. Life cycle assessment of composite materials made of recycled thermoplastics combined with rice husks and cotton linters. *The International Journal of Life Cycle Assessment* 14:73–82.

9 Modeling of Natural Fiber Composites

Liva Pupure, Janis Varna, and Roberts Joffe

CONTENTS

9.1 INTRODUCTION

Natural fibers such as flax, hemp, kenaf, jute, sisal, ramie, and coir are often considered as reinforcing polymers for enhanced performance. Indeed, the stiffness of these fibers in the axial direction is on par with that of glass fibers (for instance, 50–100 GPa stiffness is reported for flax fibers). There are well-known advantages of the use of natural fibers in composites, since they are of low density, have a smaller impact on the processing equipment (in terms of wear), are more environmentally friendly, and, of course, are renewable. Simultaneously, a number of problems hinder a wider use of these fibers in structural composites, namely sensitivity to moisture, large variation of properties, limited length, difficulty to control orientation, and poor chemical compatibility between the fibers and the matrix. In the case of non- or semistructural applications of the natural fiber, common processing methods are used (e.g., injection molding of packages, extrusion of beams for decking, and compression molding of panels for automotive use). These methods tend to produce short-fiber composites. However, recently, more advanced reinforcements based on natural fibers have been developed, such as randomly oriented fabrics, fiber rovings, and noncrimp and woven fabrics. Use of these reinforcements allowed production of composites with a much better mechanical performance. Although assembly of fibers into various types of fabrics somewhat helped solve problems associated with limited fiber length and improved control over fiber orientation, other problems mentioned earlier still remain to be solved. Therefore, the design of natural fiber composites with predefined properties for structural applications and long service life under harsh conditions is very challenging. Development of appropriate models that can describe the complex behavior of natural fiber composites would be a very useful tool for composite designers.

This chapter contains a description of modeling methods that predict the mechanical performance of natural fiber composites with inelastic behavior of constituents and their sensitivity to moisture. It has been shown in many studies that the in-plane stiffness of natural fiber composites can be reasonably well predicted by use of the rule-of-mixtures (RoM) type of models (discussed in greater detail later in the text). Even though these models are able to predict the stiffness of a wide range of short-fiber composites, they contain a number of assumptions that do not account for the actual microstructure of composites and properties of constituents. First of all, in classical application, the simple RoM is used for linear elastic materials, although nonlinearity may be introduced via the nonlinear behavior of the matrix and/or the fiber as well as through inelastic stress transfer between the fiber/matrix (e.g., elastic plastic or perfectly plastic). In this chapter, it will be demonstrated that predicting the inelastic behavior of natural fiber composites requires a model that accounts for damage, and viscoelastic and viscoplastic phenomena. Second, in a simple RoM, natural fibers are considered homogenous and isotropic and often their strength is described by a single value. It will be shown here that natural fiber is highly heterogeneous with anisotropic stiffness and its strength should be described by statistical distribution. The multiscale approach to the modeling of natural fiber composites will be used starting with the scale of the cell wall to predict fiber properties. The parameters of the fiber length and orientation efficiency will also be addressed in this chapter by showing that the orientation of fibers can be accounted for by use of more advanced micromechanical models (e.g., Fukuda and Chou) or by use of laminate analogy [and use of classical laminate theory (CLT)]. Lastly, the method that is used to account for moisture effects will be presented. Changes in the properties of natural fibers due to moisture will be discussed, as well as changes in fiber geometry due to swelling and resulting changes of volumetric composition of composites.

9.2   MULTISCALE MODELING OF NATURAL FIBER STIFFNESS

9.2.1   Hierarchical Structure of Fibers

The direct experimental measurement of fiber properties by available techniques is very complex and even impossible for some properties (e.g., transverse shear properties). Therefore, an improved understanding of the fiber behavior and development of models that are able to predict the fiber properties as dependent on the fiber ultrastructure are of high priority. In this section, we will give examples for wood fibers. Other natural fibers have the same basic features and constituents and, hence, the presented methods of analysis and observed trends are also applicable for them.

Visual inspection of fibers in the annual growth ring of a tree reveals large variations in geometrical parameters of wood fibers (tracheids) when moving from the earlywood to the latewood region. The thickness of the fiber wall increases, and the size of the cavity (lumen) decreases. The length of the softwood fiber is about 1–4 mm [1] and the diameter is 20–40 μm. Due to the large length/diameter aspect ratio (about 100), these fibers, in the context of the stress transfer in composites, may be treated as long. In spite of large variations in geometrical parameters, the fiber ultrastructure and chemical constituents remain the same.

Natural fibers have a very complex hierarchical structure covering the whole range of length scales starting from chemical structure, which is usually analyzed with nanoscale models. Fibers on the highest scale can be roughly approximated by a hollow cylinder that has an anisotropic wall material, as seen in Figure 9.1. In a composite, the hollow core of the cylinder (called lumen) may be empty or filled with resin used for impregnation. In the fiber wall, one can distinguish several anisotropic layers. Each layer of the fiber wall has a certain orientation of the material symmetry axes with respect to the fiber axis. On the next smallest scale, one can see that the microfibrils of oriented cellulosic chains surrounded by hemicellulose and lignin are the reason for anisotropy. The described length scales are sufficiently different, which is advantageous for modeling: this allows for a multiscale analysis instead of incorporating all length scales in the same model. In the multiscale modeling approach, the different length scales are linked and a model with a detailed description of the mechanisms and material and geometrical parameters on one scale, has

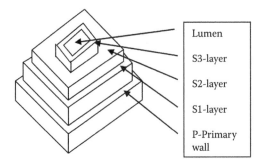

FIGURE 9.1   Schematic showing the wood fiber ultrastructure.

as output homogenized (averaged) properties, which serve as an input in the next (higher) scale modeling. Following this methodology, first, the averaged properties of each layer of the fiber wall are modeled while considering cellulose, hemicellulose, and lignin as an assembly of constituents and applying the elasticity theory. One possible representation of this heterogeneous internal structure is by the assembly of long concentric lignin, hemicellulose, and cellulose cylinders as shown in Figure 9.2. This model requires methods to calculate averaged elastic properties of concentric cylinder assemblies (CCA), consisting of cylinders with orthotropic symmetry. Similar models for two- and three-phase fiber-reinforced composite materials with isotropic or transversely isotropic fibers have been developed in [2]. On the fiber scale, the layered fiber wall with helical microfibril orientations was analyzed in [3–5]. In [6], an N-CCA model was introduced and the calculation methodology was explained in detail. A brief description of this model is presented in Section 9.2.2. In Section 9.2.3, the fiber wall is considered a laminate and the in-plane stiffness of the fiber wall is calculated using the CLT. The out-of-plane stiffness elements of the fiber wall also have to be calculated. It can be done using three-dimensional (3D) laminate theory [7], simple expressions based on rules of mixtures neglecting Poisson's interactions, or 3D finite element method (FEM) analysis. Finally, in Section 9.2.4, the fiber stiffness is modeled as a two-cylinder assembly where the central phase (lumen) may have negligible stiffness (if it is empty) or the stiffness of the matrix that is used in the composite.

9.2.2 N-PHASE CONCENTRIC CYLINDER ASSEMBLY

The unit cell of the layer in the fiber wall can be visualized as a CCA model, as shown in Figure 9.1. Since the length versus diameter ratio for the assembly is large, the average stress state in a phase is almost as in the model with infinitely long phases and the local stress perturbations at the ends have a negligible effect on the overall properties of the unit cell. Hence, infinitely long cylinders are considered in the model. A circular cross section of phases (axial symmetry) is assumed to obtain analytical solutions. The stiffness values for fibers with circular, square, and elliptical cross sections were compared in [8], showing that the shape of the cross section has a rather small effect on the results.

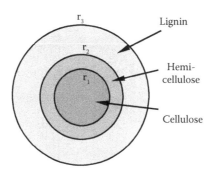

FIGURE 9.2 Schematic showing a three-phase cylindrical unit cell.

In the general case, the model consists of N infinite concentric cylinders. The material of a phase with index k is assumed to be orthotropic with orthotropic symmetry axes oriented in directions 1, r, and θ, respectively (r and θ are directions 2 and 3, respectively). The stress–strain relationship using Voigt notation is

$$\sigma_i^k = C_{ij}^k \varepsilon_j^k; \quad i,j = 1,2...,6 \tag{9.1}$$

The stiffness parameters of the assembly to be found are the longitudinal modulus E_1, Poisson's ratio ν_{12}, plane strain bulk modulus K_{23}, in-plane shear modulus G_{12}, and out-of-plane shear modulus G_{23}. Based on these, the transverse modulus E_2 and the out-of-plane Poisson's ratio ν_{23} can be found using the expressions

$$E_2 = \frac{1}{\dfrac{1}{4K_{23}} + \dfrac{1}{4G_{23}} + \dfrac{\nu_{12}^2}{E_1}} \qquad \nu_{23} = \frac{E_2}{2G_{23}} - 1 \tag{9.2}$$

The stress–strain state does not depend on the axial coordinate. The interface coordinate between the kth cylinder and $k+1$th cylinder is r_k. An empty lumen can be modeled as a cylinder with elastic properties approaching zero.

Several loading cases on the assembly have to be simulated to determine averaged elastic properties. In all cases, perfect bonding is assumed at the interface between phases and displacement and stress continuity on the interfaces between cylinders has to be satisfied:

$$u_r^k(r_k) = u_r^{k+1}(r_k), \; u_\theta^k(r_k) = u_\theta^{k+1}(r_k), \; u_1^k(r_k) = u_1^{k+1}(r_k)$$

$$\sigma_r^k(r_k) = \sigma_r^{k+1}(r_k), \; \sigma_{r\theta}^k(r_k) = \sigma_{r\theta}^{k+1}(r_k), \; \sigma_{1r}^k(r_k) = \sigma_{1r}^{k+1}(r_k), \quad k = 1,2,..,N-1 \tag{9.3}$$

Conditions at the outer boundary, $r = r_N$, of the unit cell can be either in displacements or in stresses.

To determine the average plane strain bulk modulus K_{23} of the assembly, we consider the plane strain case, $\varepsilon_{10} = 0$, and apply the nonzero radial strain, $\varepsilon_r = \varepsilon_0$, leading to $u_r^N(r_N) = \varepsilon_0 r_N$. After finding the stress state, the strain energy of the assembly is calculated. Then, the bulk modulus is found from the expression

$$U = 2K_{23}\varepsilon_0^2 V \tag{9.4}$$

The longitudinal modulus E_1 and Poisson's ratio ν_{12} of the CCA model are calculated by applying $\varepsilon_1 = \varepsilon_{10}$. The radial stress on the outer boundary $r = r_N$ is assumed to be zero, $\sigma_r^N(r_N) = 0$. The calculated radial displacement at $r = r_N$ is used to find the average strain in the radial direction, ε_0. It corresponds to free contraction of the model in the r direction. Hence, the average Poisson's ratio of the model is

$$\nu_{12} = -\frac{\varepsilon_0}{\varepsilon_{10}} \tag{9.5}$$

On calculating the average stress in direction 1, σ_1^{av}, the average longitudinal modulus E_1 of the assembly is obtained:

$$E_1 = \frac{\sigma_1^{av}}{\varepsilon_{10}} \tag{9.6}$$

To determine the in-plane shear modulus G_{12} of the composite represented by the cylinder assembly, we consider the homogenized material in Cartesian coordinates in a pure shear loading with $\gamma_{1x} = \gamma_{12}^0$, leading to $\sigma_{1x}^0 = G_{12}\gamma_{12}^0$. This strain field can be obtained by applying u_1 as a linear function of x. Then, the displacements applied on the surface of the cylindrical assembly in cylindrical coordinates are

$$u_r(r_N) = u_\theta(r_N) = 0, \ u_1(r_N) = \gamma_{12}^0 r_N \cos\theta \tag{9.7}$$

In Equation 9.7, θ is the angle between the x and r directions. The solution in this case is given in [8]. The in-plane shear modulus of the assembly can then be determined as follows. By transforming the macroscopic applied shear strain to the cylindrical system, we can write the shear stress distribution along the boundary $r = r_N$ of the homogenized material:

$$\sigma_{1r}(r_N) = G_{12}\gamma_{12}^0 \cos\theta \tag{9.8}$$

The distribution of this stress component at the outer boundary is a part of the solution and, hence, Equation 9.8 can be used to find G_{12}.

To calculate the transverse shear modulus G_{23}, shear stress $\sigma_{xy} = \tau$ is applied to the composite at infinity. The stress components in the cylindrical coordinate system are

$$\sigma_r = \tau \sin 2\theta, \ \sigma_\theta = -\tau \sin 2\theta, \ \sigma_{r\theta} = \tau \cos 2\theta \tag{9.9}$$

These stresses (or alternatively displacements) can be applied as boundary conditions to the assembly [2], obtaining upper and lower bounds of G_{23}, respectively. In the so-called generalized self-consistent scheme, suggested in [9] to analyze a transversely isotropic cylinder assembly, the assembly is embedded in an infinite domain of effective composite material. The shear modulus of the effective composite is one of the unknowns. This model was generalized for orthotropic cylinders in [8]. Since the shear modulus G_{23} of the effective composite is also an unknown, the system of equations to be solved is nonlinear with respect to G_{23} and its solution requires a numerical iterative procedure.

For N-cylinder CCA, the solution requires solving a system of algebraic equations originated from interface and boundary conditions, which is usually performed numerically. Simple analytical expressions for averaged elastic constants are available only for the two-cylinder case.

9.2.3 EXAMPLES FOR ELASTIC CONSTANTS OF A LAYER IN THE FIBER WALL

The layers of the secondary fiber wall are built of thread-like anisotropic units called microfibrils and they can be treated as unidirectional (UD) composites. The microfibrils consist of cellulose chains, hemicellulose, and lignin. The volume fractions of these chemical constituents are different in various layers of the fiber wall. The values used in this section are given in Table 9.1. The geometrical configuration of microfibrils in the layer is not fully clear. A common assumption is that cellulose fibers are surrounded by hemicellulose material and this subblock is embedded in a lignin matrix [10]. Calculation with FEM is convenient, as it uses square or rectangular shapes of constituents. Analytical modeling is possible by only assuming a concentric cylinder geometry. The CCA model in Figure 9.2 is a better representation of a homogenized material that is transversely isotropic than the model based on the square unit cell: the transverse shear modulus of the square unit cell does not obey the relationship $G_{23} = E_2/2 \, (1 + v_{23})$. The elastic properties of a layer are first calculated in its local material symmetry system of coordinates.

Elastic properties and geometrical parameters of cellulose and other constituents have been measured experimentally and also estimated theoretically [11–13]. The elastic modulus in the microfibril direction E_1^C is between 130 and 170 GPa, the transverse modulus is $E_2^C \in [17; 27]$ GPa, the axial shear modulus is $G_{12}^C \in [4.0; 5.0]$ GPa, the axial Poisson's ratio is $v_{12}^C \in [0.03; 0.2$, and the transverse Poisson's ratio is $v_{23}^C \in [0.48; 0.52]$. The variation in transverse isotropic elastic properties of hemicellulose is as large as for cellulose. Independently of the model used to calculate the properties of the cellulose + hemicellulose + lignin (C+H+L) system, this intrinsic uncertainty in input data leads to a large variation and uncertainty in the calculated averaged properties. Hence, the requirements to the model accuracy are not very high but the accuracy of the model has to be known. In simulations, we will use the elastic constants given in Table 9.2. The upper index in Table 9.2 and further in this section indicates the chemical constituent ($k = C, H, L$).

The averaged (homogenized) elastic properties of the S2 layer calculated according to the 3-cylinder CCA model are shown in Table 9.3. Alternatively, the CCA

TABLE 9.1

Volume Fractions of Constituents (Percentage) at 12% Average Moisture Content

Cell Wall Layer	Cellulose	Hemicellulose	Lignin
S2, S3	44.3	31.6	24.1
S1, primary, middle	18	17.4	64.6

TABLE 9.2

Elastic Properties of Constituents Used in the Simulations

	Constituent	E_1^k (GPa)	E_2^k (GPa)	G_{12}^k (GPa)	v_{12}^k	v_{23}^k
1	Cellulose	150.0	17.5	4.5	0.08571	0.50
2	Hemicellulose	16.0	3.5	1.5	0.4571	0.4
3	Lignin	2.75	2.75	1.034	0.33	0.33

TABLE 9.3

Elastic Properties of S2 Layer of the Fiber Wall in Material Symmetry Axes

Property	Model C+H+L	Model (C+H)+L	Model C+(H+L)	Square FEM [11]
E_1 (GPa)	72.278	72.282	72.271	72.4
E_2 (GPa)	5.964	5.973	5.947	6.70
v_{12}	0.243	0.243	0.244	0.242
G_{12} (GPa)	2.112	2.112	2.113	3.10
G_{23} (GPa)	2.056	2.060	2.049	1.36
v_{23}	0.451	0.450	0.451	0.391

model and any other analytical model may be applied in two steps. In the first step, the averaged properties of a subblock consisting of cellulose and hemicellulose cylinders are found (C+H). The calculated properties are used as input data for the next CCA model where the (C+H) subblock is embedded in the lignin (L). The result of the two-step homogenization is named (C+H)+L in Table 9.3. Another alternative is to first homogenize the (H+L) subblock and in the second step to consider the cellulose (L) cylinder embedded in the (H+L) cylinder. The result is denoted C+(H+L) in Table 9.3. This exercise was performed to evaluate the significance of geometrical details other than volume fractions. The results are very close, which proves that the sequence of the homogenization is not important.

The FEM results using a model consisting of embedded squares are also presented in Table 9.3. As expected, the axial modulus and the axial Poisson's ratio are in excellent agreement for the CCA and square model. Transverse and shear moduli are rather different. The difference in the transverse modulus reaches 10% for the S2 layer. The shear modulus values differ by almost 50%. This is because the square model cannot represent transversally isotropic materials: the relationship between G_{23} and E_2 values is not correct. Therefore, the CCA model is more reliable.

9.2.4 ELASTIC PROPERTIES OF THE FIBER WALL

The wall of the fiber is a laminate with layup [P, S1, S2, S3]. Examples of layer thicknesses and orientation of microfibrils in the layer can be found in Table 9.4. The layer thickness ratios are rather different in earlywood and in latewood fibers; particularly, the S2 layer is much thicker in the latewood. It should be noted that the primary layer has a random distribution of microfibrils and its stiffness matrix has to be calculated differently than for the rest of the layers. The in-plane elastic constants of the fiber wall may be calculated using the CLT. The calculation steps that are used to obtain the stiffness matrix of the fiber wall (laminate) are briefly explained next.

1. Using the elastic constants of layers calculated using the CCA model, we calculate the stiffness matrices of the layers in their symmetry axis. Then, the stiffness matrix Q_{ij}^k of the kth layer of the fiber wall is transformed to the global system in which z is the fiber direction, the tangential direction is φ,

TABLE 9.4

Geometrical Parameters of the Layers in the Wood Fiber Wall

Cell Wall Layer	Earlywood		Latewood	
	Thickness h^k(μm)	Microfibril Angle θ	Thickness h^k(μm)	Microfibril Angle $θ_k$
P	0.1	Random	0.1	Random
S1	0.2	±(50–70)	0.3	±(50–70)
S2	1.4	10–40	4.0	0–30
S3	0.03	60–90	0.04	60–90

Sources: Dinwoodie, J.M. 1981. *Timber—Its Nature and Behaviour*. New York, NY: Van Nostrand Reinhold; Fengel, D. 1969. The ultrastructure of cellulose from wood. Part 1: Wood as the basic material for the isolation of cellulose. *Wood Science and Technology* 3:203–17; Kollman, F.F.P. and Cote, W.A. 1968. *Principles of Wood Science and Technology, Solid Wood*. Springer-Verlag: Berlin. With permission [1,12–13].

and the radial direction (laminate thickness direction) is *r*. The P layer has in-plane random fibril orientation, and, therefore, it is an in-plane isotropic material with the stiffness matrix independent of the coordinate system.

2. Calculate the extensional stiffness matrix A_{ij} of the fiber wall (laminate). Since the laminate is unsymmetric, the coupling stiffness matrix is non-zero. However, due to the cylindrical geometry of the fiber, local curvature changes are not possible and the fiber wall behaves as if the coupling stiffness matrix is zero.

3. The elastic constants of the laminate (fiber wall) are defined through strain components as a response to applied uniaxial stress states by means of an *A* matrix that contains elements A_{16}, A_{26}. It means, for example, that fiber rotation is due to applied axial load (due to the helical fibril orientation shear strain $γ_{zϕ}$ occurs in the fiber and it will rotate). If the fiber is inside the composite, the situation is different: each fiber tends to rotate in a different direction. An interaction between the fiber and the matrix leads to a close to zero rotation. If the fiber is constrained and cannot rotate, its strain response to axial load is smaller: the apparent modulus of the fiber wall in the constrained case is higher than the real one. Obviously, the apparent modulus is more relevant and has to be used when modeling composites. The apparent modulus can be calculated assuming in CLT $γ_{zϕ} = 0$. The same result can be obtained by simply assuming $A_{16} = A_{26} = 0$. The apparent elastic properties of the earlywood and latewood fiber walls under the condition $γ_{zϕ} = 0$ are presented in Table 9.5, where true elastic constants for the rotation-free case are also presented. The difference is significant in the modulus of the wall in the fiber direction.

The apparent elastic modulus E_z^* of the fiber wall is almost 50% higher than the real modulus: if the fiber rotation is not allowed, the response becomes much stiffer. This effect using an FEM analysis was studied in [4]. Since the apparent constants are relevant for composites, all the following examples use these data. Fibril orientation in layers has an effect on the elastic properties of the fiber wall. Since the S1 and S3 layers are relatively thin as compared with the S2 layer, the most important is the fibril angle in the S2 layer. Therefore, in simulations, fixed values for fibril angles in the S1 and S3 layers (−60° and −75°, respectively) are used, leaving the fibril angle in the S2 layer as the only parameter and considering the earlywood and the latewood separately. The dependence of the apparent elastic fiber wall properties on the fibril orientation angle in the S2 layer is shown in Table 9.6.

In this chapter, we calculate the out-of-plane elastic constants of the laminate (fiber wall) using simple expressions based on constant stress or constant strain assumptions. They are

$$
\frac{1}{E_r} = \sum_k \frac{V_k}{E_r^k} \qquad \frac{1}{G_{zr}} = \sum_k \frac{V_k}{G_{zr}^k} \qquad \frac{1}{G_{r\phi}} = \sum_k \frac{V_k}{G_{r\phi}^k}
$$

$$
\nu_{zr} = \sum_k V_k \nu_{zr}^k \qquad \nu_{\phi r} = \sum_k V_k \nu_{\phi r}^k \tag{9.10}
$$

TABLE 9.5
In-Plane Elastic Constants of Wood Fiber Wall, S2 Angle 10°

Property	Earlywood		Latewood	
	Free Fiber	Apparent Constant	Free Fiber	Apparent Constant
E_z (GPa)	36.49	55.51	37.27	60.86
E_ϕ (GPa)	7.96	8.04	7.05	7.46
$\nu_{z\phi}$	0.676	0.680	0.63	0.479
$G_{z\phi}$ (GPa)	2.85	2.84	2.61	4.29

TABLE 9.6
Effect of Microfibril Angle in S2 on the Apparent In-Plane Elastic Constants of Wood Fiber Wall

Property	Earlywood			
	10°	20°	30°	40°
E_z^* (GPa)	55.40	41.93	24.00	11.68
E_ϕ^* (GPa)	8.93	8.54	8.13	8.53
$\nu_{z\phi}^*$	0.434	0.813	1.011	0.853
$G_{z\phi}^*$ (GPa)	4.45	8.47	13.04	16.02

In Equation 9.10, V_k is the volume fraction of the k layer in the laminate. The elastic constants of the layers have been recalculated to the global coordinate system $E_r^k, G_{zr}^k, G_{r\phi}^k, \nu_{zr}^k, \nu_{\phi r}^k$ using 3D stiffness transformation expressions [7]. The out-of-plane properties of the in-plane isotropic P-layer were assumed equal to the properties of the corresponding aligned (UD) material.

9.2.5   ELASTIC CONSTANTS OF THE FIBER

Macroscopic elastic constants of the fiber can be calculated in several ways. "Macroscopic" means that the lumen and the cell wall are homogenized in a fiber that has no lumen and has a radius equal to the boundary radius of the CCA model. If the fiber lumen is filled with resin, its properties should be assigned to the central cylinder. A simple way to calculate the fiber axial modulus is by using the volume fraction of the fiber wall in the fiber assembly (including the empty lumen):

$$E_z^{\text{fiber}} = E_z \frac{r_1^2}{r_N^2} \tag{9.11}$$

In another approach, a two-phase CCA model can be used: lumen (filled or empty) and the cell wall.

9.3   STIFFNESS OF COMPOSITES

9.3.1   RULE OF MIXTURES

Natural fiber composites are often attributed to the class of short fiber composites and they are treated as such when it comes to prediction of their mechanical properties. It has been demonstrated in a number of papers that axial in-plane stiffness of short fiber composites with a non-UD fiber orientation can be accurately predicted by the RoM in which corrections for finite fiber length and non-UD orientation of fibers are incorporated:

$$E_C = \left(\eta_o \eta_l V_f E_f + V_m E_m \right) \left(1 - V_p \right)^2 \tag{9.12}$$

where E_C is the Young's modulus of the composite; E_f, E_m are the stiffness of the fiber and the matrix; V_f, V_m, and V_p are the volume fractions of the fibers, matrix, and porosity, respectively; η_o is the fiber orientation factor; and η_l is the fiber length efficiency factor.

The ineffectiveness of stress transfer due to the limited length of the filaments is accounted for by the fiber length efficiency factor (η_l), which is, for example, obtained from the shear-lag model developed by Cox [14]:

$$\eta_l = 1 - \frac{\tanh(\theta)}{\theta} \quad \text{and} \quad \theta = 2 \frac{L_f}{d_f} \sqrt{G_m / E_f \ln\left(\frac{\kappa}{V_f} \right)} \tag{9.13}$$

where L_f is the average fiber length, d_f is the fiber diameter, G_m is the shear stiffness of the matrix, and κ is a constant controlled by the geometrical packing pattern of the fibers (e.g., for hexagonal packing of fibers $\kappa = \pi/2 \cdot \sqrt{3} = \sim 0.907$).

The fiber orientation factor (η_o) in Equation 9.12 accounts for the misalignment of fibers (not all fibers are aligned with the direction of loading). This factor is obtained based on a geometrical model originally developed by Krenchel: $\eta_o = \sum_{i=1}^{N} V_{f_i} \cos^4 \varphi_i$, where φ_i is the angle of fiber orientation and V_{fi} is the fraction of these fibers. For typical cases, the orientation factor is known: (1) UD composite, $\eta_o = 1$; (2) 2D random fiber orientation, $\eta_o = 3/8$ (0.375); and (3) 3D random fiber orientation, $\eta_o = 1/5$ (0.2). It should be noted that in the current form this model assumes that composite constituents are elastic and isotropic with stresses transferred between the fiber and matrix without yielding or slipping.

Some other variations for the RoM have been suggested, for example, an additional coefficient η_d is added to account for the variation of fiber diameter and the resulting changes in fiber stiffness:

$$E_C = \eta_d \eta_o \eta_l V_f E_f + V_m E_m \tag{9.14}$$

where η_d can be calculated based on average fiber diameter and slope of the curve representing dependence of fiber stiffness on diameter [15]. Most often, for convenience, it is assumed that natural fibers have a circular cross section; however, this is not the case and in reality natural fibers have a rather irregular geometry. Equation 9.14 can be modified even further by introducing the correction factor A_k, which is based on the measurements of apparent and true fiber cross-sectional areas [16]:

$$E_C = A_k \eta_d \eta_o \eta_l V_f E_f + V_m E_m \tag{9.15}$$

Both Equations 9.14 and 9.15 can be adjusted to account for the porosity, similarly to what done is for Equation 9.11.

9.3.2 SHEAR-LAG PARAMETER

The RoM model presented earlier (Equation 9.14) assumes stress transfer from the matrix to the reinforcement through interfacial shear stresses only, with maximum tensile stresses in the middle of the fiber and maximum shear stresses at the ends of the fiber. Efficiency of stress transfer is accounted for through the length efficiency factor (η_l), which is defined by the shear-lag parameter θ (see Equation 9.13). The equation for this parameter may have rather different forms depending on the accuracy of the assumptions. For example, the shear-lag parameter based on exact elasticity equations for axisymmetric stress states with transversely isotropic materials is [17]

$$\theta = 2 \frac{L_f}{d_f} \sqrt{\frac{\left(V_f E_f + V_m E_m \right)}{E_f E_m \left(\dfrac{V_m}{G_f} + \dfrac{2}{G_m} \left[\dfrac{1}{V_m} \ln \left(\dfrac{1}{V_f} \right) - 1 - \dfrac{V_m}{2} \right] \right)}} \tag{9.16}$$

where G_f is the shear modulus of the fiber. This equation can be modified even further [18] by taking into account interface imperfections and introducing the interface parameter D_S:

$$\theta = 2\frac{L_f}{d_f}\sqrt{\frac{\left(V_f E_f + V_m E_m\right)}{E_f E_m \left(\frac{V_m}{G_f} + \frac{2}{G_m}\left[\frac{1}{V_m}\ln\left(\frac{1}{V_f+\chi}\right)-1-\frac{V_m}{2}+\frac{2}{d_f D_S}\right]\right)}} \qquad (9.17)$$

where χ is a correction constant (the value of $\chi = 0.009$ seems to give good results [18] for various types of fibers).

9.3.3 Composites with In-Plane Fiber Orientation Distribution

Fibers in natural fiber composites are usually not aligned. In the next analysis, we assume that all fibers have an in-plane orientation: the fiber axis is parallel to the mid-plane of the composite plate. This assumption is justified if the fiber length is significantly larger than the thickness of the composite plate. In-plane orientation of fibers in composites may be characterized by fiber orientation angle relative frequency. The whole fiber orientation angle region $\varphi \in \left[-\frac{\pi}{2}; +\frac{\pi}{2}\right]$ may be divided into subregions of size $\Delta\varphi$, and the relative number of fibers with an orientation in the given region may be counted. Details regarding the used experimental techniques may be found in [19]. The result is presented as the relative orientation frequency distribution function (also called the fiber orientation distribution function). An example of a distribution symmetric with respect to the 0-direction is shown in Figure 9.3. Symmetric distributions result in orthotropic macroproperties of the composite.

According to the definition, the orientation distribution function $f(\varphi_i)$ multiplied by $\Delta\varphi$ is the probability that the fiber orientation angle is between $\varphi_i - \frac{\Delta\varphi}{2}$

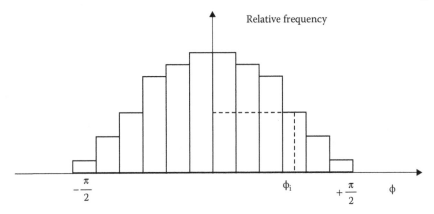

FIGURE 9.3 Example of symmetric distribution of the relative frequency of fiber orientation distribution.

and $\varphi_i + \dfrac{\Delta\varphi}{2}$. The experimentally obtained $f(\varphi_i)$ values can be fitted by a Fourier series expansion, which in the symmetric case can be written as

$$f(\varphi) = \frac{1}{\pi} \sum_{n=0}^{\infty} a_n \cos 2n\varphi \qquad (9.18)$$

Usually, two to four terms in Equation 9.18 give sufficient accuracy. If the fiber orientation in the composite is in-plane random, all orientations are represented equally and

$$f(\varphi) = \frac{1}{\pi} \qquad (9.19)$$

The effective stiffness matrix of a composite with a given in-plane fiber orientation distribution function $f(\varphi)$ may be calculated using a "laminate analogy" approach [20–21].

In this approach, the composite is replaced by an effective layered structure, which is obtained virtually by separating all fibers with a given orientation $\varphi \in [\varphi_i - d\varphi/2, \varphi_i + d\varphi/2]$ from the rest of the composite and building a UD layer with this fiber orientation. The fiber volume fraction in this layer is the same as the average volume fraction in the composite. The relative thickness of the virtual UD layer in the laminate is equal to the probability of this orientation $f(\varphi_i) d\varphi$. In this way, an effective laminate is designed with an infinite number of layers with different orientations.

The stiffness matrix of the composite using this model is equal to the extensional stiffness matrix of the effective laminate and can be written as

$$Q_{ij}^{comp} = \sum_{k=1}^{\infty} \frac{t_k}{h} \bar{Q}_{ij}^{k}(\varphi_k) = \sum_{k=1}^{\infty} \bar{Q}_{ij}^{k}(\varphi_k) f(\varphi_k) d\varphi \qquad (9.20)$$

In Equation 9.20, summation may be replaced by integration, leading to the following expression for the composite stiffness matrix elements:

$$Q_{ij}^{comp} = \int_{-\frac{\pi}{2}}^{+\frac{\pi}{2}} \bar{Q}_{ij}(\varphi) f(\varphi) d\varphi \qquad (9.21)$$

The stiffness matrix $\bar{Q}_{ij}$ of a UD layer with orientation φ may be obtained from the elastic properties of the layer in its material symmetry axes.

9.3.4 RANDOM COMPOSITES

In composites with random in-plane orientation distribution of fibers, the distribution function $f(\varphi)$ is constant and is given by Equation 9.19. In this case, the integration in Equation 9.21 has been performed analytically [21], leading to

$$Q_{11}^{comp} = Q_{22}^{comp} = \frac{1}{8}\left(3Q_{11}^a + 3Q_{22}^a + 2Q_{12}^a + 4Q_{66}^a\right)$$

$$Q_{12}^{comp} = \frac{1}{8}\left(Q_{11}^a + Q_{22}^a + 6Q_{12}^a - 4Q_{66}^a\right) \tag{9.22}$$

$$Q_{12}^{comp} = \frac{1}{8}\left(Q_{11}^a + Q_{22}^a - 2Q_{12}^a + 4Q_{66}^a\right)$$

The properties Q_{ij}^a of the corresponding aligned composite may be calculated, for example, using CCA. The in-plane elastic constants of the in-plane isotropic composite follow from

$$v^{comp} = \frac{Q_{12}^{comp}}{Q_{11}^{comp}}, \quad G^{comp} = Q_{66}^{comp}, \quad E^{comp} = 2G^{comp}\left(1 + v^{comp}\right) \tag{9.23}$$

9.3.5 NUMERICAL METHODS

Attempts have been made to predict the stiffness of the randomly oriented short fiber composites by use of numerical methods (namely FEM). Direct numerical methods using representative volume element (RVE) [22–23] have reported very accurate predictions of properties. However, the structure of these composites is rather complex and cannot be easily and accurately translated into a numerical model. Instead, statistical techniques (e.g., Monte Carlo simulation) are used to generate RVE based on available distributions of fiber length, diameter, and orientation. Due to rather intensive computational requirements, models are often smaller than RVE and contain only a small number of fibers. Moreover, in the case of natural fibers, the filaments are not straight but rather curved, which is not accounted for in the numerically generated RVE.

Some other studies combined analytical and numerical approaches, in which a simplified numerical model (e.g., single flax fiber embedded in the block of matrix) was used to compute stiffness tensor C_{ijkl}^* of the unit cell and the orientation averaging methods were then used to calculate the stiffness tensor of the natural fiber composite [24–26]:

$$C_{ijkl} = \int\int C_{ijkl}^*\left(\zeta, \vartheta\right) p\left(\zeta, \vartheta\right) \sin\zeta\,\mathrm{d}\zeta\mathrm{d}\vartheta \tag{9.24}$$

where $p(\zeta, \vartheta)$ represents the fiber orientation distribution density as a function of the azimuthal, ζ, and elevation, ϑ, angles. This approach was used to predict stress–strain curves for the flax/polypropylene composite with satisfactory accuracy, as shown in Figure 9.4.

Another study combining a numerical method with an analytical micromechanical model is presented in [8]. In this work, the authors were using the FEM to analyze the importance of geometry of wood fibers on stiffness. It is common to assume (to simplify calculations) that wood fibers have regular geometry with a circular cross section. However, in reality, fibers are noncircular and sometimes they are collapsed, as shown in Figure 9.5. In this case [8], three different configurations in the numerical model

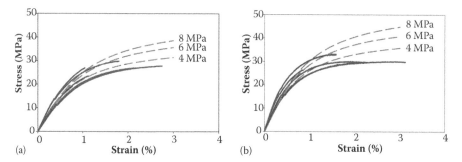

FIGURE 9.4 Stress–strain curves for flax/polypropylene composites with $V_f = 0.13$ (a) and $V_f = 0.20$ (b). Continuous lines are for experimental results and dotted for simulations (numbers near those lines show values of interfacial shear strength used in the simulations). (Courtesy of J. Modniks.)

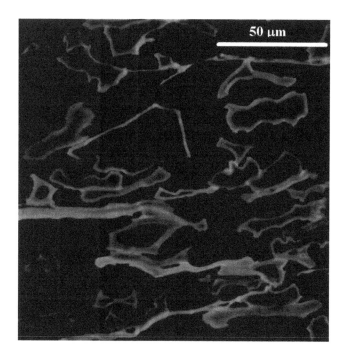

FIGURE 9.5 Confocal microscopic image of an oriented softwood kraft fiber mat. (Courtesy of Prof. K. Gamstedt.)

were used: circular and square-shaped fibers with a unit cell of square cross section and elliptic fiber in a rectangular unit cell. It was found that the fiber geometry is not important when the lumen is filled with resin and that the longitudinal elastic modulus is not affected by the shape of the fiber. However, the transverse and shear moduli, as well as the Poisson's ratio, greatly depend on the geometry of the fiber if the lumen is hollow.

9.4 STRENGTH OF COMPOSITES

9.4.1 RULE OF MIXTURES

Similar to the stiffness model (see Equation 9.12), there is a RoM expression for strength:

$$s_c^u = \left(\eta_{ls}\eta_{os}\sigma_f^u V_f + V_m \sigma_m\right)\left(1 - V_p\right)^2 \tag{9.25}$$

where σ_f^u is the fiber strength; σ_m is the stress in the matrix at the failure strain of the fiber, assuming linear elastic behavior of the fiber and the matrix $\sigma_m = \sigma_f^{u} \cdot (E_m/E_f)$; and $\eta_{ls}\eta_{os}$ are the length efficiency and orientation factors, respectively. The fiber length efficiency factor is calculated according to Kelly–Tyson model in the following manner:

$$\eta_{ls} = \begin{cases} 1 - l_c/2l & l \geq l_c \\ l/2l_c & l < l_c \end{cases}, \text{ where } l_c \text{ is the critical fiber length defined as } l_c = \frac{\sigma_f^u d_f}{2\tau} \tag{9.26}$$

and τ is the interfacial shear strength (ISS). As in case of the modulus, the strength can be also corrected [16] with respect to the error in the measurements of the fiber cross-sectional area (see A_k in Equation 9.14).

It is known that the strength of synthetic fibers (e.g., glass, carbon) can be well described by a statistical distribution, namely the Weibull distribution. The same approach works for natural fibers and the average fiber strength is given by

$$\sigma_f^u = \beta \left(\frac{l}{l_0}\right)^{-1/\alpha} \Gamma\left(1 + \frac{1}{\alpha}\right) \tag{9.27}$$

where l is the fiber length, l_0 is a normalization parameter ($l_0 = 1$ mm), α and β are parameters of the Weibull distribution, and G is the gamma function. By substituting Equation 9.27 into Equation 9.26, the critical fiber length l_c can be expressed as follows [27–28]:

$$l_c = \left(\frac{\beta\Gamma\left(1 + 1/\alpha\right)d_f}{2\tau l_0^{-1/\alpha}}\right)^{\frac{\alpha}{1+\alpha}} \tag{9.28}$$

The fiber orientation efficiency factor η_{os} can be directly obtained from optical measurements by using the Krenchel equation (see Section 9.3) or by analyzing stress–strain curves of the composite using a RoM expression for the stiffness [28–29]. It was shown that both these approaches rendered the same results for the orientation factor of short glass fiber composites with four different thermoplastic matrices [29].

9.4.2 LAMINATE ANALOGY

The laminate analogy can also be used for the calculation of strength of randomly oriented natural fiber composites. In this case, the laminate is divided into off-axis plies according to the distribution of fiber orientation (e.g., Figure 9.3). Then, CLT is used to calculate stresses (in the local coordinate system related to the fiber direction) in each layer. To apply this method, the in-plane strength of UD lamina in different directions and for different loadings (tension and compression) has to be known. The failure sequence analysis can be carried out by applying failure criteria to each of the plies to find the first ply failure. To perform such an analysis, the maximum stress criterion or more advanced quadratic criteria (e.g., Hashin [2], Tsai-Wu, or Tsai-Hill [30]) can be used.

Once the first ply in which failure is detected is found, its properties should be degraded and a new stress analysis should be performed until the next ply failure is detected. The degradation of the properties can be applied by using a ply discount model, in which the stiffness of the lamina in all directions is set to zero (or reduced almost to zero). However, it is more accurate to degrade properties according to the failure mode that has occurred. For example, if failure transverse to the fiber direction is detected, then only the transverse and shear stiffness is degraded, but the stiffness along the fibers is still unchanged. This analysis is carried out until all plies in the laminate have failed.

9.5 MOISTURE EXPANSION OF NATURAL FIBER COMPOSITES

Free hygroexpansion of materials is a well-known phenomenon. Moisture absorption leads to an increase in moisture weight content in the material and to dimensional changes (swelling). In the linear elasticity of orthotropic materials, these dimensional changes are assumed to be proportional to the moisture content change and three different constant coefficients of proportionality, called swelling or moisture expansion coefficients, characterize the relative dimensional changes in three directions of the material symmetry:

$$\beta_k^H = \frac{\varepsilon_k^H}{\Delta M}; \quad k = 1, 2, 3 \tag{9.29}$$

Here, ε_k^H is free swelling strain in the k-direction due to moisture weight content change by ΔM. Composite materials contain several connected phases with different swelling coefficients and different moisture contents. The free (unconstrained) average swelling of the composite is the result of an internal force balance and depends on the microstructure of the composite and hygroelastic properties of the constituents. In the context of the multiscale analysis used in this chapter, we performed a micromechanics analysis on one scale to obtain average properties, which serve as input moisture expansion coefficients for the next scale analysis.

While analyzing swelling of the cellulose/hemicellulose/lignin system, we first calculate the average moisture expansion coefficients of all layers in the fiber wall. Since each layer is a UD composite containing microfibrils, we can generalize the CCA model described in detail in Section 9.2.2 to account for moisture expansion.

Layers are building blocks in the fiber wall, and, therefore, the laminate theory with moisture expansion terms is the right tool to calculate the average fiber wall properties. On the next scale, we consider aligned natural fiber composites containing resin-filled or empty lumen, a cylindrical fiber wall, and the resin block surrounding it. This system may be analyzed using the CCA model with moisture terms. Finally, moisture expansion of composites with a certain fiber orientation distribution may be analyzed using the laminate analogy.

In this section, we will briefly describe the moisture expansion analysis using the CCA model and the laminate theory [31]. The CCA model may be used to determine the moisture expansion coefficients of the layer in the fiber wall. The only difference with the material model for constituents used in Section 9.2 and given by Equation 9.1 is that free moisture expansion terms are added in the elastic stress–strain relationship:

$$\sigma_i = C_{ij}\left(\varepsilon_j - \beta_j^H \Delta M\right) \tag{9.30}$$

While performing the same derivations as in [6], one can easy see that the expressions for radial displacement, and radial and axial stresses, have to be slightly modified [6] by terms that are dependent on moisture content ΔM_k. The moisture weight content (ΔM_k) in the kth constituent is not known and has to be determined experimentally or from moisture diffusion models. It is input data for the presented analysis.

The stress and displacement expressions with modified terms have to satisfy continuity conditions at interfaces. The outer boundary $r = r_N$ is stress free. The strain ε_{10} in direction 1 is constant and its value in the case of free expansion must result in zero average stress in direction 1. With these boundary conditions, the strain ε_{10} is the free moisture expansion strain of the assembly in direction 1:

$$\varepsilon_1^{H-\text{assembly}} = \varepsilon_{10} \tag{9.31}$$

The radial displacement at the free boundary $r = r_N$ can be used to calculate the radial and tangential-free moisture expansion strains of the assembly:

$$\varepsilon_r^{H-\text{assembly}} = \varepsilon_\theta^{H-\text{assembly}} = \frac{u_N\left(r_N\right)}{r_N} \tag{9.32}$$

A convenient procedure involves using ε_{10} as a numerical parameter, solving the problem, and calculating the average axial force. Then, the parameter ε_{10} is changed until zero force (with required accuracy) is obtained. The moisture expansion coefficients follow from Equation 9.29:

$$\beta_1^H = \frac{\varepsilon_1^{H-\text{assembly}}}{\Delta M_{\text{assembly}}} \quad ; \quad \beta_2^H = \beta_3^H = \frac{\varepsilon_r^{H-\text{assembly}}}{\Delta M_{\text{assembly}}} \tag{9.33}$$

The average moisture content of the analyzed assembly $\Delta M_{\text{assembly}}$ obeys the rule of mixtures:

$$\Delta M_{\text{assembly}} = \sum_k \Delta M_k \frac{\rho_k}{\rho_{\text{assembly}}} V_k \tag{9.34}$$

$$\rho_{\text{assembly}} = \sum_k \rho_k V_k \tag{9.35}$$

As discussed earlier, the fiber wall can be considered a laminate, and the composite with certain in-plane fiber orientation distribution can be analyzed by considering it a laminate with an infinite number of layers. The governing in-plane equation of the CLT for free moisture expansion of a symmetric laminate is

$$\begin{Bmatrix} N_1^H \\ N_2^H \\ N_{12}^H \end{Bmatrix} = \begin{bmatrix} A_{11} & A_{12} & A_{16} \\ A_{12} & A_{22} & A_{26} \\ A_{16} & A_{26} & A_{66} \end{bmatrix} \begin{Bmatrix} \varepsilon_1^H \\ \varepsilon_2^H \\ \gamma_{12}^H \end{Bmatrix} \tag{9.36}$$

Here, $[A]$ is the extensional stiffness matrix of the laminate. The expression for the "force" $\{N^H\}$ in Equation 9.36 written in a vector form is as follows:

$$\{N^H\} = \sum_k \Delta M_k h_k \left[\overline{Q}\right]_k \left\{\overline{\beta}^H\right\}_k \tag{9.37}$$

Here, $\left\{\overline{\beta}^H\right\}_k$ is the moisture expansion coefficient vector of the kth constituent in the global system of coordinates:

$$\overline{\beta}_1^H = m^2\beta_1^H + n^2\beta_2^H \quad \overline{\beta}_2^H = n^2\beta_1^H + m^2\beta_2^H$$

$$\overline{\beta}_{12}^H = 2\left(\beta_1^H - \beta_2^H\right)mn \quad m = \cos\theta, \quad n = \sin\theta \tag{9.38}$$

Since $\{N^H\}$ is known, the laminate strains may be found from Equation 9.36. For the laminate, these strains are free moisture expansion strains and, hence, the laminate moisture expansion coefficients can be calculated as

$$\beta_1^{H-LAM} = \frac{\varepsilon_1^H}{\Delta M_{LAM}} \quad \beta_2^{H-LAM} = \frac{\varepsilon_2^H}{\Delta M_{LAM}} \quad \beta_{12}^{H-LAM} = \frac{\gamma_{12}^H}{\Delta M_{LAM}} \tag{9.39}$$

The average moisture content in the laminate ΔM_{1AM} may be calculated using expressions such as Equations 9.34 and 9.35.

9.6 INELASTIC BEHAVIOR OF NATURAL FIBERS AND THEIR COMPOSITES

Until this point, natural fiber composites and their constituents have been treated in this chapter as linear-elastic materials. The models described earlier can fairly well predict the elastic modulus of these materials and with some success strength also can be calculated. However, a number of experimental studies [32–34] demonstrated that cellulosic fibers and composites based on them are highly nonlinear, as shown in Figure 9.6, where stress–strain curves of flax and regenerated fiber composites are presented. Thus, to obtain more accurate predictions of performance of these

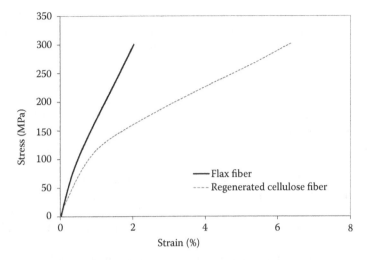

FIGURE 9.6 Stress–strain curves or natural fiber unidirectional composites.

materials, the models should include phenomena that are responsible for the nonlinear behavior. Many studies deal with this problem applied to various materials and loading conditions. Very often, however, only some of the material properties that are responsible for the nonlinear behavior are taken into account. Recently, a more general approach was presented in [35–36]. It has been shown that the model for viscoelasticity (VE) presented in [37–38] and the model for viscoplasticity (VP) by [39], modified in [40] to account for microdamage, can be used to simulate tensile tests performed in stress-controlled mode. This model can be also adopted for the case when strain is the input variable. Further within this section, this approach is discussed in detail.

9.6.1 MATERIAL MODEL IN STRESS FORMULATION

Let us look at the situation when the applied stress ramp $\sigma(t)$ causes a strain ε that can be decomposed in the VE strain, ε_{VE} and in the VP strain, ε_{VP}:

$$\varepsilon(\sigma,t) = d\left(\sigma_{max}\right)\varepsilon_{VE}\left(\sigma,t\right) + d\left(\sigma_{max}\right)\varepsilon_{VP}\left(\sigma,t\right) \qquad (9.40)$$

Strains ε_{VE} and ε_{VP} are nonlinear with respect to stress and strain evolution and this is further enhanced by accumulation of damage, characterized by $d\left(\sigma_{max}\right)$. Although in reality a more complex mathematical relationship with the interaction of all possible mechanisms is expected, the linear form (Equation 9.40) may be considered the first term in the series expansion of such a case.

It is further assumed that the damage development is an elastic process; hence, d does not depend explicitly on time, but it depends on the "maximum" (most damaging) stress state experienced during the stress ramp. In uniaxial loading considered here, it is the highest experienced axial tensile stress, σ_{max}, and hence $d = d\left(\sigma_{max}\right)$.

However, in some cases [41], it is more convenient to present d as a function of strain, ε_{max}. From the analysis of the elastic strain component, it follows [41] that the quantity of d is given by the change of the elastic compliance due to damage:

$$d\left(\sigma_{max}\right) = \frac{E_0}{E\left(\sigma_{max}\right)} \tag{9.41}$$

where E_0 is the initial elastic modulus of undamaged material and $E\left(\sigma_{max}\right)$ is the modulus of the damaged composite. For the elastic strain response, the interaction between d and VE strain in Equation 9.40 in the form of the product is exact (it has also been used in [42–43] for the transient VE strain) but there is no theoretical basis for the "product form" in the VP strain term. Of course, in general, it is reasonable to assume that damage entities are related to local stress concentrations and large local VP strains are expected, thus contributing to the macroscopic VP strain. It has been demonstrated in [40] that this form fits the test results better than a stand-alone term of ε_{VP}.

The model developed by Schapery [37–38] is used here to describe the nonlinear VE strain response, which in a one-dimensional case is

$$\varepsilon = d\left(\sigma_{max}\right) \cdot \left(\varepsilon_{el} + g_1 \int_0^t \Delta S\left(\psi - \psi'\right) \frac{d\left(g_2\sigma\right)}{d\tau} d\tau + \varepsilon_{VP}\left(\sigma,t\right) \right) \tag{9.42}$$

Here, ε_{el} represents the elastic strain in an undamaged composite, which, generally speaking, may be a nonlinear function of stress. "Reduced time" ψ is introduced as

$$\psi = \int_0^t \frac{dt'}{a_\sigma} \quad \text{and, consequently,} \quad \psi' = \int_0^\tau \frac{dt'}{a_\sigma} \tag{9.43}$$

Parameters g_1, g_2 and the shift factor a_σ in Equations 9.42 and 9.43 are stress invariant dependent (they are also affected by temperature and humidity). The $\Delta S(\psi)$ term in Equation 9.42 characterizes the transient part of the viscoelastic response, which, according to Schapery [37], does not depend on the stress and can be expressed via Prony series:

$$\Delta S\left(\psi\right) = \sum_i C_i \left(1 - \exp\left(-\frac{\psi}{\tau_i} \right) \right) \tag{9.44}$$

where C_i are constants and τ_i are retardation times. In the case $g_1 = g_2 = a_\sigma = 1$ (such a region can be found for some materials), Equation 9.42 turns into the strain–stress relationship for a linear viscoelastic, nonlinear viscoplastic material. Permanent strains that grow with time at high stresses are analyzed as viscoplastic, and it has been shown [40,43] that the development of these VP strains in natural fiber composites can be appropriately described by the following function [39]:

$$\varepsilon_{VP}\left(\sigma,t\right) = C_{VP} \left\{ \int_0^{1/t^*} \left(\frac{\sigma(\tau)}{\sigma^*} \right)^M d\tau \right\}^m \tag{9.45}$$

where C_{VP}, M, and m are constants to be determined: and t/t^* is the normalized time, where t^* is an arbitrary chosen characteristic time constant.

9.6.2 CHARACTERIZATION OF THE EFFECT OF DAMAGE

It is well known that various failure events at the microlevel (microdamage) can be observed in these materials before the final fracture of the composite. These events may include fiber–matrix debonds, fiber bundle debonding from matrix, intrabundle cracks (cracks crossing bundles with the crack plane parallel to fibers), matrix cracks bridged by several bundles, and finally, fiber and/or bundle pullouts and fiber breaks. However, in this chapter, specific damage modes are not identified or separated; instead, the overall significance of microdamage developing at high strains or stresses is characterized by the changes in the elastic modulus as defined in Equation 9.41. Experimentally, this is done through the application of multiple loading ramps in a displacement controlled test: (1) loading/unloading in a low-stress range for determination of elastic modulus, (2) viscoelastic recovery, (3) loading up to a certain high-strain level followed by unloading to almost zero stress and (4) viscoelastic recovery. Then, the same sequence (1)–(4) is repeated for a higher level of applied maximum strain in step (3) (see Figure 9.7).

There is an offset of applied stress σ_a before any significant microdamage occurs and stiffness degradation is detected; thus, a change of modulus can be expressed as a function of the applied stress:

$$\frac{E(\sigma)}{E_0} = \begin{cases} 1, & \sigma \leq \sigma_a \\ f(\sigma), & \sigma > \sigma_a \end{cases} \tag{9.46}$$

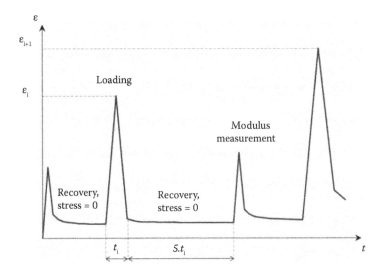

FIGURE 9.7 Example of strain dependence on time during one cycle of the displacement-controlled test.

Experimental results and a fitting curve for modulus degradation for flax fiber and lignin composite are presented in Figure 9.8. In this example, the test was carried out in the displacement controlled mode and a higher displacement was set for each subsequent step, which ensured that the applied strain increased in every step. During the first steps, a higher applied strain was always related to higher stresses, but due to the highly nonlinear behavior of this material (change of stiffness and development of VP strains), stresses never exceeded a certain value of σ_m. After reaching this threshold, the normalized modulus was still decreasing with increasing applied strain in the step even if the highest stress level reached in the current step was always lower than σ_m. In some cases, no modulus decrease was obtained before σ_m was reached. This means that at a very high stress, $d(\sigma_{max})$ is not a unique function of the maximum stress reached during the stress history: many increasing values of d correspond to σ_{max} after it has reached the value $\sigma_{max} = \sigma_m$.

Since a higher strain level is reached in each new step and the modulus is monotonously decreasing with an increase of the strain (see Figure 9.8b), it is more convenient to express the normalized modulus as a function of the applied strain ε_{max} and use a form $d(\varepsilon_{max})$. Unfortunately, the measured total strain consists of elastic, transient viscoelastic, and viscoplastic components, and it is not clear which parts of the strain and to what extent they are responsible for damage. Similar to Equation 9.46, the damage can be expressed as a function of applied strain with damage-free offset strain ε_a:

$$\frac{E(\varepsilon)}{E_0} = \begin{cases} 1, & \varepsilon \leq \varepsilon_a \\ f(\varepsilon), & \varepsilon > \varepsilon_a \end{cases} \tag{9.47}$$

9.6.3 DETERMINATION OF VISCOPLASTIC AND NONLINEAR VISCOELASTIC PARAMETERS

Similar to the damage parameters, the VP parameters are also obtained from multiple subsequent step tests, but in this case those are creep tests. Each creep stage (test with constant stress) is followed by a recovery stage (zero stress) with a duration that

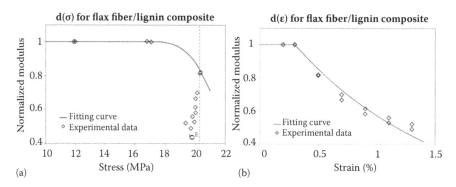

FIGURE 9.8 Elastic modulus degradation as a function of (a) stress and (b) strain for flax fiber lignin composite. The vertical dashed line in Figure 9.8a represents the σ_m value.

is long enough (~5–10 times longer than the creep test) to allow the viscoelastic strain to recover. Strains at the end of each recovery step are analyzed as viscoplastic.

Assuming $\sigma = \sigma_0$, the integration in Equation 9.45 becomes trivial and the VP strain dependence on the duration of the creep test Δt_1 is given by

$$\varepsilon_{VP}(t_1) = C_{VP}\left(\frac{\sigma_0}{\sigma^*}\right)^{Mm}\left(\frac{\Delta t_1}{t^*}\right)^{m} \tag{9.48}$$

Thus, VP strains in the constant stress creep test follow a power function with respect to time:

$$\varepsilon_{VP}(t_1) = A\left(\frac{\Delta t_1}{t^*}\right)^{m} \tag{9.49}$$

where A has power law dependence on the applied stress level in the creep test:

$$A = C_{VP}\left(\frac{\sigma}{\sigma^*}\right)^{Mm} \tag{9.50}$$

A schematic drawing of viscoplastic strains dependency on time and stress is presented in Figure 9.9. By knowing the VP strain dependence on time at a given stress, it is possible to subtract the VP strains from the creep test curve to obtain a pure viscoelastic behavior and, thus, viscoelasticity at a given stress can be characterized in a one-step creep test. Potentially, microdamage can also be present in this case and influence strain evolution. However, due to fairly low applied stress levels in the creep test, damage in the composites is negligible and does not have to be accounted for to obtain a pure viscoelastic response.

The stress dependence on time in a creep test can be represented as $\sigma = \sigma[H(t) - H(t - t_1)]$, where $H(t)$ is the Heaviside step function. This corresponds to the case when stress is applied at $t = 0$ and kept constant until time instant t_1; then, stress is suddenly removed and the strain recovery period begins. Thus, the material model (Equation 9.42) can be separately applied for time

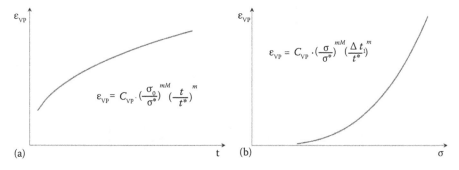

FIGURE 9.9 Viscoplastic strain dependency on (a) time and (b) stress. (Data from Pupure, L., Varna, J., and Joffe, R. 2015. Natural fiber composite: Challenges simulating inelastic response in strain-controlled tensile tests. *Journal of Composite Materials*. doi: 10.1177/0021998315579435 [46].)

intervals $t \in [0, t_1]$ (creep) and $t > t_1$ (strain recovery). The creep strain ε_{creep} and the recovery strain ε_{rec} are given by the following expressions obtained in [33,40,44]:

$$\varepsilon_{creep} = \varepsilon_0 + g_1 g_2 \sigma \sum_i C_i \left(1 - \exp\left(-\frac{t}{a_\sigma \tau_i} \right) \right) + \varepsilon_{VP}(\sigma, t) \tag{9.51}$$

$$\varepsilon_{rec} = g_2 \sigma \sum_i C_i \left(1 - \exp\left(-\frac{t_1}{a_\sigma \tau_i} \right) \right) \exp\left(-\frac{t - t_1}{\tau_i} \right) + \varepsilon_{VP}(\sigma, t_1) \tag{9.52}$$

Expressions (Equations 9.51 and 9.52) have to be used to fit the experimental creep and strain recovery data. Schematic drawings of creep and recovery tests along with equations used to fit them are presented in Figure 9.10.

9.6.4 MODEL WITH STRESS AS A FUNCTION OF STRAIN

Expressions in Section 9.6.1 allow simulation of stress-controlled tests. However, often there is a need for a model that can handle a strain-controlled experiment. For instance, codes for numerical structural analysis, most analytical micromechanics models (RoM, CCA model), the CLT, and so on require a constitutive model in which stresses are expressed as a function of strains. Moreover, most of the experiments are performed in the strain or displacement controlled mode.

Analogous to the stress-dependent model (e.g., Equation 9.42), in the one-dimensional case, stresses can be expressed through viscoelastic strains [45]:

$$\sigma = \sigma_{el} + h_1 \int_0^t \Delta E(\xi - \xi') \frac{d(h_2 \varepsilon_{VE})}{d\tau} d\tau \tag{9.53}$$

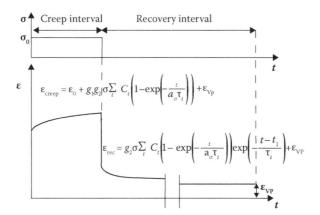

FIGURE 9.10   Schematic drawing of viscoelastic strain development.

Here, σ_{el} is the elastic component of stress, whereas the time-dependent part of the model can be described as

$$\Delta E(\xi) = -\sum_m E_m\left(1 - \exp\left(-\frac{\xi}{\tau_m}\right)\right) \tag{9.54}$$

where E_m and τ_m are stress-independent constants and the "reduced time" ξ is expressed through

$$\xi = \int_0^t \frac{dt'}{a_\varepsilon} \text{ and } \xi' = \int_0^\tau \frac{dt'}{a_\varepsilon} \tag{9.55}$$

The strain invariant-dependent functions h_1, h_2, and a_ε are equal to 1 for a material that exhibits linear viscoelastic behavior. Similar to the stress-dependent model, these parameters are also affected by temperature and humidity, but in a fixed environment they are just strain dependent.

It should be noted that for linear viscoelastic materials, the model in the form of Equations 9.42 through 9.44 can be directly inverted to Equations 9.53 through 9.55, whereas for nonlinear materials, both forms are, generally speaking, not compatible [45]. Parameters for this model are obtained from relaxation tests and, if the viscoplastic strain component is not present, these tests are simple to perform: a constant viscoelastic strain level in the sample must be maintained for a certain time. Unfortunately, in the majority of the materials, a viscoplastic strain component exists and performing a relaxation test with a constant viscoelastic strain becomes very challenging. The problem is that, even though the applied strain is maintained at the same level, the viscoplastic strain is developing with time and the viscoelastic strain decreases, as shown in Figure 9.11a. This means that, to maintain a constant viscoelastic strain during the relaxation experiment, the total applied strain must be increased to compensate for development of the viscoplastic strain component, as demonstrated in Figure 9.11b. Since viscoplastic strain development is defined by stress–time dependence (see Equation 9.45) and it is not known until the end of the

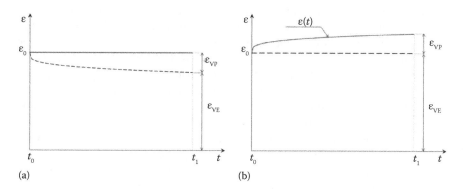

FIGURE 9.11 Schematic drawing of the relaxation test: (a) with constant applied strain and (b) strain as a function of time keeping a constant viscoelastic strain.

test, the viscoplastic strain–time dependence cannot be predicted, even if the law for viscoplastic strains is identified in creep and strain relaxation tests.

The application of the viscoelastic model in the form of $\sigma(\varepsilon)$ for any strain controlled ramp is straightforward, and the time-dependent viscoelastic strains are integrated over the defined time interval. It is not that simple (or rather impossible) with viscoplastic strain–stress dependence, since viscoplastic strains are intrinsically caused by stresses and not the opposite. A viscoplastic strain value at a certain instant of time depends on the stress history.

9.6.5 SIMULATION OF EXPERIMENTS

Simulation of the strain response in stress-controlled tests is straightforward: Equation 9.42 with parameters identified using the methodology described earlier is used to find the strain at any time instant. From the calculation efficiency point of view, it is better to rewrite Equation 9.42 in incremental form, for example, as given in [46], and to perform calculations with a small time increment. The strain at the time instant t_{k+1} is obtained from the known strain and stress at t_k.

The same model in the incremental form can be also used to simulate a strain-controlled test. The incremental expression is inverted to express the stress increment as a function of the viscoelastic strain increment (analytical inversion is not possible using the integral form of Equation 9.42). The viscoelastic strain is found by subtracting the viscoplastic strain, calculated using stresses in previous time instants, from the applied strain. If the time increment is small, the accuracy is sufficient.

In the main stress region, simulations of stress-controlled tests show good agreement with experimental data (see Figure 9.12a) for stress up to 22 MPa. At very high stresses (>22 MPa), the simulation curve does not follow the experimental results, because at this stage the material follows a different material model due to different and more complex mechanisms. Figure 9.12b shows a simulated creep test at 16 MPa for 1 hour, which is decomposed in strain components. Here, the significance of VE and VP strain components is very pronounced.

Simulations of strain-controlled quasistatic tensile tests using the inverted incremental model show two trends. The simulation showed very good agreement with the experimental data in cases when the characterization of the material in creep tests was possible until stresses were close to the maximum in the tensile test, as in the case of Figure 9.13. In cases when the material failed in creep at significantly lower stresses than reached in tensile tests, agreement was good only in the region where creep data were available. As seen in Figure 9.13b (in creep, the composite failed at 16 MPa), curves are in a good agreement until 16 MPa, whereas at higher stresses, the simulated curve is overestimating stresses. This is due to fact that the model parameters at higher stresses had to be extrapolated, which is always risky in regions where functions have high gradients. However, it was also shown in [46] that by modifying model parameters and adjusting extrapolation, it is possible to obtain a better agreement between experimental data and simulation. In Figure 9.13, two simulations with damage are shown: (a) using the damage function as a function of strain $d(\varepsilon)$ and (b) as a function of stress $d(\sigma)$. Simulations of the Schaperys model (Equation 9.53), where stresses are expressed through strains, show more realistic trends (see Figure 9.14,

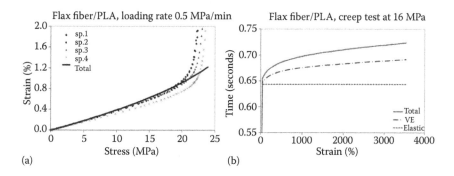

(a)

(b)

FIGURE 9.12 Data for polylactic acid (PLA). (a) Constant stress rate tensile experimental tests and modeling curve (sp.n is notation for individual specimens). (b) Simulation results for creep tests.

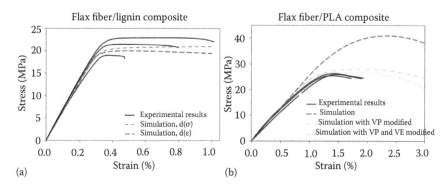

(a)

(b)

FIGURE 9.13 Simulation of stress–strain curves for (a) flax fiber/lignin and (b) flax fiber/polylactic acid (PLA) composite.

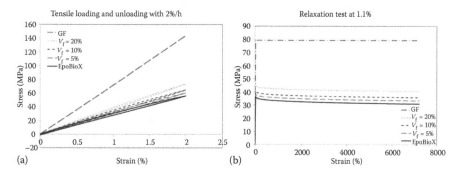

(a)

(b)

FIGURE 9.14 Simulation of (a) loading–unloading test and (b) relaxation test of glass fiber composite with different volume fractions.

where simulations of loading–unloading and relaxation tests are simulated for glass fiber and bio-based thermoset EpoBioX). This formulation of the material model is suitable for simple modeling, developing RoM type of expressions, CCA models, and laminate theory for nonlinear materials. Simulations using the isostrain assumption of

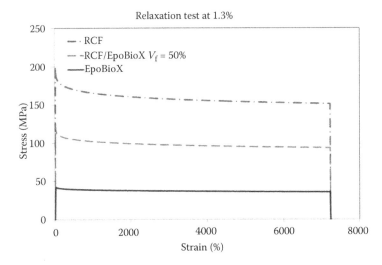

FIGURE 9.15 Simulation of relaxation test of regenerated cellulose fiber composite with a volume fraction of 50%.

the relaxation test on composites consisting of regenerated cellulose fibers, exhibiting very nonlinear behavior, with high viscoplastic and viscoelastic strains and EpoBioX matrix are shown in Figure 9.15.

9.7 CONCLUSIONS

This chapter presents a methodology for prediction of mechanical properties of natural fiber composites. The models discussed here are of different complexity and can be used for natural fiber composites at different scales. The RoM models are rather simple and often do not require very detailed information about the morphology of the composites, but they can still provide fairly accurate predictions of stiffness. However, other models are more sophisticated, require much more input information, and deliver more reliable, realistic output. For example, a laminate analogy and/or more complicated micromechanical models give more detailed information in terms of mechanical properties. The natural fiber composites exhibit nonlinear behavior and, to account for that, one must use more complicated simulation techniques. The moisture uptake and its effect on the performance of these materials must be accounted for.

There is very large potential for numerical methods to be used as modeling tools for natural fiber composites. These could be direct numerical methods or used in combination with other analytical models. Numerical simulations can provide extremely accurate predictions but only if the properties of the constituents and, more importantly, the internal structure of the materials are available as input. Such input can be obtained by the use of statistical methods (e.g., Monte Carlo simulations) or by direct 3D mapping of the composite structure by means of microcomputed tomography. The choice of the models to be used depends on the requirement set for each particular case within the design chain of natural fiber composites.

ACKNOWLEDGMENT

Funding from the strategic innovation programme, LIGHTer, provided by Vinnova is acknowledged.

REFERENCES

1. Dinwoodie, J.M. 1981. *Timber—Its Nature and Behaviour.* New York, NY: Van Nostrand Reinhold.
2. Hashin, Z. 1983. Analysis of composite materials—A survey. *Journal of Applied Mechanics* 50:481–505.
3. Jolicoeur, C. and Cardon, A. 1994. Analytical solution for bending of coaxial orthotropic cylinders. *Journal of Engineering Mechanics* 120:2556–74.
4. Neagu, R.C., Gamstedt, E.K., and Lindström, M. 2006. Modelling the effects of ultrastructural morphology on the elastic properties of wood fibres. In Proceedings of the 5th Plant Biomechanics Conference, Salmén, L. (Ed.), STFI-Packforsk AB, Stockholm.
5. Marklund, E. and Varna, J. 2009. Modeling the effect of helical fiber structure on wood fiber composite elastic properties. *Applied Composite Materials* 16:245–62.
6. Varna, J. 2008. Modelling natural fiber composites. In *Properties and Performance of Natural Fibre Composites*, ed. K.L. Pickering, 503–42. Cambridge: Woodhead Publishing Materials.
7. Joffe, R., Krasnikovs, A., and Varna, J. 2001. COD-based simulation of transverse cracking and stiffness reduction in $[S/90_n]_s$ laminates. *Composite Science and Technology* 61:637–56.
8. Marklund, E., Varna, J., Neagu, R.C., and Gamstedt, E.K. 2008. Stiffness of aligned wood fiber composites: Effect of microstructure and phase properties. *Journal of Composite Materials* 42:2377–405.
9. Christensen, R.M. and Lo, K.H. 1979. Solutions for effective shear properties in three phase sphere and cylinder models. *Journal of the Mechanics and Physics of Solids* 27:315–30.
10. Salmen, L. 1991. *Properties of Ionic Polymers: Natural and Synthetic.* Stockholm, Sweden: Swedish Pulp and Paper Research Institute, 285–94.
11. Persson, K. 2000. Micromechanical modelling of wood and fibre properties. PhD diss., Lund University.
12. Fengel, D. 1969. The ultrastructure of cellulose from wood. Part 1: Wood as the basic material for the isolation of cellulose. *Wood Science and Technology* 3:203–17.
13. Kollman, F.F.P. and Cote, W.A. 1968. *Principles of Wood Science and Technology, Solid Wood.* Springer-Verlag: Berlin.
14. Cox, H.L. 1952. The elasticity and strength of paper and other fibrous materials. *British Journal of Applied Physics* 3:72–9.
15. Summerscales, J., Hall, W., and Virk, A.S. 2011. A fibre diameter distribution factor (FDDF) for natural fibre composites. *Journal of Materials Science* 46:5876–80.
16. Virk, A.S., Hall, W., and Summerscales J. 2012. Modulus and strength prediction for natural fibre composites. *Material Science and Technology* 28:864–71.
17. Nairn, J.A. 1997. On the use of shear-lag methods for analysis of stress transfer in unidirectional composites. *Mechanics of Materials* 26:63–80.
18. Nairn, J.A. 2004. Generalized shear-lag analysis including imperfect interfaces. *Advanced Composites Letters* 13:263–74.
19. Neagu, R.C., Gamstedt, E.K., and Lindström, M. 2005. Influence of wood-fibre hygroexpansion on the dimensional instability of fibre mats and composites. *Composites Part A: Applied Science and Manufacturing* 36:772–88.

20. Chow, T-W. 1992. Microstructural design of fiber composites. In *Cambridge Solid State Science Series*, eds. R.W. Chan, E.A. Davis, and I.M. Ward, Cambridge: Cambridge University Press.

21. Megnis, M., Varna, J., Allen, D.H., and Holmberg, A. 2001. Micromechanical modeling of visco-elastic response of GMT composite. *Journal of Composite Materials* 35:849–82.

22. Lusti, H.R., Hine, P.J., and Gusev, A.A. 2002. Direct numerical predictions for the elastic and thermoplastic properties of short fibre composites. *Composite Science and Technology* 62:1927–34.

23. Hine, P.J., Lusti, H.R., and Gusev, A.A. 2002. Numerical simulation of the effects of volume fraction, aspect ratio and fibre length distribution on the elastic and thermoelastic properties of short fibre composites. *Composite Science and Technology* 62:1445–53.

24. Modniks, J. and Andersons, J. 2010. Modeling elastic properties of short flax fiber reinforced composites by orientation averaging. *Computational Materials Science* 50:595–99.

25. Modniks, J. and Andersons, J. 2013. Modeling the non-linear deformation of a short-flax-fiber-reinforced polymer composite by orientation averaging. *Composites Part B: Engineering* 54:188–93.

26. Modniks, J., Joffe R., and Andersons, J. 2011. Model of the mechanical response of short flax fiber reinforced polymer matrix composites. *Procedia Engineering* 10:2016–21.

27. Andersons, J., Joffe, R., and Sparnins, E. 2006. Stiffness and strength of flax fiber/polymer matrix composites. *Polymer Composites* 27:221–29.

28. Joffe, R., Madsen, B., Nättinen, K., and Miettinen, A. 2015. Strength of cellulosic fiber/starch acetate composites with variable fiber and plasticizer content. *Journal of Composite Materials* 49:1007–1017.

29. Tomason, J.L. 2002. Interfacial strength in thermoplastic composites—At last an industry friendly measurement method? *Composites Part A: Applied Science and Manufacturing* 33:1283–88.

30. Agarwal, B.D. and Broutman, L.J. 1990. *Analysis and Performance of Fiber Composites*. New York, NY: John Wiley and Sons.

31. Marklund, E. and Varna, J. 2009. Modeling the hygroexpansion of aligned wood fiber composites. *Composite Science and Technology* 69:1108–14.

32. Sparnins, E., Pupurs, A., Varna, J., Joffe, R., Nättinen, K., and Lampinen, J. 2011. The moisture and temperature effect on mechanical performance of flax/starch composites in quasi-static tension. *Polymer Composites* 32:2051–61.

33. Varna, J., Sparnins, E., Joffe, R., Nättinen, K., and Lampinen, J. 2012. Time dependent behavior of flax/starch composites. *Mechanics of Time-Dependent Materials* 16:47–70.

34. Nordin, L.-O. and Varna, J. 2005. Nonlinear viscoelastic behavior of paper fiber composites. *Composite Science and Technology* 65:1609–25.

35. Varna, J., Rozite, L., Joffe, R., and Pupurs, A. 2012. Nonlinear behavior of PLA based flax composites. *Plastics, Rubber and Composites* 41:49–60.

36. Pupure, L., Varna, J., Joffe, R., and Pupurs, A. 2013. An analysis of the nonlinear behaviour of lignin-based flax composites. *Mechanics of Composite Materials* 49:139–54.

37. Schapery, R.A. 1997. Nonlinear viscoelastic and viscoplastic constitutive equations based on thermodynamics. *Mechanics of Time-Dependent Materials* 1:209–40.

38. Lou, Y.C. and Schapery, R.A. 1971. Viscoelastic characterization of a nonlinear fiber-reinforced plastic. *Journal of Composite Materials* 5:208–34.

39. Zapas, L.J. and Crissman, J.M. 1984. Creep and recovery behavior of ultra-high molecular weight polyethylene in the region of small uniaxial deformations. *Polymer* 25:57–62.

40. Marklund, E., Eitzenberger, J., and Varna, J. 2008. Nonlinear viscoelastic viscoplastic material model including stiffness degradation for hemp/lignin composites. *Composite Science and Technology* 68:2156–62.

41. Varna, J. 2011. Characterization of vicoelasticity, viscoplasticity and damage in composites. In *Creep and Fatigue in Polymer Matrix Composites*, ed. R.M. Gueded, 514–42. Cambridge: Woodhead Publishing Materials.

42. Oldenbo, M. and Varna, J. 2005. A constitutive model for non-linear behavior of SMC accounting for linear viscoelasticity and micro-damage. *Polymer Composites* 26:84–97.

43. Kumar, R.S. and Talreja, R. 2003. A continuum damage model for linear viscoelastic composite materials. *Mechanics of Materials* 35:463–80.

44. Szpieg, M., Giannadakis, K., and Varna, J. 2011. Time dependent nonlinear behaviour of recycled polypropylene (rPP) in high tensile stress loading. *Journal of Thermoplastic Composite Materials* 24:625–52.

45. Schapery, R. A. 1969. On the characterization of nonlinear viscoelastic materials. *Polymer Engineering & Science* 9:295–310.

46. Pupure, L., Varna, J., and Joffe, R. 2015. Natural fiber composite: Challenges simulating inelastic response in strain-controlled tensile tests. *Journal of Composite Materials*. doi: 10.1177/0021998315579435.

10 Design of Natural Fiber-Reinforced Composite Structures

S.M. Sapuan and M.R. Mansor

CONTENTS

10.1 INTRODUCTION

Natural fiber composites have attracted the attention of many researchers due to the many good attributes that they offer. These attributes include their low cost (except silk fiber composites), light weight, abundance, being environmentally friendly, their aesthetically pleasing natural design image, comparable specific strength and stiffness, being renewable, and their formability with low investment [1]. It is reported that natural fibers such as oil palm, roselle, sisal, jute, kenaf, hemp, flax, banana pseudo-stem, pineapple leaf, and water hyacinth have been used as a reinforcement of filler in polymer composites [2]. Figure 10.1 shows an example of a roselle plant for producing roselle fiber, whereas Figure 10.2 shows an example of a stationery case made from kenaf fibers. The major problems in the use of natural fiber composites include incompatibility between hydrophilic fibers and hydrophobic polymers, which makes interfacial bonding difficult. In addition, since natural fiber composites absorb too much moisture, their applications are restricted to the indoor environment.

FIGURE 10.1   Roselle plant to produce roselle fibers.

FIGURE 10.2   Stationery case made from kenaf fibers.

However, the word "structure" in the title may not be 100% true, because in reality it is difficult to use natural fiber composites in structural components. It is reported that natural fiber composites are generally used in semi- or nonstructural components. Keeping this in mind, information in this chapter is mainly concerned with the semi- or nonstructural design of components. Hybridization seemed to be an alternative for producing structural components from natural fiber composites. Davoodi [3], Misri [4], and Mansor et al. [5] developed a bumper beam, small boat, and parking brake lever, respectively, from hybrid glass-natural fiber composites. A design with natural fiber composites is a challenging task. Literature that reports on the design of natural fiber composites is very limited, and, therefore, there is a real need to boost the publication in this subject area.

10.2   DESIGN PROCESS

In this section, a review on the design process is presented. Generally, design models discussed in this section can be used in many types of design, including designs with natural fiber composites. These are used for the benefit of readers to familiarize themselves with the design process. Pugh [6] developed a design process model called total design in product development. The total design comprises several stages such as market investigation, product design specification (PDS), conceptual design, detail design, manufacture, and sale. Dieter [7] reported that identification of customer needs, problem definition, and gathering information (including market investigation and PDS) are parts of the conceptual design activities. In the total design model by Pugh [6], detail design is the next stage to be carried out after conceptual design. But many authors such as Dieter [7], Pahl et al. [8], Wright [9], and Ashby [10] include embodiment design as an intermediate stage between conceptual and detail design. However, Ulrich and Eppinger [11] seemed to follow the total design model by Pugh with an additional stage called concept testing.

Mayer [12] used the general design methods, that is, from design brief (or general design description), to PDS, conceptual design, embodiment design, and detail design in designing composites. Mayer [12] identified one major problem in design with composites: limited availability of design data. It is because of the large permutation of fibers, matrices, and manufacturing processes. The reinforcements can be in the forms of molding compounds (intermediate materials), finished items such as pultruded or filament wound rods and bars, and raw fibers. Other variations in reinforcement include fiber length (continuous, short, and particulate), fiber arrangements (woven rovings, random chopped, and nonwoven fabrics), and fiber types such as glass, carbon, and Kevlar fibers. If one is working with natural fibers, he/she has so many types of natural fibers to choose from. It is obvious now, in composite design, that it is invariably related to the materials selection. Ashby [10] provided a good framework of product development activities that provide useful guidelines with regard to design of natural fiber composites, as shown in Figure 10.3. In his proposed framework, both geometrical design and the materials selection process are concurrently embedded in the design phase of the product. Design tools such as design for manufacturability

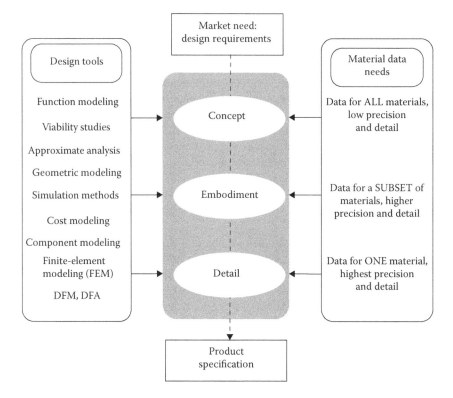

FIGURE 10.3   Product development activities [10]. (From Ashby, M. F. 2005. *Materials Selection in Mechanical Design*. 3rd Ed. Oxford: Elsevier Butterworth-Heinemann.)

(DFM) and design for assembly (DFA) are applied to accomplish geometrical design development, whereas data on the candidate materials that form the design gradually decrease as the design stage progresses from conceptual design to the detailed design stage of the product.

10.3   CONCEPTUAL DESIGN

Pugh [6] stated that conceptual design is concerned with the generation of solutions and the evaluation of these solutions to meet the PDS. It is a well-known fact that 70% of the manufacturing cost is committed during the conceptual design stage and yet, as reported by Ulrich and Eppinger [11], one of the main activities in conceptual design, that is, concept generation, consumes less than 5% of the budget and 15% of the development time of a product that they develop. Mayer [12] reported that in product design the first step is to have a design brief. Then, designers should have some design options after deciding the number of components to be developed, where these are developed by fulfilling some design requirements (specifications) such as ergonomic, functional, and aesthetic consideration. Finally, a design concept is obtained.

Two major activities in conceptual design are concept generation and concept evaluation. Both activities are briefly explained as follows.

10.3.1 CONCEPT GENERATION

After market investigation and PDS have been carried out, the next step in the design is generating the design concepts. Concept generation is also known as "initiation of concept solutions" [9]. In general, among the activities involved in concept generation are identifying the root cause of the problem, brainstorming possible solutions to address the problem, and generating design concepts by implementing the identified solution in the form of sketches. At the end of the concept generation stage, the design concepts that meet the design intent are produced in general, such as the overall product shape, style, and features associated with it.

10.3.2 CONCEPT EVALUATION

In engineering design, two of the most popular concept evaluation methods are the weighted objective method [13] and the Pugh evaluation method [6]. The weighted objective evaluation method uses a chart and in the left column of the chart design objectives are listed, as shown in Table 10.1. Then in the next column, relative weightings are assigned to the objectives. Performance parameters are established either qualitatively or quantitatively and the scores based on point scales are assigned for each concept based on how well this concept satisfies the objectives. Finally, for each design concept, the score is multiplied with weight to give a utility value. All the utility values are summed up, and the concept with the highest overall utility value is taken as the best concept. In the Pugh evaluation method, quite similar to the weighted objectives method, in the left column the design criteria (objectives) are listed. Based on the example shown in Table 10.2, with a few candidate concepts at our disposal, they are listed in the top row of the chart. However, unlike the weighted objectives method, an existing concept called "datum" is listed among all the concepts. The designer considers each concept/criterion against the datum. The legend

TABLE 10.1

Multipurpose Table Depicting Weighted Objective Evaluation Method to Select the Best Concept for Natural Fiber Composites

Concepts		1		2		3		4	
Characteristics	W	R	S	R	S	R	S	R	S
Stability of product	5	3	15	5	25	5	25	3	15
Cost of manufacture	5	3	15	2	10	2	10	2	10
Ease of manufacture	4	3	12	3	12	2	8	2	8
Reliability of service	5	4	20	4	20	4	20	4	20
Ergonomic to users	3	2	6	3	9	4	12	4	12
Low setup time	3	3	9	2	6	1	3	1	3
Ease to carry and light	3	3	9	3	9	3	9	4	12
Total Score		86		91		87		80	

Note: **W** means the weight of the characteristics. **R** means the score of the concept. **S** is the product of weight and score of the concept.

TABLE 10.2

Pugh Evaluation Method to Select the Best Concept for Sugar Palm Composite Table

	First	Second	Third	Fourth	Fifth	Sixth
Concept						
Easy Manufacture	D	+	−	+	+	+
Mass	A	+	−	+	+	+
Efficacy Cost	T	+	−	+	+	+
Ergonomics	U	+	+	+	+	−
Strength	M	−	+	−	S	+
Less Maintenance		S	−	−	S	+
Stability		−	+	+	S	+
Appearance		S	+	+	−	S
Total Score	Σ+	4	4	6	4	6
	Σ−	2	4	2	1	1
	ΣS	2	0	0	3	1

"+" is used to indicate that a particular concept is better than the datum, "−" means worse than the datum, and "S" means the same as the datum. The concept that scores the highest positive difference of "+" and "−" is chosen as the best concept.

Besides these two methods, many other methods are reported in the literature for concept evaluation, such as the Analytic Hierarchy Process (AHP) and Technique for Order Preference by Similarity to Ideal Solution (TOPSIS) [14–15]. Hambali et al. [16] demonstrated the use of the AHP method in concept design selection of a composite automotive bumper beam component. In their report, a new integrated design concept selection and material selection framework called Concurrent Design Concept Selection and Materials Selection (CDCSMS) was developed and utilized. In the integrated selection process, both concept design and material were concurrently analyzed based on the criteria obtained from the bumper beam PDS. In both activities, the AHP method was implemented to determine the weightage of the criteria and the ranking of the alternatives to obtain the final design concept as well as the best candidate composite materials. The final selection results were also validated using sensitivity analysis performed using Expert Choice software. Apart from that, Davoodi et al. [17] implemented the TOPSIS method to select the best concept design for the automotive bumper beam component using hybrid kenaf/glass fiber-reinforced epoxy composites, as shown in Table 10.3. The selection criteria involved were cost, product weight, absorption energy, deflection, manufacturing complexity,

TABLE 10.3

TOPSIS Evaluation Matrix for Bumper Beam Concept Design Selection Using Hybrid Kenaf/Glass Fiber-Reinforced Epoxy Composites [17]

No.	Concepts	Name	Materials Cost 0.15	Easy Manufacturing 0.1	Product Weight 0.2	Stain Energy 0.3	Rib Possibility 0.1	Minimum Deflection 0.15
1		RCP	24.40	2	2.44	2482.82	2	16.92
2		COP	29.00	1	2.9	43419.9	1	29.86
3		CCP	18.60	4	1.86	38825.1	5	21.34
4		DHP	25.50	3	2.55	76106.5	5	18.34
5		DCC	29.40	2	2.94	63671.6	4	25.72
6		DCP	25.60	4	2.56	44910.3	5	21.15
7		SHP	21.90	3	2.19	47231.5	4	22.92
8		SCP	22.50	5	2.25	2137.62	5	16.73

Source: Davoodi, M. M., Sapuan, S. M., Ahmad, D., Aidy, A., Khalina, A. and Jonoobi, M. 2011. Concept selection of car bumper beam with developed hybrid bio-composite material. *Materials and Design.* 32: 4857–65.

and rib design. All of the criteria were based on the bumper beam PDS document. The ranking of the alternative process was executed by using the TOPSIS method, whereas Abaqus® finite element software was used to simulate the energy absorption performance for the bumper beam under the low speed collision condition.

10.4 EMBODIMENT AND DETAIL DESIGN

Embodiment design is also called "predesign," "preliminary design," or "layout design." In composites, embodiment is done by considering materials and processes, shape and form, load path, joints, resin, reinforcement types and layup, strength and stiffness, mass, durability (corrosion, erosion, steady-state loading, and fluctuation loading), and environmental impact [12]. Embody is defined so as to give a concrete or discernible form to an idea or a concept [9]. Embodiment is defined by Pahl et al. [8] as "development of the design according to technical and economic criteria and in the light of further information, to the point where subsequent detail design can lead directly to production."

Meanwhile, in composites, detail design is done by considering shape, size and finish, dimensional tolerances, and surface finish of the geometry; identifying the fabrication route; establishing budget cost; determining relevance of code and standards; evaluating level of fire resistance; deciding on the materials; and assessing the environmental impact. However, in this chapter, both the embodiment and detail design process are briefly explained due to space limitations as well as to provide more focus on the concept design stages in the design of the natural fiber composite structures.

10.5   DESIGN OF NATURAL FIBER COMPOSITES

Design considerations of natural fiber composites are put in a different perspective by Busch [18]. He stated that design consideration in natural fiber composites includes processing consideration, selection of polymers, selection of additives, and good part design. No specific design model or design method is mentioned in the article. Ihueze et al. [19] reported their work on design and characterization for limit stress prediction in the multiaxial stress state. They used the ASTM D638-10 standard for the tensile test to design and compose composites of plantain empty fruit fiber-reinforced polyester composites. They further stated that a designer needs the estimation of failure stresses of the material for the design. They conducted a design of composites for various mechanical tests and a design for a composite modulus, and carried out design analysis using finite element analysis (FEA).

10.6   CONCEPTUAL DESIGN OF NATURAL FIBER COMPOSITES

10.6.1   Concept Generation

A design model called meanings of materials (MoM) model was reported by Taekema and Karana [1], and a creative game was organized for a group of designers in a workshop. In this meeting, the designers were asked to generate ideas using given material samples in a specified time. The purpose of this exercise is to familiarize the designers with these materials so that they will consider them in their future designs. Details of the MoM model can be found in Karana et al. [20], and Figure 10.4 shows the MoM model.

Another idea generation and concept design development method is called "Teoriya Resheniya Izobretatelskikh Zadatch" (TRIZ) in Russian, which is also termed "Theory of Inventive Problem Solving" in English. Founded by G. Altshuller,

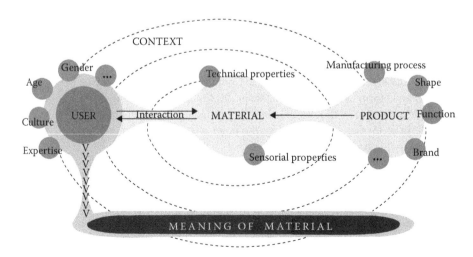

FIGURE 10.4   MoM model [20]. (From Karana, E., Hekkert, P. and Kandachar, P. 2010. A tool for meaning driven materials selection. *Materials and Design*. 31: 2932–41.)

a Russian patent officer in the 1940s, the method was developed through his early works on analyzing patents. During the process, he realized and concluded that the majority of the patents submitted were using similar groups of solution principles, and the solutions were repeated across various fields. Based on the findings, he later proposed an innovative problem-solving technique based on the solution principles synthesized from the design features of the patents. The main concept of the TRIZ method is the goal toward ideality, whereby it contradicts the mainstream thinking of having to accept compromise or trade-off in the design solution. The ideality concept was enabled by solving contradictions that occur through systematic and guided solutions, thus simultaneously engaging the process of improving features and eliminating its subsequent worsening feature [21].

The TRIZ method was reported to be an effective method in analyzing customer needs and providing innovative solutions, especially for a design purpose to satisfy the requirements [22–23]. Due to its effectiveness, the TRIZ method has also been incorporated into many design solutions and integrated with other concurrent engineering tools such as AHP and quality function deployment (QFD) to support the product development process [24–26]. In general, concept generation using the TRIZ problem-solving method consists of four main stages as illustrated in Figure 10.5, namely identification of the specific problem, converting the specific problem into the TRIZ general problem definition, applying the TRIZ tool to find a general solution based on TRIZ solution methods, and, finally, implementing the TRIZ solution suggestions in creating specific solutions for the problem to be solved [27]. Despite the unique designation used in the TRIZ problem-solving approach, the approach actually very much resembles the conventional problem-solving process, which includes problem definition, root cause identification, solution generation, and solution implementation [21]. The main difference is the tools used, whereby using the TRIZ method, specific tools have been laid out to be selected and applied. For example, several TRIZ tools can be selected for finding general solutions for the identified problem such as the contradiction matrix, cause and effect chain analysis, and substance field model. Consequently, the tools selected will also be followed by a specific TRIZ solution model in discovering the potential specific solution for the problem to be solved, such as TRIZ 40 Inventive Principles method, trimming method, and TRIZ-specific standard inventive solution method. The availability of many tools to be selected using TRIZ methods indicates the wide range of solutions that can be obtained based on the identified root cause of the problem and the flexibility of the TRIZ method to be implemented in various applications.

FIGURE 10.5 Concept design development approach using TRIZ method [27]. (From Chen, H., Tu, J. and Guan, S. 2011. Applying the theory of problem-solving and AHP to develop eco-innovative design. In *Design for Innovative Value Towards a Sustainable Society*, ed. M. Matsumoto et al., 489–494. Netherlands: Springer.)

Another innovative approach in concept generation is using biomimetics or the biomimicry concept. The biomimetics concept was inspired by nature, whereby inspiration and idea generation were developed by imitating nature's element and applying it to specific design elements to suit the intended applications [28]. Leonardo da Vinci was among the first who applied the biomimetics concept to the design process when he designed the flying contraption inspired by wings of a bat. For the design process involving a biological-based solution, Helms et al. [29] proposed several steps as a guide to apply the biomimetics concept, which includes problem definition, reframing the problem, search for an appropriate biological solution, defining the identified biological solution, and, finally, extraction of the biological principle into design ideas.

Easterling [30] reported that the scanning electron micrograph (SEM) of the cross-section of the leaf of a lily can give ideas in the design of composite sandwich panels. Li et al. [31] proposed a double helical structural design inspired by bamboo bast fiber structure to improve the compressive strength of composite laminates for structural purpose. They also found that by incorporating the interlaminar transition zone model imitating bamboo bast fiber, the interlaminar shear strength of composites can be improved by 15% of their initial value. Burns et al. [32] proposed composites joining strength improvement under bending load through the application of a bio-inspired T-joint adapted from the process of adaptive growth of a wood branch–trunk joint. The bio-inspired T-joint design was reported as being able to increase the bending failure initiation load and elastic strain energy capacity compared with the conventional baseline T-joint design without jeopardizing the composite stiffness, weight, and cost. In another report, Burns et al. [33] also reported the mimicking of similar tree branch–trunk joint characteristics into a composite T-joint design for aerospace application to improve the structural properties. The new bio-T-joint design encompassed alternating layers of composites, whereby every two plies consist of one ply that terminates as the end of the tapper whereas the other ply terminates at the end of the skin of the joint. Using the biomimetics concept, the structural joint performance was increased in terms of ductility and damage tolerance.

10.6.2 MATERIAL SELECTION

Material selection of natural fiber composites is regarded as one of the crucial elements involved in composite product development for both structural and nonstructural applications. The problem-solving process often involves a large and constant flow of information, especially for material selection of newly developed products [34]. As illustrated in Figure 10.6, conflicts often arise during the material selection process due to the involvement of many stakeholders during the product conceptual design stage with varying and conflicting requirements that need to be considered and satisfied by the best candidate material [35].

As for composite materials, particularly natural fiber composites, the task of making the optimum decision in material selection is more complex and time consuming considering the range of individual elements that make up the final natural fiber composite materials, namely the fiber that acts as the main load-bearing constituent of the whole material and the matrix that bonds the fibers together to provide the intended structural shape while simultaneously transferring and uniformly distributing the load

FIGURE 10.6 Conflicts arising from improper material selection [35]. (From Sapuan, S.M. 2010. *Concurrent Engineering for Composites*. Serdang: UPM Press.)

to the whole structure. The selection range of natural fiber and matrix resources is now increasing due to intense research being carried out nowadays as well as growing demand for sustainable raw materials to reduce the dependency on petroleum-based synthetic fibers and resins such as polyhidroxyalcanoate (PHA) and polylactic-acid (PLA) biopolymers [36]. Among the alternatives, especially to enable utilization of 100% biodegradable and renewable natural fiber composites for both the reinforcement and matrix, material was reported for natural fiber composites made from sugar palm fiber and sugar palm starch [37–38]. Figure 10.7 shows an example of a sugar palm tree as the resulting end-product available for making fully biodegradable composite materials, which are the sugar palm starch bio-resin and sugar palm fiber.

Moreover, the decision-making intricacy in selection of natural fiber composites may be further increased when hybridized natural fiber composites are included into the list of candidate materials due to an increase of different fiber types combined together to make up the final composite materials. Thus, in the material selection process, both main constituents need to be considered and analyzed before the final selection decision. Apart from that, the selection criteria for natural fiber composites also include other aspects such as final composite characteristics (which often vary from the initial individual fiber and matrix properties), general final composite performance (i.e., mechanical and thermal properties), and specific final composite performance (devised based on the specific targeted PDS, i.e., ease of maintenance and crashworthiness behavior) [39–40]. Therefore, the use of systematic material selection tools with multi-criteria analyzing capability is essential in aiding a proper decision-making process for natural fiber composite materials while simultaneously providing a quick and justified final decision for further product development work [41].

FIGURE 10.7   Sugar palm tree, sugar palm starch (biopolymer), and sugar palm fibers (Courtesy Dr. Sahari Japar, Universiti Malaysia Sabah).

In general, the material selection process may be divided into two categories, to serve screening and ranking purposes [42]. The material screening process is usually conducted in the initial stage of the overall material selection process to sort out and identify possible material candidates for the intended application. The ranking process then analyzes the candidate materials and ranks their overall importance based on the selection criteria to determine the best candidate material. Among the screening criteria used are the specific mechanical properties of the materials themselves such as density, tensile strength, and tensile modulus. The advantage of the screening process is that it helps in quickly sorting out the alternatives and narrows down the solution based on the design requirements. Hence, a suitable and manageable group of alternatives for the final decision solution can be obtained, thus resulting in faster analyzing time for the designer. One of the screening methods implemented for natural fiber composite material selection is the Ashby material selection chart. The Ashby chart was plotted to identify the possible candidate reinforcement material for plant fiber-reinforced plastic construction as well as to compare the desired properties with synthetic fiber composites [43]. In another report, the material screening process was also demonstrated to determine the optimum natural fiber composite reinforcement and matrix materials for automotive applications. A ternary diagram was used as the screening tool, which consisted of simultaneous consideration of bi-dimensional selection criteria to sort out the possible candidate materials. As opposed to the Ashby chart, which allows the comparison between two criteria simultaneously such as cost per weight or tensile strength per weight, the ternary diagram allows additional criteria to be included in the screening process. The ternary diagram helps enrich the screening capability by including more selection criteria into the process [44].

As for ranking purposes, many researchers have demonstrated the application of multi-criteria decision making (MCDM) methods in natural fiber composite materials selection. The AHP method was implemented in the material selection of final natural fiber composite materials for automotive dashboard panel application [45]. Twenty-nine types of natural fiber composites were listed as the candidate materials for the component construction, and the selection criteria were based on mechanical and physical design requirements. A sensitivity analysis was also performed to further validate the final proposed material at the end of the selection process. Ali et al. [46] also utilized the weighted-range MCDM method for material selection involving natural fiber composites for automotive application with the aid of a Java-based expert system. Three design requirements, namely tensile strength, Young's modulus, and density, were chosen as the selection criteria for the candidate materials.

Another report on natural fiber composite material selection was provided with regards to hybrid natural fiber composites for automotive structural applications. Mansor et al. [47] performed hybrid composite material selection for the design of the automotive parking brake lever component. The AHP method was utilized to select the best type of natural fiber material to be hybridized with synthetic glass fiber to formulate the hybrid composite structure. Based on the parking brake lever PDS, three selection criteria were selected, namely performance, cost, and weight for all 13 natural fiber candidates. Meanwhile, the use of the integrated AHP-TOPSIS method was also reported for natural fiber composite matrix material selection. The TOPSIS method was combined with the AHP method to perform the ranking process for the candidate thermoplastic matrix materials. Before that, the AHP method was implemented to determine the weightage of the selection criteria. In addition, the selection criteria chosen were performance, cost, weight, and service conditions [48].

In most cases, material selection related to natural fiber composites involved determination of the selection criteria to be used as the comparing reference between the list of selected material candidates in the decision-making process. Moreover, the selection criteria are often derived from the PDS used as the design guidelines for the product. Traditionally, the material selection criteria were based on technical and cost aspects of the natural fiber composite materials. However, due to growing demand to improve product sustainability, the environmental aspect has been increasingly introduced as one of the crucial elements in the material selection criteria alongside function and cost [49]. The application of life cycle assessment (LCA), for example, has enabled environmental characteristics such as energy usage and end-of-life performance to be analyzed for each of the candidate materials [50]. One example is material selection involving natural fiber composites for the electronic-command panel of an agriculture sprayer machine. The selection process involved the use of the Eco-indicator 95 LCA method to provide environmental performance data for the list of candidate natural fiber materials and later compared it with glass fiber material [51].

Furthermore, the material selection process to find the winning candidate for the intended applications may further be extended to include other selection criteria such as context and experience, apart from technical characteristics of materials and the manufacturing process. The conventional material aspect may also be elaborated to include a sensory aspect for the technical characteristics [52–53]. Inclusion of such tangible and intangible criteria into the selection process will create more complexity in the process, but, in return, will help obtain a more holistic decision of the material that is capable of functioning as per product design requirement as well as providing higher experience-centered satisfaction from the product to the user [54].

10.6.3 MANUFACTURING PROCESS SELECTION

Manufacturing process selection is another important domain in the design of natural fiber composite structures. Among the factors that influence the manufacturing process decision are the type of fiber, the fiber orientation, and volume. In addition, the final shape, size, and properties of the final natural fiber composites and their product are also crucial in the selection of the appropriate processing method. The manufacturing process capability must also suit the natural fiber processing temperature limits in order to reduce the risk of fire deterioration due to excessive heat. Major manufacturing processes available in the market for natural fiber composites are resin transfer molding, compression molding, injection molding, and sheet molding. The correct decision for manufacturing process selection will guarantee successful transformation of the natural fiber composite materials into the desired form while complying with the required product quality [55].

One example on the application of manufacturing process selection for composites was reported for the construction of the automotive bumper beam component. Five manufacturing processes were listed as the potential candidates to manufacture the component. Apart from that, based on the bumper beam PDS, six main criteria were used as the selection reference, such as production characteristics, geometry of the design, cost, and type of raw materials. The AHP method was implemented in the ranking of the candidate manufacturing processes [56].

10.6.4 CASE STUDY ON CONCEPTUAL DESIGN OF NATURAL FIBER COMPOSITES

10.6.4.1 Design Brief

A design brief is given for the design of a new portable laptop case with an embedded external cooling fan. The design intent is to create an integrated support accessory for laptop users. Market research indicates that the market demand for the product may rise to 1,000 units per month within 6 months and engineering staff would need to develop a few pre-production models and test these under service conditions with interested consumers. The design requirements for the portable laptop case with an embedded external cooling fan are shown in Table 10.4. Apart from that, the material requirements for the laptop case are also specified as shown in Table 10.5.

Based on the design brief, the TRIZ method is applied to develop the new conceptual design of the laptop case. The TRIZ method was reported as an effective method

TABLE 10.4

Design Requirements of Laptop Case with Embedded External Cooling Fan

Requirements

- Nominal laptop case internal size 15 × 11 × 1.5 inch (L × W × H) (compatible for laptops up to 14" display size)
- Internal use in room conditions as cooling fan
- External use in exposed conditions as laptop case
- Support moderate internal load
- Long life and durable
- Production target 1000 a month within 6 months
- Lightweight and low cost

TABLE 10.5
Material Requirements

Design Requirement	Reason
Lightweight	Easy to carry
Strong	To sustain the load (laptop, cooling fan, laptop accessories such as batteries, mouse, external DVD player, external hard-disk drive, etc.) while used as laptop case and external cooling fan.
Durable	Waterproof, wear resistance
Self-colored and noncorrosive	Low maintenance, long life
Electrical insulating	Safety (while in use as external cooling fan)

in analyzing customer needs and providing innovative solutions, especially for design purposes, to satisfy the requirements [22–23]. The conceptual design development of the product using natural fiber composites is explained in the next sections.

10.6.4.2 Stage 1: Specific Problem Definition (Engineering Contradiction)

According to the design brief, the external cooling fan is a device that helps reduce heat from the laptop during operation, resulting in more efficient performance and longer service life. The external cooling fan is a separate device, and thus needs to be carried along separately with the laptop case. The solution is to merge the laptop case and the cooling fan into a single product. However, this creates the problem of adding extra weight in the laptop case while carrying the laptop as well as needing to be stronger to carry the extra load. Apart from that, another problem that arises is the complexity involved in having the new product adapt according to the currently-required function

Based on the given information, several conflicting situations were found in order to achieve the overall project aim. The design intent is to merge the function of an external laptop cooling fan with the laptop case for multifunctional operation advantages. On the other hand, the merging advantage will also create several issues such as adding extra weight to the initial laptop case due to the added cooling fan function; thus, a stronger laptop case structure is required for bearing the extra load and for increasing the complexity of the product design in order to perform both casing and cooling functions with ease. Identifying the conflicting situations is the first step in applying the TRIZ method to develop the new conceptual design, and based on the TRIZ method, the conflicting situations are categorized as engineering contradictions whereby improvement of one technical aspect of the system causes deterioration of other parameters in the same system [57]. Hence, the TRIZ Contradiction Matrix and TRIZ 40 Inventive Principles Method are among the suitable methods used for solving the engineering contradiction.

10.6.4.3   Stage 2: Conversion of the Specific Problem into TRIZ General Problem

In the second stage, the specific problem to be improved and its subsequent undesired effect or trade-off is converted into TRIZ general problem classifications. In general, there are 39 system parameters according to the TRIZ general problem classifications, and the specific problem is then matched with the most appropriate TRIZ system parameters [58]. One example is the extra weight of the new product, which was found to match the system parameters definition for "weight of stationary object." Table 10.6 summarizes the overall TRIZ system parameters selected in defining the specific problem in this project.

10.6.4.4   Stage 3: Identification of General Solutions Using TRIZ Contradiction Matrix

The classified general problems identified in the previous stage were then grouped together to form the TRIZ Contradiction Matrix as shown in Table 10.7. From the contradiction matrix, the lists of general solutions were extracted to solve the contradiction using the TRIZ 40 Inventive Principles solution guidelines. The inventive principles solution was among the first methods developed in the TRIZ method, and it was derived based on studies of approximately 40,000 innovative patents by Altshuller, founder of the TRIZ method [59].

TABLE 10.6

Selected TRIZ General Problem Classification Based on the Laptop Case Design

Specific Problem	TRIZ General Problem Classification
New Design Aim	
New integrated laptop case with embedded external cooling fan	#12-shape
New Design Trade-off	
Extra weight	#2-weight of stationary object
Stronger to handle the extra load	#14-strength
Easy to be changed according to the required function	#35-adaptability or versatility

TABLE 10.7

Contradiction Matrix for Integrated Laptop Case and the Respective TRIZ General Solutions

Improving Parameter (Feature to Improve)	Worsening Parameter	TRIZ 40 Inventive Principles
No.12. Shape	No.2. Weight of stationary object	No.10. Preliminary action
		No.15. Dynamization
		No.26. Copying
		No.3. Local quality
	No.14. Strength	No.10. Preliminary action
		No.14. Curvature
		No.30. Flexible shells and thin films
		No.40. Composite materials
	No.35. Adaptability or versatility	No.1. Segmentation
		No.15. Dynamization
		No.29. Pneumatics and hydraulics

As shown in Table 10.7, 11 inventive principles are listed as the potential solution for solving the contradiction for the laptop case design. For example, to achieve improvement in the shape of the design while also solving the issue of the weight of the object, the inventive principles that can be adopted are principle number 10 (Preliminary action), number 15 (Dynamization), number 26 (Copying), or number 3 (Local quality) or a combination of these.

All the inventive principles recommended were later analyzed, and the most suitable solutions are then selected. At the end of the analysis, the best general solutions of the TRIZ 40 inventive principle method to address the contradiction were solution number 3 (Local quality), number 40 (Composite materials), number 1 (Segmentation), and number 15 (Dynamization). Using the contradiction matrix, a quick and focused general solution was able to be created in a systematic manner [60]. However, final selection of the most appropriate inventive principle to solve the contradiction may also cause some difficulty due to the overlapping and fuzzy definitions [61]. Despite the difficulty, on the other hand, the open definition also does not restrict the idea to be explored by the designer in envisioning a larger specific solution to the problem, thus encouraging a higher level of thinking to be performed in the innovative problem-solving process [62].

10.6.4.5 Stage 4: Generating Specific Solutions for the New Laptop Case Design

In the final stage, ideas for specific solution strategies related to the general solution are developed. For example, in implementing the TRIZ number 3 inventive principle (Local quality), the solution that may be adopted is making each part of the object fulfill a different and useful function. The specific solution strategy that is proposed based on the solution recommendation to the design is to make the laptop case back-support feature act as both the main support structure for the case and the structure to mount the cooling fan for the laptop. The cooling fan USB connection wire can also be stored within the laptop case back-support structure. Thus, the new structure will have multiple functions as required in the design specification for the new laptop case. A summary of the specific solution strategies for the new laptop case is listed in Table 10.8.

10.6.4.6 Conceptual Design Development

Based on the specific solution strategies obtained from the TRIZ solution principles, a new innovative conceptual design of the laptop case is developed as shown in Figure 10.8. The new laptop case is divided into two main components, which are the back-support structure and the case cover. The two components are joined together using a flexible zip for easy attachment and removal purposes, which enables it to serve as either a laptop case or an external cooling fan.

TABLE 10.8
Specific Solution Strategy Based on the TRIZ Solution Principles

TRIZ General Inventive Principles	Solution Descriptions	Specific Solution Strategy
No.3. Local quality	(a) Change an object's structure from uniform to nonuniform, change an external environment (or external influence) from uniform to nonuniform (b) Make each part of an object function in conditions most suitable for its operation (c) Make each part of an object fulfill a different and useful function	–The case back-support feature also acts as an external cooling fan for the laptop. The cooling fan USB connection wire can be easily stored within the structure. –Design antislip feature on the laptop case back-support to enable better grip with the laptop while in use. –The cooling fan can be hidden safely when used as a laptop case by a multi-segment sliding cover that can easily slide within the back-support structure (also demonstrates principle number 1, Segmentation). The sliding-cover feature also acts as a stand for the cooling fan while in use.
No.40. Composite materials	(a) Change from uniform to composite (multiple) materials	–Use NFRP as the material to fabricate the laptop case back-support feature to handle the required load and gain lighter product weight. The NFRP has higher specific strength as well as lower weight and cost compared with using 100% plastics for the same structural purpose. The NFRP is also water resistant and rigid as well as a good electrical resistor (also demonstrates principle number 3, Local quality).

(Continued)

TABLE 10.8 (*Continued*)

Specific Solution Strategy Based on the TRIZ Solution Principles

TRIZ General Inventive Principles	Solution Descriptions	Specific Solution Strategy
No.1. Segmentation	(a) Divide an object into independent parts (b) Make an object easy to disassemble (c) Increase the degree of fragmentation or segmentation (d) Transition to micro-level	−Split the laptop case into two different segments, which are the laptop case cover and the laptop case back-support. Both features are connected together using a zip connection for easy removal and attachment from each other. −Attach a strap feature to the laptop case. When the laptop case is used as a cooling fan, the laptop case cover can be neatly rolled up and stored using the strap feature (also demonstrates principle number 3, Local quality)
No.15. Dynamization	(a) Enable (design) the characteristics of an object, external environment, or process to change to be optimal or to find an optimal operating condition (b) Divide object into parts that are capable of movement relative to each other (c) If an object (or process) is rigid or inflexible, make it movable or adaptive	−The laptop case cover feature is made from flexible synthetic material such as PVC so that it can be easily removed from the back-support cooling fan feature. −The laptop case cover is also divided into several small segments for easy folding while being extracted from the back-support structure. −Using synthetic material also acts as a waterproof feature for the laptop case when being used in outdoor conditions (also demonstrates principle number 3, Local quality).

The back-support structure component performs two main functions: (1) main structural feature to support the load of the laptop case and (2) mounting structure to attach the cooling fans. The back-support structure component also performs innovative auxiliary functions to store the USB connection wire for the cooling fan to the laptop and as the stand feature for the external cooling fan. The stand is made from the cooling fan safety cover attached as the subcomponent inside the back-support structure. While in use as the laptop case, the subcomponent safely covers the cooling fan from the user's body and when in use as an external cooling fan, it can slide out from the back-support structure and be folded to form a triangular external cooling fan stand feature. Another added feature to the back-support structure is the inclusion of an antislip feature that helps improve grip with the laptop while in use as the external cooling fan. As for the case cover, it is also divided into small segments

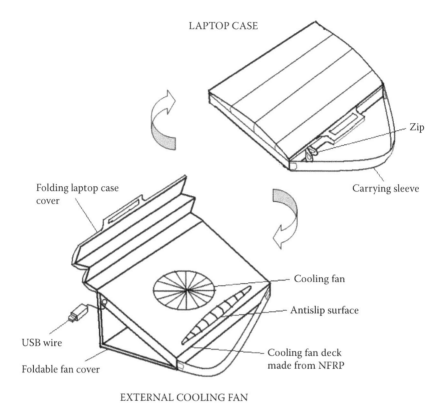

FIGURE 10.8 New conceptual design of laptop case embedded with external cooling fan developed using TRIZ method.

for easy folding and can be neatly rolled up and stored using the strap feature to the back-support structure itself.

In terms of material selection, based on the material requirements as listed in Table 10.5, the back-support structure will be made from lightweight and strong natural fiber-reinforced plastic (NFRP) materials as opposed to conventional rigid polymers such as ABS and polycarbonate. NFRP materials such as kenaf composites and hemp composites have been reported to have high-specific strength and stiffness that is comparable to synthetic glass fiber composites, due to their light weight, low material cost, good manufacturing capability with minimum harmful side effects to the user, self-colored capability due to the polymer matrix composition, and good electrical insulation ability for use as the mounting structure for an electrical component such as a cooling fan [63–65]. As for the case cover, the component will be made from flexible and durable plastic such as PVC, which also provides the water-proof capability for the laptop case for outdoor use as required in the product design requirements. Furthermore, for higher end consumers, the laptop case material can also be made from high-quality leather to satisfy the same functions while adding more elegance to the appearance of the product.

10.7 CONCLUSIONS

In this chapter, a study on the design of natural fiber-reinforced polymer composite structures is presented. The effort in bringing to life products made from natural fiber composites is different compared with conventional engineering materials whereby geometrical, materials, and manufacturing aspects need to be analyzed together for the product to suit the intended application. Thus, the concurrent engineering approach is considered more suitable to perform the task, utilizing different tools in solving each step in the overall product development stages. Among the concurrent engineering tools applicable for designers to choose from are AHP for materials and manufacturing process selection and TRIZ for concept generation. Despite the challenging task of designing natural fiber composite structures, with a correct and systematic design approach undertaken especially during the conceptual design stage of the product, the product realization using natural fiber composites may successfully be achieved, thus contributing to sustainable products for various structural applications.

ACKNOWLEDGMENTS

The authors wish to thank Dr. Campilho of Portugal for the invitation to contribute a chapter of this edited book. They are grateful to Universiti Putra Malaysia for granting sabbatical leave to the principal author spent at Universiti Malaya (October 2013 to June 2014) and to Universiti Teknikal Malaysia Melaka for granting study leave to the coauthor. This work is funded by the Ministry of Education, Malaysia, through Exploratory Research Grant Scheme (ERGS) project number ERGS/1-2013/5527190. Deep appreciation is extended to Associate Professor Dr. Nukman Yusoff of Universiti Malaya for his support.

REFERENCES

1. Taekema, J. and Karana, E. 2012. Creating awareness on natural fiber composites in design. In *Proceedings of the 12th International Design Conference (DESIGN 2012).* 1141–50.
2. Sapuan, S. M. 2014. *Tropical Natural Fiber Reinforced Polymer Composites.* Singapore: Springer.
3. Davoodi, M. M. *Development of Thermoplastic Toughened Hybrid Kenaf/Glass Fiber-Reinforced Epoxy Composite for Automotive Bumper Beam.* Ph.D. Thesis. Universiti Putra Malaysia. 2012.
4. Misri, S. *Design and Fabrication of a Boat Using Sugar Palm Glass Fiber Reinforced Unsaturated Polyester Composite.* MSc. Thesis. Universiti Putra Malaysia. 2011.
5. Mansor, M. R., Sapuan, S. M., Zainudin, E. S., Nuraini, A. A. and Hambali, A. 2014. Conceptual design of kenaf fiber polymer composite automotive parking brake lever using integrated TRIZ—Morphological chart—Analytic hierarchy process method. *Materials and Design.* 54:473–82.
6. Pugh, S. 1991. *Total Design: Integrated Methods for Successful Product Engineering.* Wokingham, UK: Addison-Wesley Publishing.
7. Dieter, G. E. 2000. *Engineering Design: A Materials and Processing Approach.* 3rd Ed. Boston, MA: McGraw-Hill.
8. Pahl, G., Beitz, W., Feldhusen, J. and Grote, K. H. 2007. *Engineering Design: A Systematic Approach.* 3rd Ed. London: Springer-Verlag London Limited.

9. Wright, I. C. 1998. *Design Methods in Engineering and Product Design*. Maidenhead, Berkshire: Mc-Graw-Hill Publishing Company.

10. Ashby, M. F. 2005. *Materials Selection in Mechanical Design*. 3rd Ed. Oxford: Elsevier Butterworth-Heinemann.

11. Ulrich, K. T. and Eppinger, S. D. 2004. *Product Design and Development*, 3rd Ed. Boston, MA: McGraw-Hill Irwin.

12. Mayer, R. M. 1993. *Design with Reinforced Plastics: A Guide for Engineers and Designers*. London: The Design Council.

13. Cross, N. 1994. *Engineering Design Methods: Strategies for Product Design*. 2nd Ed. Chichester, UK: John Wiley & Sons.

14. Saaty, T. L. 2005. *Theory and Applications of the Analytic Network Process*. Pittsburgh, PA: RWS Publications.

15. Yoon, K. P. and Hwang, C. L. 2005. *Multiple Attribute Decision Making: An Introduction to Quantitative Applications in the Social Sciences*. Thousand Oaks, CA: Sage Publications Inc.

16. Hambali, A., Sapuan, S. M., Rahim, A. S., Ismail, N. and Nukman, Y. 2011. Concurrent decisions on design concept and material using analytical hierarchy process at the conceptual design stage. *Concurrent Engineering: Research and Applications*. 19: 111–21.

17. Davoodi, M. M., Sapuan, S. M., Ahmad, D., Aidy, A., Khalina, A. and Jonoobi, M. 2011. Concept selection of car bumper beam with developed hybrid bio-composite material. *Materials and Design*. 32: 4857–65.

18. Busch, J. Plastics go on natural-fiber diet. *Machine Design*, 2012. [Online]. http://machinedesign.com/archive/plastics-go-natural-fiber-diet.

19. Ihueza, C. C., Okafor, C. E. and Okaye, C. I. 2013. Natural fiber composite design and characterization for limit stress prediction in multiaxial stress state. *Journal of King Saud University—Engineering Sciences*. 27(2): 193–206.

20. Karana, E., Hekkert, P. and Kandachar, P. 2010. A tool for meaning driven materials selection. *Materials and Design*. 31: 2932–41.

21. San, Y. T., Jin, Y. T. and Li, S. C. 2009. *TRIZ: Systematic Innovation in Manufacturing*. Selangor: Firstfruits Sdn Bhd.

22. Wang, F. K. and Chen, K. S. 2010. Applying lean six sigma and TRIZ methodology in banking services. *Total Quality Management and Business Excellence*. 21: 301–15.

23. Albiñana, J. C. and Vila, C. 2012. A framework for concurrent material and process selection during conceptual product design stages. *Materials and Design*. 41: 433–46.

24. Yamashina, H., Ito, T. and Kawada, H. 2002. Innovative product development process by integrating QFD and TRIZ. *International Journal of Production Research*. 40: 1031–50.

25. Li, T. S. and Huang, H. H. 2009. Applying TRIZ and Fuzzy AHP to develop innovative design for automated manufacturing systems. *Expert Systems with Applications*. 36: 8302–12.

26. Yeh, C. H., Huang, J. C. Y. and Yu, C. K. 2011. Integration of four-phase QFD and TRIZ in product R&D: a notebook case study. *Research in Engineering Design*. 22: 125–41.

27. Chen, H., Tu, J. and Guan, S. 2011. Applying the theory of problem-solving and AHP to develop eco-innovative design. In *Design for Innovative Value Towards a Sustainable Society*, ed. M. Matsumoto et al., 489–494. Netherlands: Springer.

28. Benyus, J. M. 1997. *Biomimicry: Innovation Inspired by Nature*. New York: HarperCollins Publisher Inc.

29. Helms, M., Vattam, S. S. and Goel, A. K. 2009. Biologically inspired design: process and products. *Design Studies*. 30: 606–22.

30. Easterling, K. E. 1990. *Tomorrow's Materials*. 2nd Ed. London: The Institute of Materials.

31. Li, S. H., Zeng, Q. Y., Xiao, Y. L., Fu, S. Y. and Zhou, B. L. 1995. Biomimicry of bamboo bast fiber with engineering composite materials. *Materials Science and Engineering:C*. 3: 125–30.

32. Burns, L. A., Mouritz, A. P., Pook, D. and Feih, S. 2012. Strength improvement to composite T-joints under bending through bio-inspired design. *Composites Part A: Applied Science and Manufacturing*. 43: 1971–80.

33. Burns, L. A., Mouritz, A. P., Pook, D. and Feih, S. 2012. Bio-inspired design of aerospace composite joints for improved damage tolerance. *Composite Structures*. 94: 995–1004.

34. van Kesteren, I. E. H. 2008. Product designers' information needs in materials selection. *Materials and Design*. 29: 133–45.

35. Sapuan, S.M. 2010. *Concurrent Engineering for Composites*. Serdang: UPM Press.

36. Marques, T., Esteves, J. L., Viana, J., Loureiro, N. and Arteiro A. 2012. Design for sustainability with composite systems. In *15th International Conference on Experimental Mechanics (ICEM15)*. 1–2.

37. Sahari, J., Sapuan, S. M., Zainudin, E. S. and Maleque, M. A. 2013. Mechanical and thermal properties of environmentally friendly composites derived from sugar palm tree. *Materials and Design*. 49: 285–89.

38. Sahari, J. and Salit, M.S. 2012. The development and properties of biodegradable and sustainable polymers. *Journal of Polymer Materials*. 29: 153–165.

39. Al-Oqla, F.M. and Sapuan, S. M. 2014. Natural fiber reinforced polymer composites in industrial applications: Feasibility of date palm fibers for sustainable automotive industry. *Journal of Cleaner Production*. 66: 347–54.

40. Dicker, M. P. M., Duckworth, P. F., Baker, A. B., Francois, G., Hazzard, M. K. and Weaver, P. M. 2014. Green composites: A review of material attributes and complementary applications. *Composites Part A: Applied Science and Manufacturing*. 56: 280–89.

41. Sapuan, S. M. and Mansor, M. R. 2014. Concurrent engineering approach in the development of composite products: A review. *Materials and Design*. 58: 161–67.

42. Jahan, A., Ismail, M. Y., Sapuan, S. M. and Mustapha, F. 2010. Material screening and choosing methods—A review. *Materials and Design*. 31: 696–705.

43. Shah, D. U. 2014. Natural fiber composites: Comprehensive Ashby-type materials selection charts. *Materials and Design*. 62: 21–31.

44. Koronis, G., Silva, A. and Fontul, M. 2013. Green composites: A review of adequate materials for automotive applications. *Composites Part B: Engineering*. 44: 120–27.

45. Sapuan, S. M., Kho, J. Y., Zainudin, E. S., Leman, Z., Ahmed Ali, B. A. and Hambali, A. 2011. Materials selection for natural fiber reinforced polymer composites using analytical hierarchy process. *Indian Journal of Engineering and Materials Sciences*. 18: 255–67.

46. Ahmed Ali, B. A., Sapuan, S. M., Zainudin, E. S. and Othman, M. 2013. Java based expert system for selection of natural fiber composite materials. *Journal of Food Agriculture and Environment*. 11: 1871–77.

47. Mansor, M. R., Sapuan, S. M., Zainudin, E. S., Nuraini, A. A. and Hambali, A. 2013. Hybrid natural and glass fibers reinforced polymer composites material selection using analytical hierarchy process for automotive brake lever design. *Materials and Design*. 51: 484–92.

48. Mansor, M. R., Sapuan, S. M., Zainudin, E. S., Nuraini, A. A. and Hambali, A. 2014. Application of integrated AHP-TOPSIS method in hybrid natural fiber composites materials selection for automotive parking brake lever component. *Australian Journal of Basic and Applied Sciences*. 8: 431–39.

49. Lindahl, P., Robèrt, K. H., Ny, H. and Broman, G. 2014. Strategic sustainability considerations in materials management. *Journal of Cleaner Production*. 64: 98–103.

50. Milani, A. S., Eskicioglu, C., Robles, K., Bujun, K. and Hosseini-Nasab, H. 2011. Multiple criteria decision making with life cycle assessment for material selection of composites. *Express Polymer Letters*. 5: 1062–74.

51. Alves, C., Ferrão, P. M. C., Freitas, M., Silva, A. J., Luz, S. M. and Alves, D. E. 2009. Sustainable design procedure: The role of composite materials to combine mechanical and environmental features for agricultural machines. *Materials and Design*. 30: 4060–68.

52. Wastiels, L. and Wouters, I. 2012. Architects' considerations while selecting materials. *Materials and Design*. 34: 584–93.

53. Karana, E., Hekkert, P. and Kandachar, P. 2009. Meanings of materials through sensorial properties and manufacturing processes. *Materials and Design*. 30: 2778–84.

54. Karana, E., Hekkert, P. and Kandachar, P. 2008. Material considerations in product design: A survey on crucial material aspects used by product designers. *Materials and Design*. 29: 1081–89.

55. Faruk, O., Bledzki, A. K., Fink, H.-P. and Sain, M. 2014. Progress report on natural fiber reinforced composites. *Macromolecular Materials and Engineering*. 299: 9–26.

56. Hambali, A., Sapuan, S. M., Ismail, N. and Nukman, Y. 2009. Composite manufacturing process selection using analytical hierarchy process. *International Journal of Mechanical and Materials Engineering*. 4: 49–61.

57. Kim, Y. S. and Cochran, D. S. 2000. Reviewing TRIZ from the perspective of axiomatic design. *Journal of Engineering Design*. 11: 79–94.

58. Kremer, G. O., Chiu, M.-C., Lin, C.-Y., Gupta, S., Claudio, D. and Thevenot, H. 2012. Application of axiomatic design, TRIZ, and mixed integer programming to develop innovative designs: A locomotive ballast arrangement case study. *The International Journal of Advanced Manufacturing Technology*. 61: 827–42.

59. Li, T. 2009. Applying TRIZ and AHP to develop innovative design for automated assembly systems. *The International Journal of Advanced Manufacturing Technology*. 46: 301–13.

60. Sheu, D. D. and Lee, H. -K. 2011. A proposed process for systematic innovation. *International Journal of Production Research*. 49: 847–68.

61. Cascini, G., Rissone, P., Rotini, F. and Russo, D. 2011. Systematic design through the integration of TRIZ and optimization tools. *Procedia Engineering*. 9: 674–79.

62. Luttropp, C. and Lagerstedt, J. 2006. EcoDesign and the ten golden rules: Generic advice for merging environmental aspects into product development. *Journal of Cleaner Production*. 14: 1396–1408.

63. Aji, I., Zainudin, E. S., Abdan, K., Sapuan, S. M. and Khairul, M. D. 2012. Mechanical properties and water absorption behavior of hybridized kenaf/pineapple leaf fiber-reinforced high-density polyethylene composite. *Journal of Composite Materials*. 47: 979–90.

64. Ishak, M. R., Sapuan, S. M., Leman, Z., Rahman, M. Z. A., Anwar, U. M. K. and Siregar, J. P. 2013. Sugar palm (Arenga pinnata): Its fibers, polymers and composites. *Carbohydrate Polymers*. 91: 699–710.

65. El-Shekeil, Y. A., Salit, M. S., Abdan, K. and Zainudin, E. S. 2011. Development of a new kenaf bast fiber-reinforced thermoplastic polyurethane composite. *Bioresources*. 6: 4662–72.

11 Joint Design in Natural Fiber Composites

R.D.S.G. Campilho and Lucas F.M. da Silva

CONTENTS

11.1 INTRODUCTION

The use of composite materials, in general, is increasing in the aerospace, aeronautical, automotive, and sports industries due to their attractive specific characteristics, which enables weight to be reduced. Composite substrates in the form of fiber-reinforced plastics are not isotropic, and several tests are necessary to determine all the mechanical properties of the material. One common form of manufacturing synthetic fiber composites is using prepregs. A prepreg consists of a combination of a matrix (or resin) and fiber reinforcement. It is ready for use in the component manufacturing process. It is available in unidirectional (UD) form (one direction of reinforcement) or in fabric form (several directions of reinforcement). For example, the new BOEING 787

Dreamliner airplane has 50% composite materials in its structure. This higher use of composites combined with other technological improvements enables this airplane to have fuel consumption that is 20% lower than that of other airplanes with a similar size. The behavior of composites is highly anisotropic in respect to both stiffness and strength. In the fiber direction, UD composites can be very strong and stiff whereas the transverse and shear properties are much lower. Natural fiber composites have been studied more recently because of environmental issues such as recyclability, energy saving during the entire life cycle, and emissions during processing and disposal [1]. Because of this global awareness, car manufacturers such as Toyota, Ford, Mercedes Benz, and BMW are introducing natural fiber-reinforced parts for nonstructural components such as panels and insulation [2]. Because of the design complexity and dimensions of products, and also limitations in the fabrication processes, many times a joining method is required to produce a product or structure. The joining methods can basically be divided into two categories: mechanical joining and adhesive bonding. Mechanical methods for composite materials, in general, include fastening, bolting, or riveting, but these have a number a limitations that currently make adhesive bonding more attractive: stress concentrations at the vicinity of the holes that cancel the fiber continuity, adherend weakening, weight penalty, and large shape disturbance of the joined parts (with aesthetic and possibly functional disadvantages) [3]. Adhesive joints surpass these limitations and, in addition, can be designed as a function of the load typology. Adhesively bonded joints provide several benefits, such as more uniform stress distribution than conventional techniques such as fastening or riveting, high fatigue resistance, and the possibility of joining different materials [4]. A structure bonded at several regions is much more flexible in terms of assembly and allows a cost reduction of the entire process. However, it is likely that the weaker parts of the structure become the bonded joints instead of the composites. The main drawback of adhesive joints is that disassembly is usually not possible [5]. It should also be emphasized that despite the geometry continuity, bonded joints still suffer from length-wise stress concentrations. In this context, the single-lap joint (SLJ) configuration is one of the most affected geometries, because it undergoes bending under load due to its asymmetry, which results in the development of considerable peel stresses at the overlap ends [6]. Other geometries, such as the double-lap joint (DLJ), are less affected by peel stresses. In addition, shear stresses also peak at the same locations because of gradual adherend straining along the overlap [7]. Other joint configurations such as the DLJ or scarf enable reduction of shear stress gradients, although these cannot be completely eliminated. Peak stresses are highly detrimental to the joints, especially for brittle adhesives, in which cracks tend to initiate at the overlap ends when peak stresses are attained, since these adhesives do not allow plasticity. Because of this handicap of bonded joints, techniques used to reduce peak stresses were largely studied. Thus, the careful design and fabrication quality are very important aspects to be accounted for in bonded joints. Derived from all these issues and the improvements in structural adhesives, bonded joints have begun enhancing or replacing the use of traditional mechanical fasteners in composite and metallic structures. Because of this, this chapter deals only with adhesively bonded joints. Many bonded joint configurations can be applied in composites, and the most common are the SLJ and DLJ, single- and double-strap, and stepped and scarf joints [3]. Other joint configurations that are not so widely used are joggle-lap joints (mainly for aircraft fuselage halves, doublers, and repair patches) and L-section joints (to bond the internal structure to outer skins of aircraft wings). In terms of joining methods for natural fiber composites, literature is very scarce. Thus, in this chapter, most of the references and described works pertain to synthetic composites, which behave in a similar manner and whose results can be extrapolated with some degree of confidence to natural fiber composites.

11.2   APPROACHES FOR STRENGTH PREDICTION

Information about the stress distributions inside the adhesive layer of an adhesively bonded joint is essential for joint strength prediction and joint design. In the case of composites (either synthetic or natural), this data can also help to assess the possibility of delamination, which is recurrent when addressing composite joints [8]. Two big groups of techniques are used for the stress analysis of bonded joints, namely analytical and numerical methods [9]. Analytical methods using closed-form algebra employ classical linear theories in which some simplifications are used. Volkersen [10] was the precursor of the application of these methods to bonded joints by using a closed-form solution that considers the adhesive and adherends as elastic and that the adhesive deforms only in shear. The equilibrium equation of an SLJ led to a simple governing differential equation with a simple algebraic equation. However, the analysis of adhesive joints can be highly complex if composite adherends are used, the adhesive is deformed plastically, or there is an adhesive fillet. In those cases, several differential equations of high complexity might be obtained (nonlinear and nonhomogeneous). There are many analytical models in the literature for obtaining stress and strain distributions. Many closed-form models are based on modified shear-lag equations, as proposed originally by Volkersen [10]. Reviews of these closed-form theories and their assumptions can be found in da Silva et al. [11–12] and Tong and Luo [13]. With the inclusion of factors such as stress variation through the adhesive thickness, plasticity, thermal effects, composite materials, and others in the formulations, the analytical equations become more complex and there is a greater requirement to use computing power to attain a solution. Hart-Smith [14] developed a number of methods for stress analysis of adhesive joints. Versions of this method have been prepared as Fortran programs and were largely used in the aerospace industry. Formulations by different authors were implemented in spreadsheets or as a program to be run on personal computers (PC). Currently, software packages that assist in the design of efficient joints are available. Table 11.1 reviews the available commercial PC-based analysis/design software packages. The main characteristics and field of application of each software package are compared. As shown in Table 11.1, the available software packages are tailored to specific geometries and most of them only cover few of them. It is also visible that only a few of these find application in the field of composites. A common limitation of this software, apart from those of da Silva et al. [15] and Dragoni et al. [16], is that the choice of process is dependent on the previous experience of the designer.

The finite element (FE) method is a well-established numerical technique that can handle complex structures and nonlinear material properties where classical theories generally fail to work. The FE method is, by far, the most common numerical

TABLE 11.1

Software Packages Available in the Market

Name	Supplier	Application	Features
BOLT	G.S. Springer, Stanford University, Stanford, CA	Design of pin-loaded holes in composites	Prediction of failure strength and failure mode Three types of bolted joints: joints with a single hole, joints with two identical holes in a row, joints with two identical holes in tandem Applicable to uniform tensile loads and symmetric laminates
BISEPSLOCO	AEA Technology, UK	Closed-form computer code for predicting stresses and strains in adhesively bonded SLJ	Tensile/shear/bending moment loading Adhesive peel and shear stress predictions Allowance for plasticity in adhesive layer Thermal stress analysis
BISEPSTUG	AEA Technology, UK	Closed-form computer code for predicting stresses and strains in adhesively bonded coaxial joints	Stepped and profiled joints Orthotropic adherends Torsional and axial loading Allowance for plasticity in adhesive layer Thermal stress analysis
CoDA	National Physical Laboratory, UK	Preliminary design of composite beams and panels, and bolted joints	Synthesis of composite material properties (lamina and laminates for a range of fiber formats) Parametric analyses Panel and beam design Bonded and bolted double shear joints Bearing, shear-out, pin shear, and by-pass tensile failure prediction
DLR	DLR-Mitteilung, Germany	Preliminary design of composite joints	Adhesively bonded and bolted joints Linear-elastic and linear-elastic/plastic behavior Tension and shear loading Symmetric and asymmetric lap joints Bearing, shear-out, pin shear, and by-pass tensile failure prediction (washers and bolt tightening)

(Continued)

TABLE 11.1 (Continued)
Software Packages Available in the Market

Name	Supplier	Application	Features
FELOCO	AEA Technology, UK	FE module computer code for predicting stresses and strains in adhesively bonded SLJ	Stepped and profiled joints Tensile/shear/bending moment/pressure loading Linear and nonlinear analysis Peel, shear, and longitudinal stress predictions in adhesive layer and adherends Thermal stress analysis for adherend and adhesive
PAL	Permabond, UK	"Expert" system for adhesive selection	Joined systems include: lap and butt joints, sandwich structures, bushes/gears/bearings/shafts/pipes/threaded fittings Elastic analysis Creep/fatigue effects on joint stiffness (graphical) Joint strength
RETCALC	Loctite, UK	Interactive Windows-based software, general purpose	Correction factors (temperature and fatigue)
ESDU	Engineering Science Data Unit, UK	Software for use in structural design	ESDU 78042 Shear stresses in the adhesives in bonded joints. Single-step DLJ loaded in tension ESDU 79016 Inelastic shear stresses and strains in the adhesives bonding lap joints loaded in tension or shear (computer program) ESDU 80011 Elastic stresses in the adhesive in single-step double-lap bonded joints ESDU 80039 Elastic adhesive stresses in multistep lap joints loaded in tension (computer program) ESDU 81022 Guide to the use of Data Items in the design of bonded joints
Joint Designer [15]	Faculty of Engineering, University of Porto, Portugal	Closed-form computer code for predicting stresses and strength in adhesively bonded joints	Accessible to nonexperts Lap joints Adhesive peel and shear stress predictions Joint strength prediction Allowance for plasticity in the adhesive layer and adherends Orthotropic adherends Thermal stress analysis
JointCalc [16]	Henkel AG, Germany	Closed-form computer code for predicting stresses and strength in adhesively bonded joints	Accessible to nonexperts SLJ and DLJ, single- and double-strap joints, peel joints, and cylindrical joints Adhesive peel and shear stress predictions Joint strength prediction

technique used in the context of adhesively bonded joints. Adams and coworkers are among the first to have used the FE method for analyzing adhesive joint stresses [17–18]. One of the first reasons for the use of the FE method was for evaluating the influence of the spew fillet on the stresses in the adhesive layer. Other aspects that can be analyzed by FE are the joint rotation, the adherends, and adhesive plasticity. The study by Harris and Adams [19] is one of the first FE analyses taking into account these three aspects. Different numerical methods are available for bonded joints, but these are not widespread. The boundary element method is still very limited in the analysis of adhesive joints. One of the few exceptions is the work of Gonçalves et al. [20]. The finite differences method is particularly suited for solving complex governing differential equations in closed-form models. In this chapter, only the FE is discussed, and several approaches to failure analysis are accounted for: continuum mechanics, fracture mechanics, and the more recent damage mechanics and extended finite element method (XFEM). Examples of application of damage mechanics and XFEM are given to illustrate the most recent advances in numerical modeling of adhesively bonded joints. Cohesive zone models (CZM) are being increasingly used in FE models [9]. This technique gives the complete response of structures, including their bonded joints, up to final failure, modeled in a single analysis without additional postprocessing of FE analysis results being necessary. Numerical methods for bonded joints are an emerging field, and the techniques used to model damage can be divided into continuum or local approaches. The continuum approach refers to damage modeling over a finite region, and it usually acts by variation of the material properties when damage occurs. In the local approach, a possibility is given for damage to grow along zero volume lines and surfaces in two and three dimensions, respectively, and this is often referred to as the CZM approach. The CZM approach is based on the traction–separation relationship for the interfaces. In these laws, with an increasing interfacial separation, the traction across the interface reaches a maximum (damage initiation), then decreases (softening), and, finally, the crack propagates, thus permitting a total debond of the interfaces. It is thus possible to simulate the whole crack propagation process and failure. CZM models the fracture process, extending the concept of continuum mechanics by including a zone of discontinuity modeled by cohesive zones and thus using both strength and fracture mechanics concepts. CZM may be used for adhesive debonding or for composite delaminations. Campilho et al. [6] evaluated by CZM the strength of carbon-epoxy bonded SLJ and the influence of the CZM shape used to model the adhesive, for different geometry/adhesive combinations. The SLJ were bonded with either a brittle or ductile adhesive and tensile strength was tested for different values of overlap length, which enabled the testing of different solutions (short overlaps give small shear stress gradients, while large overlaps give rise to large stress concentrations). Guidelines were discussed to achieve the best predictions depending on the most suited CZM shape for each bonded system, and the authors concluded that CZM analyses are powerful tools that are used to describe the failure behavior of composite bonded joints. Chen et al. [21] used a continuum mechanics approach to predict crack initiation and propagation in SLJ and also the failure load. Two adhesives were used in the study, Ciba® MY750 and the same adhesive rubber-toughened with carboxyl terminated butadieneacryl (CTBN). The damage process in the joints was simulated by the FE method by an incremental analysis in which, after a converged solution was obtained after a given load increment, a check was made to see whether the failure condition had been attained anywhere in the joint. If the condition had been attained at a given element, the values of Young's modulus (E) and Poisson's ratio (ν) of the material within the failed region were reset to zero (or near zero), such that it could deform almost freely without transferring any load. Concurrently, the stresses were also reset to zero and then the whole system of equations (stiffness) was reassembled and solved. The proposed method accounted for damage by voiding the cracked elements. A new load increment was not applied until the damage process had stopped. The numerical failure loads compared very well with the experimental results with or without a fillet of adhesive. XFEM is a more recent technique and its application in bonded joints is still very limited. Campilho et al. [22] evaluated the capabilities of the XFEM formulation embedded in Abaqus® for the simulation of SLJ and DLJ between aluminum adherends bonded with a brittle adhesive (Araldite® AV138). A comparison was also performed with CZM modeling, and there was a detailed discussion regarding the applicability of both methods. Overlap lengths between 5 and 20 mm were considered in this comparative study. Figure 11.1 shows the progressive failure of a SLJ with a 20 mm overlap (detail at the overlap edge) using the principal strain criterion for the initiation of damage and estimation of crack growth direction [22]. It is visible that the direction of maximum strain led to crack growth in the direction of the adherend. This is not consistent with the real behavior of SLJ. The authors concluded that the simulation of damage propagation along the adhesive bond with XFEM is unfeasible with this technique. However, the sites of damage initiation and respective loading can still be predicted with accuracy. On the other hand, CZM accurately predicted the failure loads.

By comparing the different techniques, although the closed-form solutions have their limitations, they are easy to use, especially for parametric studies. The FE method is highly accurate for any joint geometry but it requires significant computer

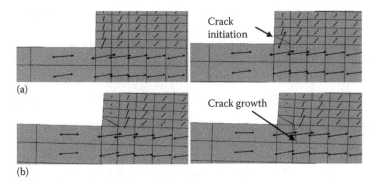

(a)

(b)

FIGURE 11.1   Progressive failure of a SLJ with a 20 mm overlap using the XFEM (the arrows represent the directions of maximum principal strain): damage initiation within the adhesive at the overlap edges (a) and damage growth to the aluminum adherend (b). (From Campilho, R.D.S.G., Banea, M.D., Pinto, A.M.G., da Silva, L.F.M., and de Jesus, A.M.P., Strength prediction of single- and double-lap joints by standard and extended finite element modelling, *International Journal of Adhesion & Adhesives*, 31, 363–72, 2011. With permission [22].)

power and experienced personnel who know how to deal with the software. Thus, closed-form solutions are widely used for joint design and FE for research, complex geometries, or elaborate material models. The design process of adhesive joints is radically different from that of other traditional methods of joining such as bolts or rivets. Therefore, it is not advised to take a joint initially designed for another type of joining method and to modify it for adhesive bonding. The first task is to choose a suitable adhesive, and this choice depends on the type of loading, the adherends to bond, and the environment (temperature and humidity). Adhesives are not as strong as the adherends they are joining such as metals or composites. However, when they are loaded over a large area such as in an SLJ, they can provide a high strength that is sufficient to plastically deform the metal in some cases or to break the composite. This is the main reason that, when designing an adhesive joint, the load must be spread over a large area and not concentrated in one point. Peel loads should also be avoided because of the low peel strength of adhesives [23]. This is particularly true for composite materials, which add the possibility of delaminations. Because of these issues, the adhesive layer should be primarily loaded in shear whenever possible, and peel and cleavage stresses should be minimized. However, these considerations should be balanced with limitations in terms of manufacturing process, cost, consequences of failure, and desired final appearance that may complicate the designing process.

The strength of an adhesive joint that is not subjected to external environmental factors is ruled by the mechanical properties of the adhesive and adherends, the joint geometry, the stress concentrations, and the residual internal stresses. Actually, residual stresses may occur as a result of different thermal expansion between the adhesive and adherends. Another cause is the adhesive shrinkage during cure. All of these factors are addressed in this chapter, and simple design guidelines are given to improve the joint load-bearing properties. The joint strength can also be improved by geometrical modifications such as fillets and adherend tapering. All these aspects are discussed in detail. Hybrid joints, combining adhesive bonding with riveting or bolting, are increasingly being used for adhesive joints. The advantages of these solutions are explained in this chapter. Bonded repair methods are also discussed, because pristine structures or their joints may be damaged in some way and it is important to discuss methods that guarantee an efficient repair design.

11.3   STRENGTH PREDICTION IN NATURAL FIBER COMPOSITES

The strength prediction of bonded joints for synthetic composites is widely documented in the literature and works are available regarding the most common predictive methods, such as continuum mechanics, fracture mechanics, and advanced methods such as CZM. Natural fiber composites are a more recent field of research, triggered by environmental reasons and ecological purposes and, thus, research works are very scarce. In this section, a few of these are listed, together with relevant results and with emphasis on the predictive methods. Campilho et al. [24] experimentally evaluated the tensile fracture toughness of adhesive double-cantilever beam joints between natural fiber composites of jute-reinforced epoxy, by bonding with a ductile adhesive and co-curing. To estimate the fracture toughness, conventional methods were used for the co-cured specimens, whereas for the adhesive within the bonded joint, the *J*-integral was considered. As an output of the work, the fracture behavior in tension of bonded and co-cured joints in jute-reinforced natural fiber composites was characterized for subsequent strength prediction. In addition, for the adhesively bonded joints, the tensile cohesive law of the adhesive was also derived by the direct method. Regarding the two bonding methods, the authors found that, for the co-curing method, crack propagation was accompanied by large-scale fiber-bridging events between the adherends, which resulted in crack initiation for a smaller value of toughness than the bonded joints. However, with the continuing crack growth and appearance of fiber bridging, the toughness for crack propagation became significantly higher than that for the adhesively bonded specimens, turning the co-cured joints more resistant to crack propagation. Figure 11.2 shows a co-cured specimen at crack initiation, before fiber bridging (a), and during the steady state (with fiber bridging) (b) [24].

Gonzalez-Murillo and Ansell [3] studied unidirectionally reinforced natural fiber composites with henequen and sisal fibers as reinforcement with overlaps from 5 to 50 mm. Different bonding methods/solutions were compared: epoxy-bonded SLJ and co-cured joints with two solutions, intermingled fiber joints (by interlacing the fibers in the overlap before embedding the resin) and laminated fiber joints (by superimposing the fibers of both adherends and embedding the resin). Figure 11.3 shows an SLJ with an overlap of 50 mm (a) and fiber interactions for an intermingled (b) and

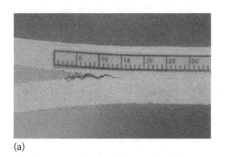
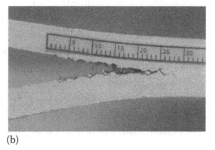

(a)   (b)

FIGURE 11.2 Co-cured specimen at crack initiation before fiber bridging (a) and during the steady state (with fiber bridging) (b). (From Campilho, R.D.S.G., Moura, D.C., Gonçalves, D.J.S., da Silva, J.F.M.G., Banea, M.D., and da Silva, L.F.M., Fracture toughness determination of adhesive and co-cured joints in natural fiber composites, *Composites: Part B*, 50, 120–26, 2013. With permission [24].)

FIGURE 11.3 SLJ with an overlap of 50 mm (a) and fiber interactions for an intermingled (b) and laminated (c) joint. (From Gonzalez-Murillo, C., Co-cured in-line joints for natural fiber composites, *Composites Science and Technology*, 70, 442–49, 2010. With permission [3].)

laminated (c) joint [3]. The results showed that the intermingled and laminated solutions have much higher lap shear strengths than the SLJ, even reaching a very fair proportion of the pristine composite strength. For the laminated configuration, 92% of the undamaged composite was attained with an overlap length of 50 mm. The intermingled joints revealed a strength reduction with the overlap length because of the increasing misalignment of the bonding fibers, which negatively affected the bond strength. The SLJ were largely affected by out-of-plane rotation and differential shear effects, and this reflected in lower failure strengths. The joints were also modeled using an FE analysis, enabling the interpretation of the experimental observations. A simplified linear and elastic FE model of the joints confirmed the significant stress concentrations at the overlap edges of the bonded joints. The models for the intermingled joints were built assuming that the bonding region had doubled the value of E of the adjacent composite, because of the doubling of the fiber volume fraction at that region. For these joints, failure was experimentally observed to be close to the ends of the overlapping region, which was in close agreement with the highest predicted strains in the joint.

Bajpai et al. [25] studied the joining between natural fiber composites made of nettle fibers (NF) and grewia optiva fibers (GOF) embedded in polylactic acid or polypropylene bases. Two joining techniques were tested to produce SLJ: microwave heating with a suitable susceptor and adhesive bonding. In the microwave heating process, the composites are heated with microwave radiations that penetrate and simultaneously heat the inside of the materials. By this process, microwaves supply energy in a rapid manner even to thick materials. The composites were made of GOF mat and NF mat reinforcements. The polyactic acid and polypropylene matrices were converted into 1 mm thick films by compression molding. The composites were fabricated by manual layup, by alternatively stacking polymer and fiber plies and heating at 170°C, to produce 4 mm thick composite plates. Testing of the joints revealed that, for the microwave bonded joints (Figure 11.4a, [25]), the maximum loads were obtained by the polylactic acid/GOF adherend joints, whereas the polypropylene/NF joints behaved the worst. For these joints, adherend failure near the overlap edges occurred, which the authors attributed to unwanted heating and weakening of the specimens near the overlap. The failure loads could not attain the composite strength, but the authors discussed the possibility to further increase the strength by incorporating the joint interface with matrix material that slightly dopes with susceptor material. With this modification, the bonding process takes less time and damage to the composite is minimized. The experimental results of the adhesively bonded joints revealed a much lesser failure load than the microwave bonded joints (Figure 11.4b, [25]), because of poor chemical bonding between the adhesive and matrices. The failure loads were between 4.9% and 13.2% of the pristine strength of the composites, which is clearly low, and was justified by adhesive failures detected by visual inspection after failure. The microwave heating process was also simulated with a standard multiphysics FE analysis to describe the microwave heating mechanism. The results of the experimental study were in close agreement with the FE investigation.

Ferreira et al. [26] performed a fatigue study for composite SLJ. Two setups were considered: bidirectional woven E-glass fibers and polypropylene composites; and hybrid stacked composites. This second solution aimed at improving the fatigue strength by using hybrid fiber composites with a polypropylene/hemp natural fiber layer adjacent to the bond interface, which was expected to produce more uniform stresses in the transient regions (Figure 11.5, [26]). The adhesive considered to fabricate the joints was a cyanoacrylate (Bostik 7452). Several results were presented

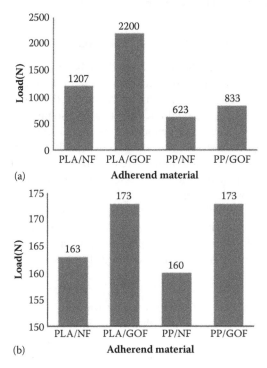

FIGURE 11.4 Maximum load sustained by the joints formed with microwave heating (a) and adhesive bonding (b). (From Bajpai, P.K., Singh, I., and Madaan, J. Joining of natural fiber reinforced composites using microwave energy: Experimental and finite element study, *Materials and Design*, 35, 596–602, 2012. With permission [25].)

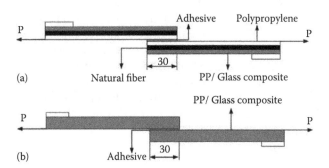

FIGURE 11.5 Hybrid (a) and regular (b) composite SLJ. (From Ferreira, J.M., Silva, H., Costa, J.D., and Richardson, M., Stress analysis of lap joints involving natural fiber reinforced interface layers, *Composites: Part B*, 36, 1–7, 2005. With permission [26].)

(S–N curves, relating stresses with the cyclic count to failure), together with the failure mechanisms. Conversely to what was hypothesized in the beginning, the hybrid stacked joints showed a lower fatigue strength than that for the original thermoplastic composite joints, which was justified by the higher peak stresses at the overlap and lower adhesive strength measured in double-cantilever beam tests.

11.4   FACTORS AFFECTING JOINT STRENGTH

The main factors that affect the joint strength of bonded joints are the material properties (adherends and adhesive) and the stress concentrations that, in turn, depend on the joint geometry (adherend and adhesive thickness, and overlap). If thermal effects are present, residual internal stresses also play an important role in the joint strength. It is known that neither shear nor peel stresses in adhesive joints are uniform, and that the average shear stress (i.e., load divided by the bonded area) can be much lower than the maximum stress in the bondline. If the adhesive is ductile, it will enable redistribution of stresses before failure occurs. On the other hand, when using a brittle adhesive, damage propagates as soon as the adhesive-allowable stresses are attained anywhere in the bondline. However, damage initiation always takes place at the regions of stress concentrations, notwithstanding the type of adhesive and, thus, it is of the upmost importance to decrease these stress peaks if a joint strength improvement is required. To achieve this goal, a few guidelines can be considered:

- Use an adhesive with low stiffness and ductile behavior
- Use identical adherends (E and thickness) or, if not feasible, balance the stiffness
- Minimize residual thermal stresses
- Avoid peel stresses
- Use a thin adhesive layer
- Use a large bonded area

All these factors are mentioned in detail and means to employ them are suggested. When considering composite joints, peel loadings should be avoided at all costs; otherwise, the composite may fail in transverse tension (interlaminar failure) before the adhesive fails [23]. In the occurrence of an interlaminar failure, the composite separates longitudinally at the bonding region and the shear transfer capacity between the two adherends is cancelled, although the adhesive remains unharmed. To prevent this occurrence, peel stresses can be smoothed by using adhesive fillets, adherend tapers, and adhesive bands along the overlap or hybrid joints. These issues are discussed in detail in Section 11.5.

11.4.1   ADHESIVE PROPERTIES

The adhesive properties highly influence the joint strength, but one should always bear in mind that a stronger adhesive does not necessarily give a higher joint strength. Actually, a strong but brittle adhesive locally achieves high stress in the corners of adhesive joints, but it does not allow stress redistributions to the low stressed areas. As a result, the average shear stress at failure is very low [6]. On the other hand, adhesives with high ductility and a low value of E generally have a low strength. However, they are able to distribute the stresses more uniformly along the bondline (because of the low stiffness) and to deform plastically, which makes the joints much stronger than with strong but brittle adhesives [27]. The adhesive stiffness has a marked effect on stresses: a low stiffness adhesive provides a more uniform stress distribution compared with a stiff adhesive, which gives rise to major stress concentrations at the overlap edges (Figure 11.6). Moreover, if the adhesive is ductile, it will be able to redistribute the load and make use of the less stressed parts of the overlap, whereas a brittle adhesive concentrates the load at the ends of the overlap without giving the possibility of plasticization, giving a low average shear stress at failure (Figure 11.7). Usually, adhesives are strong, brittle and stiff or weak, ductile, and flexible. Although ideally an adhesive should be strong, ductile, and flexible, this is very difficult to achieve with modern adhesives. In view of this, the recommendation goes to using less strong but ductile adhesives, although this also depends

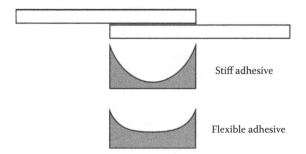

FIGURE 11.6 Influence of the adhesive's value of E on the shear stress distributions along the bondline.

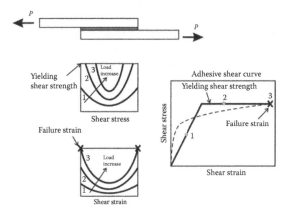

FIGURE 11.7 Influence of the adhesive ductility on the bondline stress distribution along the overlap.

on the overlap, as will be explained later. This is because ductile adhesives also resist crack propagation more than brittle ones, which is reflected in a much higher toughness. The fatigue strength of bonded joints is typically lower for brittle adhesives. This difference is justified by more uniform stress distributions and higher damping energy of ductile adhesives. For nonuniform loadings such as peel, cleavage, or thermal internal stresses, a joint with a ductile adhesive will also behave better.

11.4.2 Adherend Properties

The adherend properties (e.g., E, strength, and ductility) have a huge impact on the joint strength. Adherends with a high value of E promote more uniform stress distributions along the bondline, while those with a low value give rise to increased stress gradients in the adhesive layer [10]. The adherend strength and ductility are also essential and can explain many joint failures. Actually, for metallic adherends, yielding can trigger premature joint failure because with the increasing applied load, the stress reaches the yield point of the steel, and large plastic strains appear, giving rise to a plastic hinge (Figure 11.8). Although the stresses do not overcome the material strength, the associated strains become very large. When the nearby adhesive allowable strain is achieved, it fails. As a result, it is the adherend yielding that controls failure [28–29]. For laminated composite adherends, the outer plies of the adherends should be parallel to the loading direction to avoid intralaminar failure of these plies. However, the major problem of joints between composite adherends is the low transverse strength of composites, which is of the same order or lower than the adhesive tensile strength because it is governed by the matrix properties. As a result, the composite tends to fail in an interlaminar manner due to the high peel stresses at the overlap edges, as shown in Figure 11.9.

da Silva and Adams [30] addressed this problem by studying a bismaleimide reinforced with a carbon fiber fabric. The strength in the fiber direction was measured with a four-point flexure test as a function of temperature, as shown in Figure 11.10 [30]. The failure occurred between the inner rollers. Tests were performed at –55, 22, 100, and 200°C. It is interesting to note that the strength increases with temperature. This is because the resin becomes tougher and more ductile as the temperature approaches the glass transition temperature (T_g) and is less sensitive to defects. It is widely acknowledged that the through-thickness strength of composites is a difficult property to measure. In this study [30], the composite plate was cut and bonded to two steel blocks

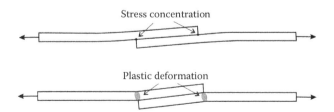

FIGURE 11.8 Adherend yielding and creation of plastic hinges in a SLJ.

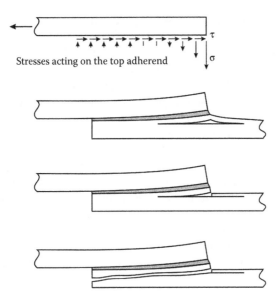

Stresses acting on the top adherend

FIGURE 11.9 Interlaminar failure of the composite in adhesive joints.

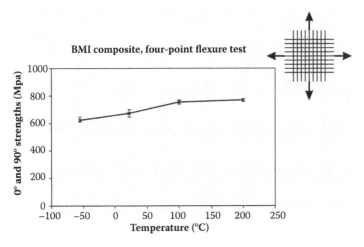

FIGURE 11.10 0° and 90° strength in flexure of a bismaleimide composite. (From da Silva, L.F.M., and Adams, R.D., Effect of temperature on the mechanical and bonding properties of a carbon reinforced bismaleimide, *Proceedings of the Institution of Mechanical Engineers, Part L, Journal of Materials: Design and Applications*, 222, 45–52, 2008. With permission [30].)

waisted to 12 × 15 mm². The steel and composite surfaces to bond were grit-blasted. However, to avoid damaging the fibers, the composite was grit-blasted with the gun of the shot blaster well away from the surface so as to decrease the blast pressure, and for a short period of time (approximately 5 seconds). Several designs were tested to make sure that failure occurs in the composite (design 3 in Figure 11.11, [30]). The adhesive used to bond the composite to the steel blocks was the epoxy AV119 from Huntsman. Its T_g is 120°C, so the composite is not tested at 200°C. A jig was used to guarantee alignment during manufacture. The specimen was loaded normal to the plane of the fibers. It is necessary to ensure that the load is perfectly aligned and that there is no bending effect that would lead to a premature failure. This was achieved by loading the steel blocks through precisely aligned pins. Figure 11.12 shows that the strength obtained in the thickness direction is one order of magnitude lower than the longitudinal strength, comparable to the strength of an adhesive [30]. If failure occurs in the composite, a failure criterion in the adhesive (maximum stress and maximum strain) overestimates the joint strength. The use of dissimilar adherends decreases the joint strength due to a nonuniform stress distribution (Figure 11.13). To reduce this problem, the joint should be designed so that the longitudinal stiffness of the adherends to be bonded is equal, that is, $E_1 t_1 = E_2 t_2$, where t is the thickness, and the subscripts (1, 2) refer to adherend 1 and adherend 2, respectively.

The composite material itself can be modified in order to increase its through-thickness strength. Many novel techniques have been developed. The most common

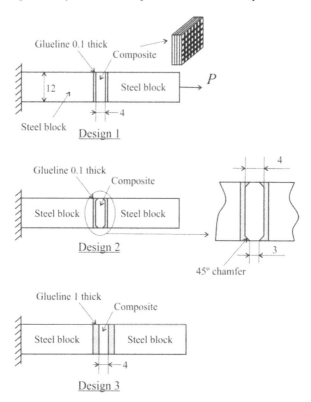

FIGURE 11.11 Designs studied for the determination of the transverse bismaleimide composite properties (dimensions in mm, not to scale). (From da Silva, L.F.M., and Adams, R.D., Effect of temperature on the mechanical and bonding properties of a carbon reinforced bismaleimide, *Proceedings of the Institution of Mechanical Engineers, Part L, Journal of Materials: Design and Applications*, 222, 45–52, 2008. With permission [30].)

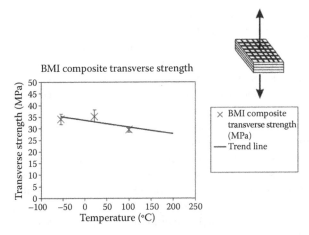

FIGURE 11.12 Transverse (through the thickness) strength of the bismaleimide composite. (From da Silva, L.F.M., and Adams, R.D., Effect of temperature on the mechanical and bonding properties of a carbon reinforced bismaleimide, *Proceedings of the Institution of Mechanical Engineers, Part L, Journal of Materials: Design and Applications*, 222, 45–52, 2008. With permission [30].)

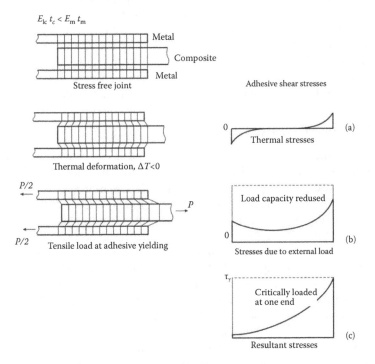

FIGURE 11.13 Adhesive shear stresses in a metal/composite DLJ for the case where the composite has a lower longitudinal stiffness than the metal ($E_{lc}t_c < E_m t_m$) under a tensile load and a thermal load, at adhesive yielding (τ_y). (a) Thermal stresses, (b) stresses due to external load, and (c) resultant stresses.

through-thickness reinforcement techniques are three-dimensional (3D) weaving [31], stitching [32], and braiding [33]. However, the only technique that is capable of reinforcing prepreg laminates in the through-thickness direction in large commercial quantities is z-pinning. Z-pin-reinforced composites are composed of short small-diameter rods inserted in the through-thickness direction (called the "z-axis" in analytical models and hence the name "z-pins") of the composite laminates. However, there are several concerns about the z-pinning process, particularly the accurate insertion of the z-pins in the orthogonal direction, swelling of the laminate that reduces the fiber volume fraction, and fiber damage [34].

Hybrid materials are a rather new type of material that have been developed in the past 20–30 years [35]. One type, the fiber metal laminates, has been developed at the Delft University of Technology, in cooperation with a number of partners. These material systems are created by bonding composite laminate plies to metal plies. It was found that the fatigue crack growth rates in adhesive bonded sheet materials can be reduced, if they are built up by laminating and adhesively bonding thin sheets of the material, instead of using one thick monolithic sheet. The concept is usually applied to aluminum with aramid (ARALL) and glass fibers (GLARE) but can also be applied to other constituents [36–37]. In particular, the GLARE material has been intensively investigated and has become one of the new materials used in the large Airbus A380 aircraft. Two large sections of the fuselage and the leading edges of the horizontal tail planes are made of GLARE. Since metals and fiber-reinforced polymers have characteristic properties and features with respect to manufacturing, the manufacturing of hybrid materials has properties and features related to both material groups. This multilayer composition of the hybrid laminates also offers the opportunity to mix and combine constituent materials with the aim of optimizing the component or substructure and, therefore, have a tailor-made material.

11.4.3 Adhesive Thickness

The effect of the bondline thickness on SLJ is well documented in the literature. Most of the results are for typical structural adhesives and show that the lap joint strength decreases as the bondline increases [17,27]. Experimental results show that for structural adhesives, the optimum joint strength is obtained with thin bondlines, in the range of 0.1–0.2 mm. However, the classical analytical models such as those of Volkersen [10] or Goland and Reissner [38] predict the opposite. Many theories attempt to explain this fact, and this subject is still controversial. Adams and Peppiatt [17] explained that an increase in the bondline thickness increases the probability of having internal imperfections in the joint (voids and microcracks), which will lead to premature failure of the joints. Crocombe [39] shows that thicker SLJ have a lower strength by considering the plasticity of the adhesive. An elastic analysis shows that the stress distribution of a thin bondline is more concentrated at the ends of the overlap than a thicker bondline that has a more uniform stress distributions. A thin bondline will therefore reach the yielding stress at a lower load than a thick bondline. However, when yielding occurs in a thicker joint, there is less "elastic reserve" to sustain further loading and, thus, yielding spreads more quickly (Figure 11.14). Gleich et al. [40] showed with an FE analysis on SLJ that increases in the interface stresses

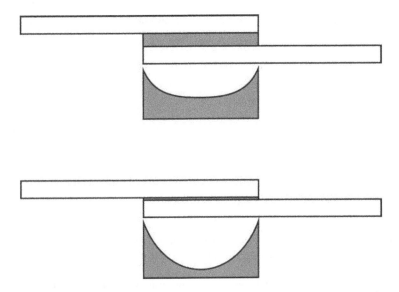

FIGURE 11.14 Stress distribution in a thick (top) and in a thin (bottom) adhesive layer.

(peel and shear) as the bondline gets thicker cause the failure load of a bonded joint to decrease with increasing bondline thickness. They found that for the low bondline thickness range, an optimum distribution of stresses along the joint interface exists for maximum joint strength. Grant et al. [41] found a reduction in joint strength with increasing bondline thickness when testing SLJ for the automotive industry with an epoxy adhesive. The strength reduction was attributed to the higher bending moments for the lap joints with thick bondlines due to the increase in the loading offset. The analyses described earlier are, in general, for metal adherends. Recent unpublished results of the authors have shown that a similar effect is obtained when composites are used. A stiff polyurethane was used with carbon fiber-reinforced plastics (CFRP) adherends. Two overlaps were studied (20 and 60 mm), and the adhesive thickness was varied from 0.2 to 3 mm.

11.4.4 OVERLAP

Increasing the joint width increases the strength proportionally. However, the effect of the overlap length depends of the type of adhesive (i.e., ductile or brittle) and on the type of adherend. For metal adherends, three cases should be considered: elastic adherends (e.g., high-strength steel) and ductile adhesive, elastic adherends and brittle adhesive, and adherends that yield. For elastic adherends and ductile adhesives (more than 20% shear strain to failure), the joint strength is approximately proportional to the overlap. This is because ductile adhesives can deform plastically, redistribute the stress as the load increases, and make use of the whole overlap. In this case, the failure criterion is the global yielding of the adhesive. For adhesives with intermediate ductility, the adhesive fails because the strain in the adhesive at the ends of the overlap reaches the adhesive shear strain to failure (Figure 11.15).

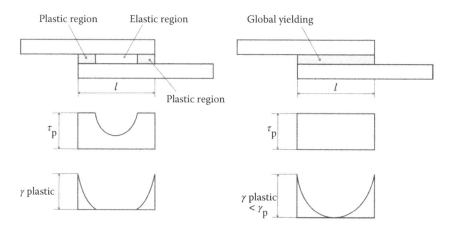

FIGURE 11.15 Failure due to adhesive shear strain (left) and due to global yielding (right). τ_p is the yielding shear stress, and γ_p is the plastic shear strain at failure.

For elastic adherends and brittle adhesives, the joint strength is not proportional to the overlap and a plateau is reached. This is because the stress is concentrated at the ends of the overlap, and a longer overlap does not alter the stress distribution along the overlap. For adherends that yield, failure is dictated by the adherend yielding and again a steady-state value is reached corresponding to adherend yielding.

The effect of the overlap for composite adherends is mainly dictated by the composite transverse strength. Recently, Neto et al. [42] studied this matter with carbon fiber-reinforced epoxy composites. The main objectives of this work were the characterization of the failure process and strength of adhesive joints with composites, bonded with different adhesives and from short to long overlaps, and the validation of different predicting methods. This work allowed for the conclusion that for SLJ with composites bonded with a ductile adhesive (SikaForce® 7888) the failure load increases as the overlap increases (Figure 11.16, [42]) and the failure was cohesive in the adhesive for all overlaps (Figure 11.17a, [42]). In the case of the brittle adhesive (AV138), as the overlap increased a plateau was reached (Figure 11.16, [42]), since for the overlap of 30 mm, the experimental failure observed was interlaminar and therefore it was dictated by the composite (Figure 11.17b, [42]). Analytical and numerical methods were used to predict the strength of the joints. For the joints with the brittle adhesive, the analytical model of Hart-Smith predicted the failure load in the composite using peel stresses. For the ductile adhesive, the best prediction was obtained with the global yielding criterion (Figure 11.15):

$$P_{GY} = \tau_y bl \tag{11.1}$$

where P_{GY} is the failure load of the adhesive due to global yielding, τ_y is the yield strength of the adhesive, b is the joint width, and l is the overlap length. The numerical methods used (CZM) returned satisfactory values with the brittle adhesive but with the ductile adhesive these models did not work with the same precision. This

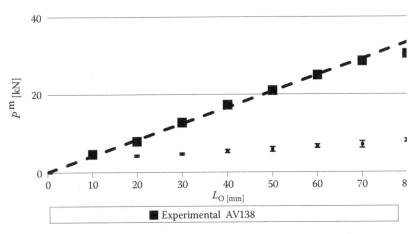

FIGURE 11.16 Experimental failure loads for composite SLJ with a brittle adhesive AV138 and a ductile adhesive SikaForce 7888. (From Neto, J.A.B.P., Campilho, R.D.S.G., and da Silva, L.F.M., Parametric study of adhesive joints with composites, *International Journal of Adhesion & Adhesives*, 37, 96–101, 2012. With permission [42].)

FIGURE 11.17 (a) Cohesive failure of joints with the ductile adhesive SikaForce 7888 for an overlap of 50 mm. (b) Composite failure for joints with the brittle adhesive AV138 for an overlap of 40 mm. (From Neto, J.A.B.P., Campilho, R.D.S.G., and da Silva, L.F.M., Parametric study of adhesive joints with composites, *International Journal of Adhesion & Adhesives*, 37, 96–101, 2012. With permission [42].)

probably happened because a triangular law was used and the behavior of the ductile adhesive is closer to a trapezium shape. The determination of the cohesive law properties can also be improved with an inverse method or a direct determination method. Nevertheless, the numerical models were capable of simulating the failure initiation and propagation observed in the tests.

11.4.5 Residual Stresses

One of the main advantages of using adhesive bonding is the possibility of bonding dissimilar materials, such as CFRP to aluminum in many aeronautical applications. However, dissimilar adherends may have very different coefficients of thermal expansion (CTE). Thus, temperature changes may introduce thermal stresses in addition to the externally applied loads. The adhesive curing and the resulting thermal shrinkage may also introduce internal stresses. Deformations or even cracks can appear. It is important to consider thermal effects, because these generally lead to a joint strength reduction, even though in some cases the opposite happens [43]. Several authors have found that the stresses caused by adhesive shrinkage have much less effect on the lap joint strength than those generated by the adherend thermal mismatch. Thermal loads are especially important when bonding adherends with different CTE [14]. For metal/composite joints, for example, the metal tends to shrink as the temperature is decreased from the cure value (generally a high temperature) and this is partially resisted by the composite (lower CTE), thereby inducing residual bond stresses, especially at the ends of the joint. One end has positive residual shear stresses, and the other end has negative residual shear stresses (Figure 11.13a). The thermal stresses are beneficial at one end of the joint but have a reverse effect at the other side of the joint. The thermal load ΔT is given by

$$\Delta T = T_O - T_{SF}, \tag{11.2}$$

where T_O is the operating temperature and T_{SF} is the stress-free temperature. It is reasonable to consider the stress-free temperature as the normal cure temperature of the adhesive for most cases.

11.5 METHODS TO INCREASE JOINT STRENGTH

11.5.1 Fillets

Various authors have shown that the inclusion of a spew fillet at the ends of the overlap reduces the stress concentrations in the adhesive and the adherends [17–18,44–49]. The load transfer and shear stress distribution of an SLJ with and without fillet are schematically represented in Figure 11.18. It can be seen that there is a stress concentration at the ends of the overlap for the SLJ with a square end. Modification of the joint end geometry with a spew fillet spreads the load transfer over a larger area and gives a more uniform shear stress distribution. The fillet not only gives a smoother load transfer but also alters the stress intensity factors, as shown in Figure 11.19. Adams and Peppiatt [17] found that the inclusion of a 45° triangular spew fillet decreases the magnitude of the maximum principal stress by 40% when compared with a square-end adhesive fillet. Adams and Harris [45] tested aluminum/epoxy SLJ with and without fillet and found an increase of 54% in joint strength for the filleted joint. Adams et al. [44] tested aluminum/CFRP SLJ and found that the joint with a fillet is nearly two times stronger than the joint without a fillet. Crocombe and Adams [18] did similar work but included geometric (overlap length, adhesive thickness, and adherend thickness) and material (E ratio) parameters. The reduction in peel and shear stresses was the greatest for a low

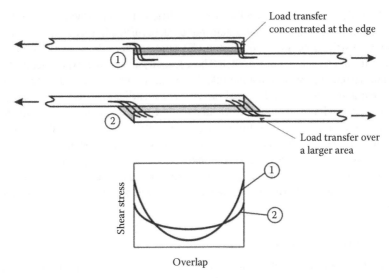

FIGURE 11.18 Load transfer and shear stress distribution in SLJ both with and without fillet.

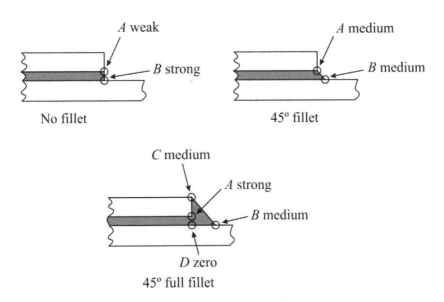

FIGURE 11.19 Stress intensity factors in adhesive lap joints with different spew fillet geometry.

E ratio (adhesive with low E), a high adhesive thickness, and a low adherend thickness. Dorn and Liu [46] investigated the influence of the spew fillet on plastic/metal joints. The study included an FE analysis and experimental tests, and it was concluded that the spew fillet reduces the peak shear and peel adhesive stresses and decreases stress and strain concentrations in the adherends at the most critical regions. They also studied the influence of different adhesives and different metal adherends. A ductile adhesive and a more balanced joint (aluminum/plastic instead of steel/plastic) give a better stress distribution. Tsai and Morton [47] studied the influence of a triangular spew fillet in laminated composite SLJ. The FE analysis and the experimental tests (Moiré interferometry) proved that the fillet helps carry a part of the load, thus reducing the shear and peel strains. The analyses described earlier are limited to triangular geometry. Lang and Mallick [48] investigated eight different geometries: full and half triangular, full and half rounded, full rounded with fillet, oval, and arc. They showed that shaping the spew to provide a smoother transition in joint geometry significantly reduces the stress concentrations. Full-rounded fillets and arc spew fillets give the highest percentage of reduction in maximum stresses, whereas half-rounded fillets give a lower percentage. Furthermore, increasing the size of the spew also reduces the peak stress concentrations.

11.5.2 ADHEREND SHAPING

Adherend shaping is a powerful way to decrease the stress concentration at the ends of the overlap. Figure 11.20 presents typical geometries used for that purpose. Some analytical models were proposed to have a more uniform stress distribution along the overlap [50]. However, the FE method is more appropriate for the study of adherend shaping. The concentrated load transfer at the ends of the overlap can be more uniformly distributed if the local stiffness of the joint is reduced. This is particularly relevant for adhesive joints with composites due to the low transverse strength of composites. Adams et al. [44] addressed this problem. They studied various configurations of DLJ where the central adherend is CFRP and the outer adherends are made of steel. They found with FE and experiments that the inclusion of an internal taper and an external fillet can triplicate the joint strength. Later, da Silva and Adams [49] tested joints with an internal taper and an adhesive fillet (Figure 11.20) that were manufactured and tested with the epoxy adhesive Supreme 10HT (Master Bond) at 22°C. The failure load is higher than for the joint without a taper and an adhesive fillet (basic design), but the increase is very small (Figure 11.21). The increase in

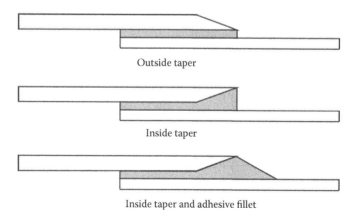

Outside taper

Inside taper

Inside taper and adhesive fillet

FIGURE 11.20 Adherend shaping.

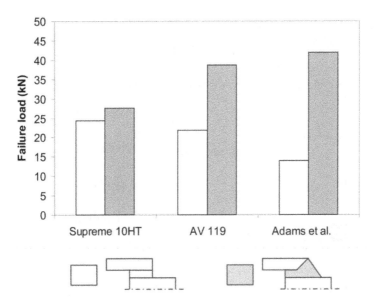

FIGURE 11.21 Basic design vs. design with taper and adhesive fillet: failure loads for various adhesives at 22°C (titanium/composite DLJ with the composite as the inner adherend).

strength obtained by Adams et al. [44] was three times. They also showed that, with an adhesive fillet and an internal taper, the loading is predominantly tensile in the adhesive and that the locus of failure is at the outer surface of the adhesive fillet close to the outer adherend corner. The Supreme 10HT joints also failed at the outer adherend corner, as shown in Figure 11.22. The tensile strength of the Supreme 10HT at room temperature is 46 MPa [51], whereas that of the adhesive used by Adams et al. [44] was 82 MPa. Therefore, another adhesive with a higher tensile strength than Supreme 10HT was used to check whether a joint with an internal taper and an adhesive fillet can give a higher strength increase. The epoxy AV119 (Huntsman) was selected, as its tensile strength at room temperature is 67 MPa. The strength increase compared with the basic design is now 1.8 times (Figure 11.21). The locus of failure was also at the outer adherend corner (Figure 11.22b). These results show that the use of an internal taper and adhesive fillet is not necessarily beneficial and will depend on the adhesive properties.

Hildebrand [52] did similar work on SLJ between fiber-reinforced plastic and metal adherends. The optimization of the SLJ was done by modifying the geometry of the joint ends. Different shapes of adhesive fillet, reverse tapering of the adherend, rounding edges, and denting were applied in order to increase the joint strength. The results of the numerical predictions suggest that, with a careful joint-end design, the strength of the joints can be increased by 90%–150%. The use of internal tapers in the adherends in order to minimize the maximum transverse stresses at the ends of bonded joints has also been studied by Rispler et al. [53]. An evolutionary structural optimization method (EVOLVE) was used to optimize the shape of adhesive fillets.

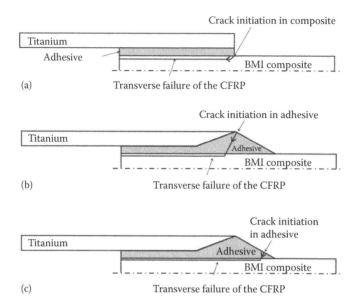

FIGURE 11.22 Failure mode of titanium/composite DLJ: (a) in basic design; (b) with an internal taper and an adhesive fillet at 22°C; (c) with an internal taper and an adhesive fillet at –55°C (schematic representation).

EVOLVE consists of an iterative FE analysis and a progressive removal of elements in the adhesive that are low stressed. Other examples of joint-end modifications for joint transverse stress reduction but using external tapers are those of Sancaktar and Nirantar [54] and Kaye and Heller [55]. Kaye and Heller [55] used numerical optimization techniques in order to optimize the shape of the adherends. This is especially relevant in the context of repairs using composite patches bonded to aluminum structures (Section 11.7) due to the highly stressed edges.

Tapers (internal or external) or more complex adherend shaping are excellent methods to reduce peel stresses at the ends of the overlap, and therefore to increase the joint strength. Internal tapers with a fillet seem to be a more efficient method to have a joint strength increase, especially with brittle adhesives and when composites are used. The FE method is a convenient technique for the determination of the optimum adherend geometry; however, the complexity of the achieved geometry is not always possible to realize in practice.

11.6 HYBRID JOINTS

Joints with different methods of joining are increasingly being used. The idea is to gather the advantages of the different techniques while leaving out their problems. Another possibility is to use more than one adhesive along the overlap or vary the adhesive and/or adherend properties. All these cases have been grouped here under a section called "hybrid joints" (Figure 11.23). Such joints are particularly

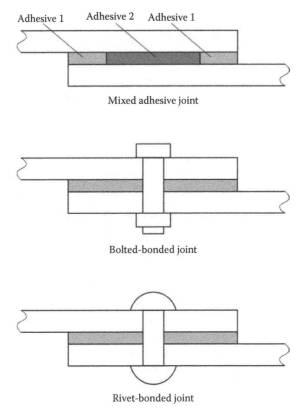

Adhesive 1   Adhesive 2   Adhesive 1

Mixed adhesive joint

Bolted-bonded joint

Rivet-bonded joint

FIGURE 11.23   Hybrid joints.

difficult to simulate using analytical models for obvious reasons. The FE method is the preferred tool to investigate the application of such techniques and find design guidelines.

11.6.1   MIXED ADHESIVE JOINTS

Mixed modulus joints have been proposed in the past to improve the stress distribution and to increase the joint strength of adhesives with a high value of E. The stiff and brittle adhesive should be at the middle of the overlap, whereas an adhesive with a low value of E is applied at the edges that are prone to stress concentrations. Sancaktar and Kumar [56] used rubber particles to toughen the part of the adhesive located at the ends of the overlap and to increase the joint strength. The concept was studied with the FE method and proved experimentally. Pires et al. [57] also proved with an FE analysis and experimentally with two different adhesives that the mixed adhesive method gives an improvement in joint performance. Temiz [58] used an FE analysis to study the influence of two adhesives in

DLJ under bending and found that the technique greatly decreased the stresses at the ends of the overlap. Bouiadjra et al. [59] used the mixed modulus technique for the repair of an aluminum structure with a composite patch. The use of a more flexible adhesive at the edge of the patch increases the strength performance of the repair. das Neves et al. [60–61] have developed an analytical model that takes into account two adhesives along the overlap and permits one to determine the best combination of adhesives and the optimum geometric factors (e.g., overlap) to have the maximum joint strength. The authors showed that the technique is more efficient for SLJ than for DLJ. da Silva and Lopes [62] have studied SLJ maintaining the same brittle adhesive in the middle of the overlap and using three different ductile adhesives of increasing ductility at the ends of the overlap. A simple joint strength prediction is proposed for mixed adhesive joints. The mixed adhesive technique gives joint strength improvements in relation to a brittle adhesive alone in all cases. For a mixed adhesive joint to be stronger than the brittle adhesive and the ductile adhesive used individually, the load carried by the brittle adhesive must be higher than that carried by the ductile adhesive. Marques and da Silva [63] studied mixed adhesive joints with an internal taper and a fillet. They showed that the use of a taper and a fillet have little effect on the strength of mixed adhesive joints. The ductile adhesive at the ends of the overlap is sufficient to have an improved joint strength in relation to a brittle adhesive alone. The taper and the fillet are only useful when the brittle adhesive is used alone. One of the problems associated with the mixed adhesive technique is the proper adhesive separation. The best way to control the process is to use film adhesives. However, it is difficult to find compatible adhesives in film form. This is a problem for the manufacturers to solve. Meanwhile, adhesive separation can be done with the use of silicone strips, even though a small portion of the load bearing area is reduced.

The technique of using multimodulus adhesives has been extended to solve the problem of adhesive joints that need to withstand low and high temperatures by da Silva and Adams [64–65]. At high temperatures, a high temperature adhesive in the middle of the joint retains the strength and transfers the entire load whereas a low temperature adhesive is the load-bearing component at low temperatures, making the high temperature adhesive relatively lightly stressed. The authors studied various configurations with the FE method and proved experimentally that the concept works, especially with dissimilar adherends (titanium/CFRP).

11.6.2 ADHESIVE JOINTS WITH FUNCTIONALLY GRADED MATERIALS

The mixed adhesive joint technique can be considered a rough version of a functionally graded material. The ideal would be to have an adhesive functionally modified with properties that vary gradually along the overlap, allowing a true uniform stress distribution along the overlap. Ganesh and Choo [66] and Apalak and Gunes [67] have used functionally graded adherends instead of functionally graded adhesives. Ganesh and Choo [66] used a braided preform with continuously varying braid angle to realistically evaluate the performance of adherend E grading in an SLJ. An increase of 20% joint strength was obtained due to a more uniform stress distribution. Apalak and Gunes [67] studied the flexural behavior of an adhesively bonded SLJ with adherends composed of a functionally gradient layer between a pure ceramic (Al_2O_3) layer and a pure metal (Ni) layer. The study is not supported with experimental results, and the adhesive stress distribution was not hugely affected. To our knowledge, the works by Ganesh and Choo [66] and Apalak and Gunes [67] are the only studies that deal with the application of functionally graded materials to adhesive joints. There has never been an attempt to modify the adhesive, which might be easier and more logical than modifying the adherends. Only Sancaktar and Kumar [56] used rubber particles to locally modify the adhesive at the ends of the overlap, but that is not a gradually modified adhesive. This is an area that is being intensively studied and where modeling at different scales is essential. More recently, Stapleton et al. [68] used the same idea by strategically placing glass beads within the adhesive layer at different densities along the joint with composite adherends. Kumar and Scanlan [69] numerically studied the joint behavior of a functionally graded joint with different degrees of grading in the adhesive stiffness along the overlap. However, several practical concerns impede the actual use of such adhesives. Carbas et al. [70–71] developed an apparatus that allows gradually varying the adhesive properties (stiffness, strength, and ductility) along the overlap by using a differential cure. The graded cure ensures gradual adhesive properties along the bondline, and an adhesive with ductile behavior where there is greater stress concentrations and a strong adhesive where high strength is required. Two adhesives were tested as a function of the cure temperature and then applied in a joint with a graded cure obtained by induction heating [72–73]. The graded joints give significant improvement in terms of strength in relation to the joints cured isothermally (Figure 11.24). The failure mode of the graded joints is also safer with a progressive failure. The analytical simulations were able to predict the behavior of the joints, including the specimens repaired with a graded bondline [74].

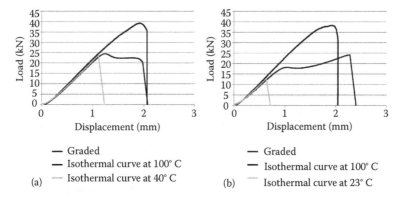

FIGURE 11.24 Typical load-displacement curves of joints with isothermal cure and graded cure for adhesives Araldite 2011 (a) and Loctite Hysol 3422 (b).

11.6.3 RIVET-BONDED JOINTS

Liu and Sawa [75] investigated, using a 3D FE analysis, rivet-bonded joints and found that for thin substrates, riveted joints, adhesive joints, and rivet-bonded joints gave similar strengths whereas for thicker substrates the rivet-bonded joints were stronger. They proved this experimentally. Later, the same authors [76] proposed another technique similar to rivet-bonded joints: adhesive joints with adhesively bonded columns. Strength improvements are also obtained in this case. The advantage of this technique of that the appearance is the joint is maintained in relation to an adhesive joint. Grassi et al. [77] studied through-thickness pins for restricting debond failure in joints. The pins were simulated by tractions acting on the fracture surfaces of the debond crack. Pirondi and Moroni [78] found that the adhesive layer strongly increases the maximum load and the initial stiffness in comparison with a joint with a rivet alone. When failure of the adhesive occurs, the joint behaves as with only a rivet.

11.6.4 BOLTED-BONDED JOINTS

Chan and Vedhagiri [79] studied the response of various configurations of SLJ, namely bonded, bolted, and bonded-bolted joints, using the 3D FE method, and the results were validated experimentally. The authors found that, for the bonded-bolted joints, the bolts help to reduce the stresses at the edge of the overlap, especially after the initiation of failure. The same type of study was carried out by Lin and Jen [80].

11.7 REPAIR TECHNIQUES

Adhesively bonded repairs usually imply complex geometries, and for a detailed design and optimization of these repairs, the FE method has been extensively used, especially with composites. The literature review of Odi and Friend [81] about repair techniques clearly illustrates this point. Typical methods and geometries are presented in Figure 11.25. Among the various techniques available, single- and double-strap repairs are the easiest to perform, but they lack the capacity to restore the undamaged strength of materials. Scarf or stepped repairs are particularly attractive because a flush surface is maintained, which permits good aerodynamic behavior. Tong and Sun [82] developed a pseudo 3D element to perform a simplified analysis of strap-bonded repairs to curved structures. The analysis is supported by a full 3D FE analysis. The authors found that external patches are preferred when the shell is under an internal pressure whereas internal patches are preferred when under an external pressure. Soutis and Hu [83] studied numerically and experimentally bonded composite patch repairs to repair cracked aircraft aluminum panels. The authors concluded that the bonded patch repair provides a considerable increase in the residual strength. Vallée et al. [84] tested and studied with the help of FE different stress reduction methods in double-strap bonded repairs with fiber-reinforced composite and wooden adherends. Strength improvements were found by patch tapering, although the large stress reductions were not followed by identical strength improvements. Marques and da Silva [63] studied aluminum

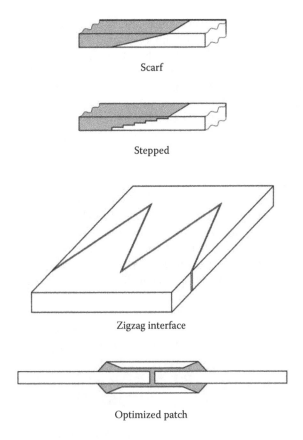

Scarf

Stepped

Zigzag interface

Optimized patch

FIGURE 11.25   Repair techniques for composite structures.

double-strap bonded repairs with a taper and a spew fillet at the ends of the overlap, and they also considered the mixed adhesive technique (ductile adhesive at the overlap ends). The experimental results were validated with an FE stress analysis, and it was found that the taper angle is only recommended for the brittle adhesive. On the other hand, the mixed adhesive technique only has a real advantage for the taper-less configuration.

Campilho et al. [85] addressed with the help of FE and CZM the tensile strength optimization of carbon-epoxy single- and double-strap bonded repairs by includ-ing geometric modifications at the bonding region. Since the optimization work was purely numerical, an initial validation of the numerical method was accomplished with experimental results from the literature. Different strength improvement tech-niques and respective combinations were tested: chamfering the patch outer and inner faces at the overlap ends, plug filling the adherend gap with adhesive, use of fillets of different shapes and dimensions at the patch ends, and chamfering of the outer and inner adherend edges. It was found that the strength of the single-strap repairs was highly improved by a 45° fillet (accounting for the entire patch thickness), as well as adherend inner and outer chamfering. By combining these three modifications

(Figure 11.26, [85]), an overall strength improvement of nearly 27% was found relative to the unmodified repair. The strength improvement of double-strap repairs reached 12% by using a 45° fillet and plug filling the adherend gap. Additional studies were also performed to assess the adherend/patch layups and the adherend gap effects on the predicted strength. It was found that the single-strap repair strength is highly dependent on the adherend layup because of its influence on the transverse flexure. Layups with load-oriented plies at the adherend-free surfaces were recommended to minimize peel stresses. Figure 11.27 shows repairs with stiff (a) and compliant (b) adherends and patches under a similar applied load, showing crack growth only for the compliant repair because of higher peel stresses [85]. Opposed to this behavior, the double-strap repairs showed little variations of strength because of the flexure elimination of the adherends. The adherend gap affected the strength for values smaller than 5 mm. The strength was gradually reduced below gaps of 5 mm because of localized stress concentrations at the overlap edges. In the end, design principles were proposed to fabricate adhesively bonded repairs in composite structures. Hu and Soutis [86] presented a numerical analysis of composite double-strap repairs under a compressive load, concluding that localized stresses act at the end zones of the overlap. Results showed that peak shear strains can be markedly reduced, thus increasing the adhesive thickness at the patch edges. Therefore, a joint with patches tapered from inside was hypothetically considered to reduce stress concentrations in the adhesive layer and, consequently, to increase the joint strength. No experimental analyses were performed to corroborate this fact.

FIGURE 11.26   Single-strap repair combining a 45° fillet with adherend inner and outer chamfering. (From Campilho, R.D.S.G., de Moura, M.F.S.F., and Domingues, J.J.M.S., Numerical prediction on the tensile residual strength of repaired CFRP under different geometric changes, *International Journal of Adhesion & Adhesives*, 29, 195–205, 2009. With permission [85].)

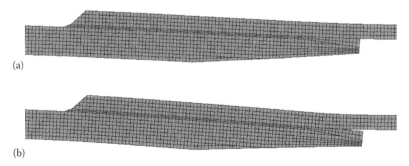

(a)

(b)

FIGURE 11.27   Deformed configuration of the repairs for stiff (a) and compliant (b) adherends and patches for the same value of prescribed load. (From Campilho, R.D.S.G., de Moura, M.F.S.F., and Domingues, J.J.M.S., Numerical prediction on the tensile residual strength of repaired CFRP under different geometric changes, *International Journal of Adhesion & Adhesives*, 29, 195–205, 2009. With permission [85].)

Gunnion and Herszberg [87] studied scarf repairs and carried out an FE analysis to assess the effect of various parameters. They found that the adhesive stress is not much influenced by mismatched adherend layups and that there is a huge reduction in peak stresses with the addition of an over-laminate. Odi and Friend [88] performed a bidimensional FE study regarding scarf repairs with carbon-epoxy laminates and scarf angles of 1.1°, 1.9°, 3.0°, 6.2°, and 9.0°. A $[0_2/\pm45/90/\pm45/0_2]_S$ layup and an adhesive thickness of 0.25 mm were considered. The tensile failure loads of the repairs were obtained with the Tsai-Wu and the maximum stress criterion for the laminates and the average stress criterion for the adhesive layer. The laminates were modeled ply by ply, allowing applying failure criteria to be developed specifically for composite laminates. Figure 11.28 shows shear stress distributions in the adhesive layer, normalized by the remote applied stress. Figure 11.28a compares results of a

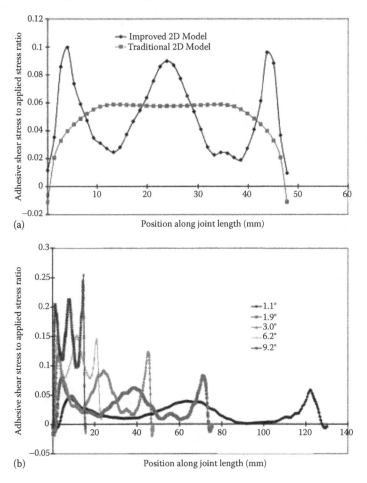

FIGURE 11.28 Shear stresses in the adhesive for a 3° scarf repair (a) and comparison between five scarf angles (b). (From Odi, R.A., and Friend, C.M., An improved 2D model for bonded composite joints, *International Journal of Adhesion & Adhesives*, 24, 389–405, 2004. With permission [88].)

traditional model (using equivalent elastic properties for the laminates) and the proposed methodology, for a 3° scarf angle, whereas Figure 11.28b equates the five scarf angles evaluated [88]. The traditional approach results in a practically uniform shear stress distribution. The improved approach captures the shear stress gradients along the bond length caused by the compliance of the distinct plies in the load direction. It was found that shear stresses are highest near the 0° plies. In addition, shear stresses in the adhesive layer increase with the scarf angle, diminishing the repair strength. It was observed that the failure load increased, reducing the scarf angle. With a 1.1° scarf angle, a reduction of the failure load was observed numerically, due to element distortion. The comparison between the test results and simulations resulted in the conclusion that the proposed methodology accurately predicted the failure load of these repairs.

Campilho et al. [7] numerically studied scarf repairs in UD composites with a triangular CZM and concluded that the strength of the repair increased exponentially with the reduction of the scarf angle. In a subsequent work, Campilho et al. [89] experimentally and numerically tested carbon-epoxy scarf repairs in tension, using scarf angles between 2° and 45°, considering CZM for the strength prediction. The selected adhesive allowed to perform the repairs was very ductile and, to account for this behavior, a trapezoidal damage law including the adhesive plasticity was used. The numerical models allowed failure in the adhesive layer and also composite interlaminar and composite intralaminar failures (in the transverse and fiber direction)—local approach—that are necessary to numerically replicate the experimental failure paths. Validation of the proposed model with experimental data was accomplished by comparing the global stiffness, strength and corresponding value of displacement, and failure mode. The layout of the cohesive elements in the models is shown in Figure 11.29 [89]. Two failure modes were experimentally observed: cohesive failure of the adhesive for the repairs with 15°, 25°, and 45° scarf angles, and mixed cohesive and interlaminar/intralaminar failure for the repairs with 2°, 3°, 6°, and 9° scarf angles. The CZM simulations accurately reproduced the experimental fracture modes and the failure mode variations. Figure 11.30 compares the numerical and experimental failure loads as a function of the scarf angle, showing an exponential increase with the reduction of the scarf angle, related to the increase of available

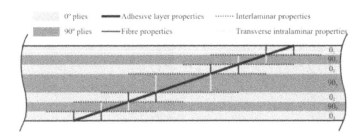

FIGURE 11.29 Layout of cohesive elements with different CZM laws in the numerical models. (From Campilho, R.D.S.G., de Moura, M.F.S.F., Pinto, A.M.G., Morais, J.J.L., and Domingues, J.J.M.S., Modelling the tensile fracture behaviour of CFRP scarf repairs, *Composites: Part B*, 40, 149–57, 2009. With permission [89].)

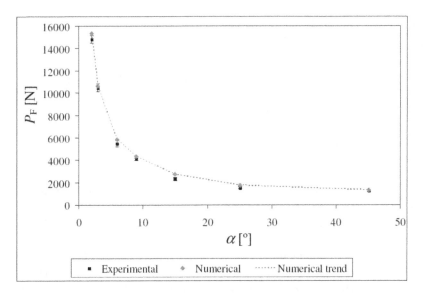

FIGURE 11.30 Comparison of the numerical and experimental failure loads as a function of the scarf angle. (From Campilho, R.D.S.G., de Moura, M.F.S.F., Pinto, A.M.G., Morais, J.J.L., and Domingues, J.J.M.S., Modelling the tensile fracture behaviour of CFRP scarf repairs, *Composites: Part B*, 40, 149–57, 2009. With permission [89].)

shear area for load transfer between the adherends and the patch, together with a reduction of peel stresses along the bonding length [89]. Overall, the experimental and FE results showed a good agreement on the global elastic stiffness, strength, and respective value of displacement, although the displacement showed some scatter, which the authors imputed to difficulties in obtaining the desired adhesive thickness during the fabrication of the specimens. On account of the described results, the authors concluded that the CZM can be successfully applied to predict the mechanical behavior of bonded structures. Bahei-El-Din and Dvorak [90] proposed new design concepts for the repair of thick composite laminates. The regular butt joint with a patch on both sides was modified by the inclusion of pointed inserts or a zigzag interface to increase the area of contact and improve the joint strength.

Recently, Carbas et al. [91] applied the graded cure technique (Section 11.6.2) to the repair of wood beams with defects with CFRP patches. The idea was to use a graded joint to reduce the stress peaks at the edges and improve the strength of the repair. Two common types of defects of wood beams under bending loadings, cross-grain tension and compression failure, were analyzed. The CFRP patches were bonded to the damaged area using the Loctite® Hysol 3422 adhesive. The effect of the cure processes of the adhesive in the effectiveness of the repair was assessed. Two different cure processes were studied: isothermal cure and graded cure. The results regarding the bending tests on the wood beams showed that graded joints can be used to improve the strength and reliability of repaired beams. The numerical simulations were able to predict the behavior of the wood specimens, including the specimens repaired with a graded bondline.

11.8 CONCLUSIONS

Simple design rules for SLJ were proposed as a function of the main variables that influence joint strength: adhesive and adherend properties, adhesive thickness, overlap, and residual stresses. The following main rules should be considered:

* An adhesive with a low value of E and high ductility
* Similar adherends whenever possible
* A thin adhesive layer
* A large bonded area

Several methods that increase the joint strength were discussed for lap joints: fillets and adherend shaping. The designer should always bear in mind that when designing an adhesive joint, the load must be spread over a large area and not concentrated in one point. Peel loads are the greatest enemy of the designer of bonded joints, especially with composites, due to the low transverse strength. Adhesives in conjunction with other methods of joining (rivet, bolt) were explained. The idea is to get a synergetic effect and combine the advantages of the two methods. Mixed adhesive joints and functionally graded joints are also very promising techniques. Finally, various types of repairs were discussed to obtain the maximum strength recovery. The scarf joint is particularly efficient and aesthetically attractive.

REFERENCES

1. Mohanty, A.K., Misra, M., Drzal, L.T., Selke, S.E., Harte, B.R., and Hinrichsen, G. 2005. Natural fibers, biopolymers, and biocomposites: an introduction. In *Natural Fibers, Biopolymers, and Biocomposites*, (ed.) Mohanty, A.K., Misra, M., and Drzal, L.T., 1–38. Boca Raton: CRC Press.
2. Scuh, T.G. 2005. *Renewable Materials for Automotive Applications*. Stuttgart: Daimler-Chrysler AG.
3. Gonzalez-Murillo, C. 2010. Co-cured in-line joints for natural fibre composites. *Composites Science and Technology* 70:442–49.
4. da Silva, L.F.M., Öchsner, A., and Adams, R.D. (ed.) 2011. *Handbook of Adhesion Technology*. Heidelberg: Springer.
5. Feraren, P., and Jensen, H.M. 2004. Cohesive zone modelling of interface fracture near flaws in adhesive joints. *Engineering Fracture Mechanics* 71:2125–42.
6. Campilho, R.D.S.G., Banea, M.D., Neto, J.A.B.P., and da Silva, L.F.M. 2013. Modelling adhesive joints with cohesive zone models: effect of the cohesive law shape of the adhesive layer. *International Journal of Adhesion & Adhesives* 44:48–56.
7. Campilho, R.D.S.G., de Moura, M.F.S.F., and Domingues, J.J.M.S. 2007. Stress and failure analyses of scarf repaired CFRP laminates using a cohesive damage model. *Journal of Adhesion Science and Technology* 21:855–70.
8. Campilho, R.D.S.G., de Moura, M.F.S.F., and Domingues, J.J.M.S. 2005. Modeling single and double lap repairs on composite materials. *Composites Science and Technology* 65:1948–58.
9. da Silva, L.F.M., and Campilho, R.D.S.G. 2012. *Advances in Numerical Modelling of Adhesive Joints*. Heidelberg: Springer.
10. Volkersen, O. 1938. Die nietkraftoerteilung in zubeanspruchten nietverbindungen mit konstanten loschonquerschnitten. *Luftfahrtforschung* 15:41–7.

11. da Silva, L.F.M., das Neves, P.J.C., Adams, R.D., and Spelt, J.K. 2009. Analytical models of adhesively bonded joints—Part I: Literature survey. *International Journal of Adhesion & Adhesives* 29:319–30.

12. da Silva, L.F.M., das Neves, P.J.C., Adams, R.D., Wang, A., and Spelt, J.K. 2009. Analytical models of adhesively bonded joints—Part II: Comparative study. *International Journal of Adhesion & Adhesives* 29:331–41.

13. Tong, L., and Luo, Q. 2011. Analytical approach. In *Handbook of Adhesion Technology*, (ed.) da Silva, L.F.M., and Öchsner, A., and Adams, R.D. Heidelberg: Springer.

14. Hart-Smith, L.J. 1973. Adhesive-bonded double-lap joints, NASA CR-112235.

15. da Silva, L.F.M., Lima, R.F.T., and Teixeira, R.M.S. 2009. Development of a computer program for the design of adhesive joints. *The Journal of Adhesion* 85:889–918.

16. Dragoni, E., Goglio, L., and Kleiner, F. 2010. Designing bonded joints by means of the Jointcalc software. *International Journal of Adhesion & Adhesives* 30:267–80.

17. Adams, R.D., Peppiatt, N.A. 1974. Stress analysis of adhesive-bonded lap joints. *Journal of Strain Analysis* 9:185–96.

18. Crocombe, A.D., and Adams, R.D. 1981. Influence of the spew fillet and other parameters on the stress distribution in the single lap joint. *The Journal of Adhesion* 13:141–55.

19. Harris, J.A., and Adams, R.D. 1984. Strength prediction of bonded single lap joints by non-linear finite element methods. *International Journal of Adhesion & Adhesives* 4:65–78.

20. Gonçalves, D.J.S., Campilho, R.D.S.G., da Silva, L.F.M., and Fernandes, J.L.M. 2014. The use of the boundary element method in the analysis of single lap joints. *The Journal of Adhesion* 90:50–64.

21. Chen, Z., Adams, R.D., and da Silva, L.F.M. 2011. Prediction of crack initiation and propagation of adhesive lap joints using an energy failure criterion. *Engineering Fracture Mechanics* 78:990–1007.

22. Campilho, R.D.S.G., Banea, M.D., Pinto, A.M.G., da Silva, L.F.M., and de Jesus, A.M.P. 2011. Strength prediction of single- and double-lap joints by standard and extended finite element modelling. *International Journal of Adhesion & Adhesives* 31:363–72.

23. Adams, R.D., Comyn, J., and Wake, W.C. 1997. *Structural Adhesive Joints in Engineering*. 2nd edn. London: Chapman & Hall.

24. Campilho, R.D.S.G., Moura, D.C., Gonçalves, D.J.S., da Silva, J.F.M.G., Banea, M.D., and da Silva, L.F.M. 2013. Fracture toughness determination of adhesive and co-cured joints in natural fibre composites. *Composites: Part B* 50:120–26.

25. Bajpai, P.K., Singh, I., and Madaan, J. 2012. Joining of natural fiber reinforced composites using microwave energy: Experimental and finite element study. *Materials and Design* 35:596–602.

26. Ferreira, J.M., Silva, H., Costa, J.D., and Richardson, M. 2005. Stress analysis of lap joints involving natural fibre reinforced interface layers. *Composites: Part B* 36:1–7.

27. da Silva, L.F.M., Rodrigues, T., Figueiredo, M.A.V., de Moura, M., and Chousal, J.A.G. 2006. Effect of adhesive type and thickness on the lap shear strength. *The Journal of Adhesion* 82:1091–115.

28. Karachalios, E.F., Adams, R.D., and da Silva, L.F.M. 2013. Single lap joints loaded in tension with high strength steel adherends. *International Journal of Adhesion & Adhesives* 43:81–95.

29. Karachalios, E.F., Adams, R.D., and da Silva, L.F.M. 2013. Single lap joints loaded in tension with ductile steel adherends. *International Journal of Adhesion & Adhesives* 43:96–108.

30. da Silva, L.F.M., and Adams, R.D. 2008. Effect of temperature on the mechanical and bonding properties of a carbon reinforced bismaleimide. *Proceedings of the Institution of Mechanical Engineers, Part L, Journal of Materials: Design and Applications* 222:45–52.

31. Mouritz, A.P. 2007. Review of z-pinned composite laminates. *Composites: Part A* 38:2383–97.
32. Dransfield, K., Baillie, C., and Mai, Y.W. 1994. Improving the delamination resistance of CFRP by stitching—A review. *Composites Science and Technology* 50:305–17.
33. Tong, L., Mouritz, A.P., and Bannister, M.K. 2002. *3D Fibre Reinforced Polymer Composites*. Boston: Elsevier.
34. Chang, P., Mouritz, A.P., and Cox, B.N. 2006. Properties and failure mechanisms of z-pinned laminates in monotonic and cyclic tension. *Composites: Part A* 37:1501–13.
35. Vlot, A., and Gunnink, J.W. 2001. *Fibre Metal Laminates—An Introduction*. Dordrecht: Kluwer Academic Publishers.
36. Alderliesten, R. 2009. On the development of hybrid material concepts for aircraft structures. *Recent Patents on Engineering* 3:25–38.
37. Vermeeren, C.A.J.R. 2003. An historic overview of the development of fibre metal laminates. *Applied Composite Materials* 10:189–205.
38. Goland, M., and Reissner, E. 1944. The stresses in cemented joints. *Journal of Applied Mechanics* 66:A17–27.
39. Crocombe, A.D. 1989. Global yielding as a failure criteria for bonded joints. *International Journal of Adhesion & Adhesives* 9:145–53.
40. Gleich, D.M., van Tooren, M.J.L., and Beukers, A. 2001. Analysis and evaluation of bond line thickness effects on failure load in adhesively bonded structures. *Journal of Adhesion Science and Technology* 15:1091–101.
41. Grant, L.D.R., Adams, R.D., da Silva, L.F.M. 2009. Experimental and numerical analysis of single lap joints for the automotive industry. *International Journal of Adhesion & Adhesives* 29:405–13.
42. Neto, J.A.B.P., Campilho, R.D.S.G., and da Silva, L.F.M. 2012. Parametric study of adhesive joints with composites. *International Journal of Adhesion & Adhesives* 37:96–101.
43. da Silva, L.F.M., Adams, R.D., and Gibbs, M. 2004. Manufacture of adhesive joints and bulk specimens with high temperature adhesives. *International Journal of Adhesion & Adhesives* 24:69–83.
44. Adams, R.D., Atkins, R.W., Harris, J.A., and Kinloch, A.J. 1986. Stress analysis and failure properties of carbon-fibre-reinforced-plastic/steel double-lap joints. *The Journal of Adhesion* 20:29–53.
45. Adams, R.D., and Harris, J.A. 1987. The influence of local geometry on the strength of adhesive joints. *International Journal of Adhesion & Adhesives* 7:69–80.
46. Dorn, L., and Liu, W. 1993. The stress state and failure properties of adhesive-bonded plastic/metal joints. *International Journal of Adhesion & Adhesives* 13: 21–31.
47. Tsai, M.Y., and Morton, J. 1995. The effect of a spew fillet on adhesive stress distributions in laminated composite single-lap joints. *Composite Structures* 32:123–31.
48. Lang, T.P., and Mallick, P.K. 1998. Effect of spew geometry on stresses in single lap adhesive joints. *International Journal of Adhesion & Adhesives* 18:167–77.
49. da Silva, L.F.M., and Adams, R.D. 2007. Joint strength predictions for adhesive joints to be used over a wide temperature range. *International Journal of Adhesion & Adhesives* 27:362–79.
50. Cherry, B.W., and Harrison, N.L. 1970. The optimum profile for a lap joint. *The Journal of Adhesion* 2:125–28.
51. da Silva, L.F.M., and Adams, R.D. 2005. Measurement of the mechanical properties of structural adhesives in tension and shear over a wide range of temperatures. *Journal of Adhesion Science and Technology* 19:109–42.
52. Hildebrand, M. 1994. Non-linear analysis and optimization of adhesively bonded single lap joints between fibre-reinforced plastics and metals. *International Journal of Adhesion & Adhesives* 14:261–7.

53. Rispler, A.R., Tong, L., Steven, G.P., and Wisnom, M.R. 2000. Shape optimisation of adhesive fillets. *International Journal of Adhesion & Adhesives* 20:221–31.

54. Sancaktar, E., and Nirantar, P. 2003. Increasing strength of single lap joints of metal adherends by taper minimization. *Journal of Adhesion Science and Technology* 17:655–75.

55. Kaye, R., and Heller, M. 2005. Through-thickness shape optimisation of typical double lap-joints including effects of differential thermal contraction during curing. *International Journal of Adhesion & Adhesives* 25:227–38.

56. Sancaktar, E., and Kumar, S. 2000. Selective use of rubber toughening to optimize lap-joint strength. *Journal of Adhesion Science and Technology* 14:1265–96.

57. Pires, I., Quintino, L., Durodola, J.F., and Beevers, A. 2003. Performance of bi-adhesive bonded aluminium lap joints. *International Journal of Adhesion & Adhesives* 23:215–23.

58. Temiz, S. 2006. Application of bi-adhesive in double-strap joints subjected to bending moment. *Journal of Adhesion Science and Technology* 20:1547–60.

59. Bouiadjra, B.B., Fekirini, H., Belhouari, M., Boutabout, B., and Serier, B. 2007. Fracture energy for repaired cracks with bonded composite patch having two adhesive bands in aircraft structures. *Computational Materials Science* 40:20–6.

60. das Neves, P.J.C., da Silva, L.F.M., and Adams, R.D. 2009. Analysis of mixed adhesive bonded joints—Part I: Theoretical formulation. *Journal of Adhesion Science and Technology* 23:1–34.

61. das Neves, P.J.C., da Silva, L.F.M., and Adams, R.D. 2009. Analysis of mixed adhesive bonded joints—Part II: Parametric study. *Journal of Adhesion Science and Technology* 23:35–61.

62. da Silva, L.F.M., and Lopes, M.J.C.Q. 2009. Joint strength optimization by the mixed adhesive technique. *International Journal of Adhesion & Adhesives* 29:509–14.

63. Marques, E.A.S., and da Silva, L.F.M. 2008. Joint strength optimization of adhesively bonded patches. *The Journal of Adhesion* 84:917–36.

64. da Silva, L.F.M., and Adams, R.D. 2007. Adhesive joints at high and low temperatures using similar and dissimilar adherends and dual adhesives. *International Journal of Adhesion & Adhesives* 27:216–26.

65. da Silva, L.F.M., and Adams, R.D. 2007. Techniques to reduce the peel stresses in adhesive joints with composites. *International Journal of Adhesion & Adhesives* 27:227–35.

66. Gannesh, V.K., and Choo, T.S. 2002. Modulus graded composite adherends for single-lap bonded joints. *Journal of Composite Materials* 36:1757–67.

67. Apalak, M.K., and Gunes, R. 2007. Elastic flexural behaviour of an adhesively bonded single lap joint with functionally graded adherends. *Materials and Design* 28:1597–617.

68. Stapleton, S.E., Waas, A.M., and Arnold, S.M. 2011. Functionally Graded Adhesives for Composite Joints. NASA/TM—2011-217202.

69. Kumar, S., and Scanlan, J.P. 2011. Modeling of modulus graded axisymmetric adhesive joints. *Mechanics of Materials* (submitted for publication).

70. Carbas, R.J.C., da Silva, L.F.M., and Critchlow, G. 2014. Adhesively bonded functionally graded joints by induction heating. *International Journal of Adhesion & Adhesives* 48:110–18.

71. Carbas, R.J.C., da Silva, L.F.M., and Critchlow, G. 2014. Effect of post-cure on adhesively bonded functionally graded joints by induction heating. *Journal of Materials: Design and Applications* doi:10.1177/1464420714523579.

72. Carbas, R.J.C., da Silva, L.F.M., Marques, E.A.S., and Lopes, A.M. 2013. Effect of post-cure on the physical and mechanical properties of epoxy adhesives. *Journal of Adhesion Science and Technology* 27:2542–57.

73. Carbas, R.J.C., Marques, E.A.S., da Silva, L.F.M., and Lopes, A.M. 2014. Effect of cure temperature on the glass transition temperature and mechanical properties of epoxy adhesives. *The Journal of Adhesion* 90:104–19.

74. Carbas, R.J.C., da Silva, L.F.M., Madureira, M.L., and Critchlow, G. 2014. Modelling of functionally graded adhesive joints. *The Journal of Adhesion* 90:698-716.

75. Liu, J., and Sawa, T. 2001. Stress analysis and strength evaluation of single-lap adhesive joints combined with rivets under external bending moments. *Journal of Adhesion Science and Technology* 15:43–61.

76. Liu, J., Liu, J., and Sawa, T. 2004. Strength and failure of bulky adhesive joints with adhesively-bonded columns. *Journal of Adhesion Science and Technology* 18:1613–23.

77. Grassi, M., Cox, B., and Zhang, X. 2006. Simulation of pin-reinforced single-lap composite joints. *Composites Science and Technology* 66:1623–38.

78. Pirondi, A., and Moroni, F. 2009. Clinch-bonded and rivet-bonded hybrid joints: application of damage models for simulation of forming and failure. *Journal of Adhesion Science and Technology* 23:1547–74.

79. Chan, W.S., and Vedhagiri, S. 2001. Analysis of composite bonded/bolted joints used in repairing. *Journal of Composite Materials* 35:1045–61.

80. Lin, W.H., and Jen, M.H.R. 1999. Strength of bolted and bonded single-lapped composite joints in tension. *Journal of Composite Materials* 33:640–66.

81. Odi, R.A., and Friend, C.M. 2002. A comparative study of finite element models for the bonded repair of composite structures. *Journal of Reinforced Plastics and Composites* 21:311–32.

82. Tong, L., and Sun, X. 2003. Nonlinear stress analysis for bonded patch to curved thin-walled structures. *International Journal of Adhesion & Adhesives* 23:349–64.

83. Soutis, C., and Hu, F.Z. 1997. Design and performance of bonded patch repairs of composite structures. *Proceedings of the Institution of Mechanical Engineers, Part G* 211:263–71.

84. Vallée, T., Tannert, T., Murcia-Delso, J., and Quinn, D. J. 2010. Influence of stress-reduction methods on the strength of adhesively bonded joints composed of orthotropic brittle adherends. *International Journal of Adhesion & Adhesives* 30:583–94.

85. Campilho, R.D.S.G., de Moura, M.F.S.F., and Domingues, J.J.M.S. 2009. Numerical prediction on the tensile residual strength of repaired CFRP under different geometric changes. *International Journal of Adhesion & Adhesives* 29:195–205.

86. Hu, F.Z., and Soutis, C. 2000. Strength prediction of patch repaired CFRP laminates loaded in compression. *Composites Science and Technology* 60:1103–14.

87. Gunnion, A.J., and Herszberg, I. 2006. Parametric study of scarf joints in composite structures. *Composite Structures* 75:364–76.

88. Odi, R.A., and Friend, C.M. 2004. An improved 2D model for bonded composite joints. *International Journal of Adhesion & Adhesives* 24:389–405.

89. Campilho, R.D.S.G., de Moura, M.F.S.F., Pinto, A.M.G., Morais, J.J.L., and Domingues, J.J.M.S. 2009. Modelling the tensile fracture behaviour of CFRP scarf repairs. *Composites: Part B* 40:149–57.

90. Bahei-El-Din, Y.A., and Dvorak, G.J. 2001. New designs of adhesive joints for thick composite laminates. *Composites Science and Technology* 61:19–40.

91. Carbas, R.J.C., Viana, G.M.S.O., da Silva, L.F.M., and Critchlow, G. 2014. Functionally graded adhesive patch repairs of wood beams in civil applications. *Journal of Composites for Construction* doi: 10.1061/(ASCE)CC.1943-5614.0000500.

State-of-the-Art
Applications of Natural
Fiber Composites
in the Industry

Dipa Ray

CONTENTS

12.1 INTRODUCTION

A new surge of interest in utilizing "greener" materials such as natural fibers (NFs) in place of synthetic fibers has again put natural fiber composite (NFC) into the spotlight. The growing consumer awareness along with increased commercial demand to use greener materials has led to new innovations and new applications. The use of NF in a range of industrial applications has increased considerably in recent years. The NFs are used in combination with synthetic as well as biodegradable polymers, both thermoplastic and thermosetting type, for various applications. NFCs have found application in a number of diverse sectors ranging from automotive and construction industries, to packaging and leisure-based products. In 2012, the European Union (EU) reported production of 350,000 t of wood and NFC. The most important application sectors are construction (decking, siding, and fencing) and automotive interior parts. Between 10% and 15% of the total European composite market are covered by wood-plastic composites (WPCs) and NFC [1]. In recent times, many companies have shifted their focus to using NFs that weigh less, are durable and efficient, have high specific strength properties, cost less, are neutral to CO_2, are recyclable and biodegradable, can be separated easily, have low density, and contain desirable physical properties. These features give NF great advantages compared with traditional fibers, such as carbon or glass.

12.2 HISTORICAL APPLICATIONS OF NATURAL FIBER COMPOSITES

The NFs have fulfilled the requirements of fibers in human life in various ways. The first utilization of NFC made with clay in Egypt can be dated back to 3000 years ago. The ancient Egyptians realized the usefulness of NF long ago and they used straw-reinforced clay as their building material. But the advancement of industrialization gradually enhanced the use of fossil fuels such as coal, petroleum, and natural gases in all sectors. In addition to energy and heat production, these fossil fuels have been and still are used as the basic raw materials for industries.

It was only from the second half of the twentieth century that a new social awareness started happening to protect the environment and to reduce the depletion of petroleum resources. The main reasons that sharpened this awareness are (1) a critical increase in the release of "greenhouse" gases into the atmosphere due to burning of fossil fuels, (2) alarming environmental pollution due to a severe increase in the volume of wastes, and (3) depletion of petroleum resources of our planet. The search began for renewable natural resources to develop new industrial products that offered a two-way benefit: reduce consumption of petroleum resources and trim down the burden for the environment. Although it cannot be expected that renewable resources fully replace petroleum products, new emerging techniques explored by the scientific community all over the world can definitely contribute in a positive way to this problem. Thus, the reinvention of NF and combining them with polymers give rise to a new class of materials that can compete with and replace synthetic materials for some structural, but mostly for nonstructural and semistructural, applications in many sectors such as automotives, construction, packaging, biomedical, and so on. NFs such as flax, hemp, jute, sisal, bamboo, and so on have already been in use in many industrial sectors. In this chapter, we will discuss various state-of-the-art applications of NFC as industrial products.

12.3 STATE-OF-THE-ART APPLICATIONS OF NATURAL FIBER COMPOSITES

The NFC market is divided into two broad segments: wood fibers and nonwood fibers. Wood fiber is the most common fiber, used for building and construction industries, whereas nonwood fibers, such as flax, kenaf, hemp, and jute, are the materials of choice for automotive applications. Wood flour is obtained as sawmill waste and wood fibers are produced from sawmill waste by a wet thermomechanical process [2]. The industrial applications of NFC include window and door frames, furniture, railroad sleepers, automotive panels, fencing and decking items, packaging materials, and general products that do not require very high mechanical properties.

The role of the automotive industry in bringing NFC into real-life application is of primary importance. The largest areas in which NFC are being currently used are the automotive and construction industries. North America is big in building and construction applications, and Europe is the largest region for automotive applications. Europe has the highest total NFC consumption [3]; the Asian market for NFC is emerging due to the rapidly increasing demand in China and India. Due to increased environmental awareness, the use of NFC in other sectors such as structural, sports and leisure, marine, wind turbines, consumer goods, packaging, and medicine is increasing rapidly. Future markets are expected to be highly competitive, and companies with innovative ideas can succeed and gain market share. The relative costs of NF such as jute, coir, sisal, and hemp are significantly lower than that of glass, and they vary depending on various agronomic factors between regions [4]. However, flax fibers of good quality, which are suitable for high-end applications, are more expensive than lower grades of flax fibers that are derived from linseed oil production [4]. Conventional processing equipment used for glass fiber composites are used for NFC with equivalent processing conditions. This offers a practical advantage of replacing glass fibers fully or partially with NF. Production costs are reported to be lowered by 10%–30% [4] when NFs are used in place of glass fibers in addition to reduced health hazards.

General-purpose synthetic thermoplastics such as polyethylene (PE), polypropylene (PP), polyvinyl chloride (PVC), and so on are the most common candidate matrices used with NF. Among thermoset resins, unsaturated polyester, epoxy, vinylester, and phenolic resins are the most commonly used. The biodegradable polymer matrices are more desired to keep the green credential of such products, but they are still considerably higher priced than the traditional thermoplastics, although the costs have fallen in the past few years as industrial production has scaled up.

The use of NFC is predicted to be a growing market. According to Lucintel's report in 2010, *Natural Fiber Composites Market Trend and Forecast 2011–2016: Trend, Forecast, and Opportunity Analysis*, the global NFC market had reached US\$289.3 million in 2010, with a compound annual growth rate (CAGR) of 15% from 2005 onward. By 2016, the market is expected to reach US\$531.3 million with a CAGR of 11 % over the next 5 years. NFCs have shown steady growth in the past 6 years [3].

In the next sections, some state-of-the-art applications of NFC in different industrial sectors will be discussed while highlighting what has been showcased as the most recent developments in the field of NFC products in the market.

12.3.1 Structural Applications of Natural Fiber Composites

NFs are a useful reinforcement for the sustainable construction industry. The call for sustainable, safe, economical, and secure building material is a global problem and numerous challenges are present to produce environmentally friendly construction products that are structurally safe and durable. NFCs are used in a large variety of building materials such as fiber/cement composites, fencing, decking, siding, doors, windows, bridges, and so on.

Fiber-cement composites (fiber-reinforced cement composites) were originally developed by James Hardie (1980). Recently, fiber cement-based materials have found increasing applications in residential housing construction. Fiber-cement composite products are found in the exterior and interior of a building such as in siding, roofing, external cladding, internal lining, floors, walls, building boards, bricks, bracing, fencing, and so on. Fiber cement is also used for constructing dams, bridge decks, road building, and so on. One of the main ingredients of fiber-cement products is cellulose fibers from wood or nonwood sources that are added to reinforce the cement composite. Small amounts of additives are also utilized to aid the process or provide products with particular characteristics. To enhance the durability of such fiber-cement composites and to reduce the chemical interactions between the cements and the bio fibers, additives such as fly ash, slag, and silica fume are used.

Hempcrete is a commercially available product of the NF/cement composite that is becoming popular as a green building material [5]. Hempcrete is a combination of chopped hemp shiv (inner woody core of hemp plant) and a binder comprising natural hydraulic lime and cement. The hemp core or shiv has a high silica content that enables it to bind well with lime. The product is a lightweight cementitious insulating material weighing about a seventh or an eighth of the weight of concrete. Fully cured hempcrete blocks float in water (Figure 12.1).

It is not a load-bearing or structural element, but it is a firm and self-insulating material which is suitable for uses such as timber frame infill, insulation and, with the addition of aggregate, as floor slabs. Hemp is a renewable biomaterial, and lime is an abundant quarried material. Hempcrete regulates the temperature and humidity of a building, in some cases completely eliminating the need for heating and cooling systems, resulting in energy savings. Hempcrete is a carbon-negative material. The carbon trapped in the hemp offsets not only the carbon from the hemp production but also the residual carbon from the lime production after reabsorption of carbon as the lime cures [6–7].

Biocomposite-based particleboards are extensively used as building materials in many parts of the world. The bagasse fiber obtained from sugar cane is utilized in

FIGURE 12.1 Hempcrete building materials. (From Hemp materials, http://www.hempbuilding .com/, accessed April 1, 2014. With permission [5].)

the United States to make bagasse composition panels. Bagasse fibers are used for particleboards, fiber boards, and composition panel production in North America. In addition to bagasse, other NFs such as wheat and ryegrass straw fibers, hulls, sunflower, and soybean stalks are also currently used in composites for various applications [8].

Cereal straw is the second most common agro-based fiber that is used in panel production. The high silica content in cereal straw makes it naturally flame retardant, and the low density of straw panels makes them resilient and more resistant to earthquakes. Straw is also used in particleboards. The presence of rice husks in building products helps increase the acoustic and thermal properties.

In Asia, bamboo fibers are used to a large extent for construction. In addition, researchers have developed a variety of building materials utilizing industrial and agricultural wastes that combine well with cement as binders. These combinations are utilized to make composition boards, flooring tiles, roof sheathing, and so on. Rice straw is used to produce fiber boards. In comparison to wood fibers, rice straw is of lower quality because of its high content of nonfibrous materials. European researchers have also examined a multitude of agricultural waste fibers for fiberboards such as beech fibers, hemp, tobacco, vines, cotton, raspberry, maize, or sunflower stalks. Among these waste fibers, promising results were obtained with hemp and tobacco fibers [8]. In exterior construction, wood fiber polymer composites are made in long-profile cross-section dimensions. These bioproducts are utilized as dock surface boards, decks, picnic tables, landscape timbers, and industrial flooring. Many industries are now manufacturing biocomposite decking, cladding, fencing, and furniture products that are easily available in the market [9–12]. Biocomposite products are becoming popular in industries such as sports and leisure, transportation, marine, electronics, acoustics, and interior design. Bcomp [13] is offering different products such as flax fabrics, core materials, and composite reinforcements that are finding applications in different industries (Figure 12.2).

Composites Evolution [14] manufactures lightweight, innovative, and sustainable materials, including fiber reinforcements, resins, and intermediates based on natural, bio-derived, recycled, and recyclable feedstock. Along with jute and flax fiber reinforcements, polyfurfuryl alcohol (PFA)-based (PFA bioresin) ecopregs have recently been introduced in the market for fire-resistant composite applications (Figure 12.3). They have similar performance to phenolic prepregs but with a significantly lower environmental impact and improved health and safety. PFA is a thermosetting bioresin derived from crop waste and is similar in nature to a phenolic resin. The prepregs can be consolidated by vacuum bagging, autoclave, or press moulding and are designed for a range of applications in mass transport, aerospace, furniture, and construction. These are used as lightweight stiff door modules, sustainable biocomposite cabinets, turbine blades, acoustic guitars, and so on.

Amber Composites [15–16] has introduced a prepreg based on a Biotex flax fabric made by Composites Evolution [14]. Biotex flax fabrics provide the high performance and easy processing that are usually associated with glass fiber composites but with lower weight and environmental impact. These prepregs can be used for semistructural and decorative applications in automotive, marine, sports, and consumer goods. This fabric is only available with the epoxy resin system.

FIGURE 12.2 Different applications of biocomposite products. (From Bcomp, http://www
.bcomp.ch/, accessed August 13, 2014. With permission [13].)

GreenCore NCell™ Natural Fiber-Reinforced Thermoplastics are a family of
high-performance, biocomposite materials available in the market for injection
molding and extrusion [17]. High-strength parts made from NCell™ are green, lightweight, manufactured with less energy, and have unique surface aesthetics. NCell™
consists of a PP or PE matrix reinforced with approximately 40% natural cellulosic
microfibers. NCell™ compounds have high-performance mechanical properties and
are able to replace glass-reinforced thermoplastics (Figure 12.4).

A new range of biocomposite products have been introduced in the market by
NPSP that combine NF with mineral oil-based as well as natural resource-based
polyester resins (Nabasco N-3010 and Nabasco N-5010, respectively) and are suitable
for use indoors as well as outdoors (Figure 12.5) [18].

NFC products are now seen in many new applications, which indicates promising
growth in this sector. LINEO, France is providing new solutions for the use of natural flax fibers in composite applications. They are supplying FlaxPly and FlaxPreg
products for manufacturing of sporting goods [19]. There are some more companies
supplying hemp- and flax-based semicomposite products such as dry NF preforms
and wet prepregs [20–24]. Various long and short fiber-reinforced products are available that are suitable for different processing techniques such as extrusion, injection
molding, and so on.

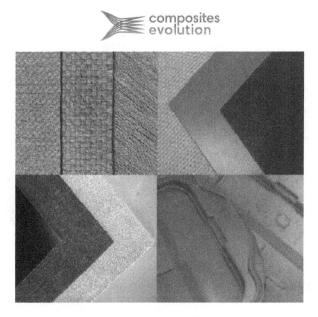

FIGURE 12.3   Biotex fabric, PFA ecopreg, and biocomposite products. (From Composites Evolution, http://www.compositesevolution.com/, accessed August 12, 2014. With permission [14].)

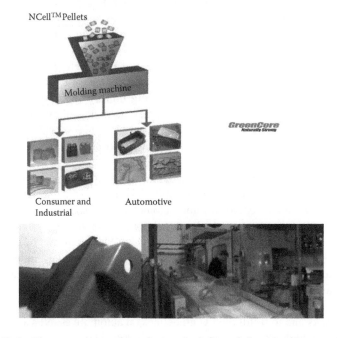

FIGURE 12.4   Biocomposite products from polyolefins reinforced with natural cellulosic microfibers. (From GreenCore: Naturally Strong, http://greencorenfc.com/products.htm, accessed July 8, 2014. With permission [17].)

FIGURE 12.5 Sustainable biocomposite products. (From NPSP: Solution in composites, http://www.npsp.nl/, accessed August 10, 2014. With permission [18].)

12.3.2 APPLICATIONS IN THE AUTOMOTIVE INDUSTRY

NFCs with thermoplastic and thermoset matrices have been embraced by car manufacturers for dashboards, door panels, seatbacks, headliners, package trays, and some other interior applications in automotives. NFs such as kenaf, hemp, flax, jute, and sisal have been used, which offer advantages such as weight reduction, cost reduction, lower carbon footprint, and less dependence on petroleum products. However, several important technical challenges must be resolved before the car manufacturers gain the confidence to accept NFC on a large scale for exterior automotive applications. The major challenges are achieving uniform fiber properties, strong adhesion between the fiber and the matrix, flame retardancy, ultraviolet (UV) resistance, and, most importantly, moisture repellence [25]. The variation in quality is also one of the limitations of NF due to the natural processes affecting growth with harvests varying from year to year.

The EU and some Asian countries have introduced stringent guidelines on automotives. EU legislation implemented in 2006 has accelerated the introduction of natural fiber-reinforced composites for automotive applications. The legislation stated that by 2006, 80% of a vehicle must be reused or recycled and by 2015, it must be 85% [26]. Japan required 88% of a vehicle to be recovered by 2005, rising to 95% by 2015. As a result, the search for new, recyclable, and lighter materials is on for future automotives. Glass fiber-reinforced polymers have already proved themselves as suitable candidates for automotives. But they have some drawbacks such as high fiber density, poor recyclability, and hazards in handling. To overcome these drawbacks and to fulfill the EU guidelines on automotives, NFs are becoming more popular candidates that can replace glass fibers to a large extent. NFs are nearly 40% lighter in weight than glass fibers and are renewable, eco-friendly, hazard-free materials with low carbon footprint and high recyclability [27].

The first motorcar using plant-based fibers in the body was made by Henry Ford way back in 1941 and, with an early picture, the hammer-wielding Mr. Ford showed the world how strong plant fibers were in reinforcing man-made materials [28]. The frame of the car, made of tubular steel, had 14 plastic panels attached to it. The car weighed 2000 lb., which was 1000 lb. lighter than a steel car. The exact ingredients of the plastic panels are unknown, because no record of the formula exists these days. One article claims that they were made from a chemical formula, which among many other ingredients included soybeans, wheat, hemp, flax, and ramie; however, Lowell E, the man who was instrumental in creating the car, claimed that phenolic resin with formaldehyde was used for impregnating soybean fibers [29].

Major car manufacturers such as Volkswagen, BMW, Mercedes, Ford, Toyota, Audi, DaimlerChrysler, and Opel currently use NFC in applications, as shown in Table 12.1 [30]. BMW has been using NF in the 3, 5, and 7 series models of their cars since the early 1990s. In 2001, BMW used 4000 t of NF in the 3 series alone. The combination here is 80% flax with 20% sisal blend for enhanced strength and impact resistance. The major applications include the interior door linings and paneling.

TABLE 12.1
Application of Natural Fiber Composites by Vehicle Manufacturers

Manufacturer	Model	Application
BMW	3, 5, and 7 series	Seatback, headliner panel, boot lining, door panels
Volkswagen	Passat Variant, Golf A4, Bora	Seatback, door panel, boot-lid finish panel, boot-liner
Mercedes-Benz	C, S, E, and A classes	Door panels (flax/sisal/wood fibers with epoxy resin/unsaturated polyester resin matrix), trunk panel (cotton with PP/PET fibers), seat surface/backrest (coconut fiber/natural rubber)
Mercedes-Benz	Trucks	Internal engine cover, engine insulation, bumper, wheel box, roof cover
DaimlerChrysler	A, C, E, S class, EvoBus (exterior)	Door panel, windshield/dashboard, pillar cover panel
Audi	A2, A3, A4, A4 Avant, A6, A8, Roadster, Coupe	Hat rack, boot lining, spare tire lining, side and back door panel, seatback
Toyota	Raum, Brevis, Harrier, Celsior	Floor mats, spare tyre cover, door panels
Ford	Mondeo CD162, Focus	Door panels, boot-liner

Source: Suddell, B.C., Industrial fibers: Recent and current developments, Proceedings of the Symposium on Natural Fibers, 2009, ftp://ftp.fao.org/docrep/fao/011/i0709e/i0709e10.pdf, accessed January 10, 2014. With permission [4].

Wood fibers are used to enclose the rear side of seat backrests, and cotton fibers are used as a soundproofing material.

In May 2003, Toyota used eco plastics for the first time in the Raum and developed new plastic materials for interior parts, such as scuff plates, roof headliners, and seat cushions, in 2008. Toyota has expanded the use of eco plastics to 60% of the total area in interior parts for the Toyota Sai electric hybrid car launched in October 2009. Furthermore, in extension of the technology developed in recycling bumpers for repair services, a new recycled material based on end-of-life vehicles was created. Toyota applied it first to deflectors and so on after developing a process to assure the quality by eliminating foreign particles, such as dirt, adhering to bumpers, and fine-tuning the compound [31]. Several Toyota vehicles, including their hybrid models, such as the Toyota Prius and Lexus CT 200h, contain plant-based eco plastics. Other Lexus models such as the GS and CT use bamboo for components such as the steering wheels and speakers. Toyota plans to replace 20% of the conventional plastics in its vehicles with bioplastics by 2015. Another percentage will be based on synthetic polymers but reinforced with materials from biomass such as wood fibers. A variety of wood fibers are being tested. Wood fiber is stronger, greener, recyclable, and cheaper than synthetic fiber. Cars can be delivered at a lower cost in comparison to other cars made with conventional materials. Being lighter than traditional petroleum-based materials, greenhouse gas emissions will be cut and the Corporate Average Fuel Economy (CAFE) standards will be met. For each 100 kg saved, fuel consumption decreases by 0.5 L [32].

Ford has been a pioneer in bio-based car parts since Henry Ford used soybean and hemp in the body of his 1941 vehicle, known as both the Soybean car and the Hemp body car. These days, Ford uses parts made of soy, rice, and wood in many of its vehicles. Soy-based foam is used in all seatbacks and cushions in every Ford vehicle in North America and in 75% of all headrests. One of the other innovations is rice hulls used in a bracket in a plastic part in the 2014 F-150. Ford began using cellulose fiber-based-reinforced PP in the armrest reinforcement of the 2014 Lincoln MKX. The wood fiber is 7% lighter than petroleum-based fiber, with a weight savings of about half a kilogram. Ford plans to use the bio fiber for the entire center console, which would mean an additional weight savings of 3.6 kg. It also plans to replace all of the plastic and polymeric materials—about 10% or 136 kg—in Ford vehicles with bio fibers by 2015 [32].

In spite of so many promising applications, there are still some roadblocks. The main challenges include moisture absorption, odor, fogging, aging due to sun exposure, and variability in performance of raw materials. But research is going on to overcome those challenges and to provide solutions that will be acceptable to the industry [32]. A very recent report mentioned that the Volkswagen Golf, one of the most popular cars on the road, is newly introducing seating, door panels, and a front-end module made of a flax-reinforced PP that will feature on cars sold from 2015 onward. A new injection molding press has been installed for production of 50% PP and 50% flax composite parts, which will enable reduced cycle times essential for automotive production of high volume [28]. The consumption of NFC in the European automotive industry is shown in Figure 12.6 [1]. Figure 12.6 shows the total volume of 80,000 t of wood and NF used in 150,000 t of composites in

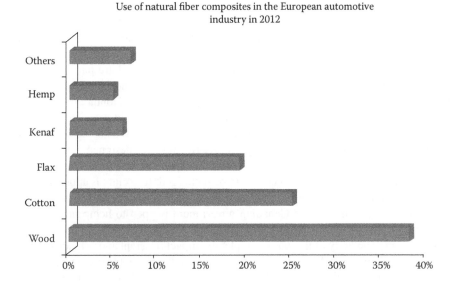

Use of natural fiber composites in the European automotive industry in 2012

FIGURE 12.6 Use of natural fibers in composites in the European automotive industry in 2012, including cotton and wood (total volume 80,000 t); others include jute, coir, sisal, and abaca fibers [1].

passenger cars and lorries produced in Europe in 2012. Compression molding of wood and NFC is an established and proven technique for manufacturing lightweight and high-class interior parts in mid-range and luxury cars that offer advantages such as crash resistance and price. However, there are well-known disadvantages as well, such as limited shapes and designs, scraps, and cost disadvantages with regard to part integration [1]. Since 2009, new improved compression molding techniques have shown impressive properties in weight reduction, which has increased interest in new car models. Today, with the new technologies, an area weight of down to 1500 g/m^2 (with thermoplastics) and even down to 1000 g/m^2 (with thermosets) is possible—which are real advantages compared with glass fiber composites. In the EU, 15.7 million passenger cars were produced in 2011 and 2 million other motor vehicles such as trucks, transporters, motor bikes, and so on were manufactured. Considering that 30,000 t of NF and another 30,000 t of wood fibers were used in 15.7 million passenger cars, every passenger car in Europe contains 1.9 kg of NF and 1.9 kg wood fibers on average; so, in total, that makes almost 4 kg of these fibers per car. From a technical point of view, much higher volumes are possible. Vehicles with considerably larger amounts such as 20 kg of natural and wood fibers have been successfully produced in series for years. The market development also depends on the political framework. Any incentive for the use of natural and wood fibers in the European automotive industry could help extend the existing amount of 30,000 t/ year for natural and wood fiber each. The vision could be an increase of up to five times, which means up to 150,000 t/year for each fiber type and the technologies are ready to use [33].

DaimlerChrysler's sustainability efforts have undertaken unique initiatives involving the use of bio-based materials in the Philippines, South America, and South Africa. As a part of their strategy to create a global sustainability network, DaimlerChrysler has actively taken part in identifying bio-based materials and also has used renewable energies to replace conventional fuels, and are pursuing a bio-based automotive supply chain that includes a network from the farmer to the automotive distributor [34]. Global automotive suppliers such as Honda embarked on using wood-fiber parts in the floor area of the Pilot sport utility vehicle (SUV), a decision that was driven by engineering considerations as well as by corporate philosophy. The world's most eco-friendly car, the Kestrel, was designed in Canada by Calgary-based Motive Industries [35]. It has a top speed of 90 km/h and a range of approximately 100 miles before needing to be recharged. Its weight is approximately 2500 lb., and it has a very affordable price given the fact that hemp is very easy to grow and is inexpensive. The Canadian government is open to hemp farming and is actively supporting the industrial hemp industry and its potential benefit to the environment. The body of the car is completely impact resistant and is made entirely out of hemp. The Kestrel's hemp composite body shell passed its crash test in strong form, and unlike steel, the panels bounce back into shape after impact. Hemp also has the same mechanical properties as glass. It is even lighter than glass, and these properties help boost fuel efficiency [36].

Another biocomposite-based concept car is designed and manufactured in partnership with Helsinki Metropolia University of Applied Sciences, the Finnish Funding Agency for Technology and Innovation, and several other partner companies. This Biofore Concept Car showcases the use of innovative biomaterials in the automotive industry (Figure 12.7). The majority of parts traditionally made from plastics are replaced with high-quality, safe, and durable biomaterials, UPM Formi and UPM Grada, which can significantly improve the overall environmental performance of car manufacturing [37].

Driving real
sustainable change
The Biofore Concept Car

FIGURE 12.7   Biofore Concept Car. (From Biofore, http://www.upm.com/upmcc-en/Pages/default.aspx, accessed August 5, 2014. With permission [37].)

12.3.3 APPLICATIONS IN MARINE ENVIRONMENTS

Boat industries generate large amounts of waste of glass fiber-reinforced polyester composites that are very difficult to break down. The development of NFC is a suitable alternative combined with low environmental impact compared with the glass/polyester composites. However, there are some negative aspects of using NF reinforcement in the marine environment, as they are sensitive to water. When they are exposed to severe environmental conditions (immersion in sea water, for example), biocomposites undergo degradation related to high water uptake, causing hydrolysis, swelling, matrix cracking, and fiber–matrix debonding [38]. The retention of biocomposite properties in such environments is thus a major issue if they are to be used in external applications. The first ever racing boat prototype was built with Huntsman Advanced Materials, which incorporated up to 50% natural flax fiber in the composite structure [39]. Lineo [19] used their technology to coat flax fibers with epoxy resins in such a way that absorption of water from the flax was prevented and strong bonds between the flax and the epoxy resin were created, guaranteeing the quality of the laminate.

However, in spite of these challenges, some biocomposites are appearing as products in marine industries. E-Tech Boards (Hawthorne) has used BioMid cellulose-fiber woven fabric in place of glass to fabricate several surfboards. E-Tech boards use up to 70% less fiberglass than industry standard surfboards. All boards are made using USDA-certified bio epoxy from Entropy Resins. Bio-plastic leash plugs, handles, recyclable foam, sustainably harvested wooden stringers, or bamboo stringers and bamboo veneers are some other biomaterials used in these biocomposite surf boards [40]. Ashland formulated a laminating grade of renewable material-based green resin, "Envirez," for Canadian boat builder Campion Marine (Kelowna, British Columbia) that is now used in the manufacture of Campion's entire boat line [41]. Bio fibers such as flax and a bio resin called EcoComp® UV-L have recently been used to make eco-friendly marine products such as kayaks, canoes, and surfboards (Figure 12.8) [42]. Further research is needed to develop water-resistant biocomposite products that can perform well in marine environments.

12.3.4 PACKAGING

Synthetic packaging materials are nonrenewable, nondegradable in nature, and are depleting petroleum resources. They are causing environmental problems with regard to their disposal, and disturbing the environmental ecosystem, water supplies, and sewage systems. Hence, the search is continuing for alternative packaging materials that can effectively replace their nonrenewable and nonbiodegradable counterparts in terms of cost and properties. Biodegradable polymers such as poly(lactic acid) (PLA), polycaprolactone (PCL), polyhydroxyalkanoates, aliphatic biodegradable polyesters, and copolyesters are commercially available [43–44] and their use as packaging materials could reduce the environmental problems. However, the costs of these biodegradable polymers are higher than those of the synthetic ones. An economically viable approach involves producing low-cost degradable packaging materials from renewable resources by using natural fillers as load-bearing constituents

FIGURE 12.8 Eco-friendly marine products. (From Sustainable Composites Ltd, http://www .suscomp.com/, accessed July 30, 2014. With permission [42].)

in the composite materials. A very useful combination is nanoclay and NF as hybrid reinforcement in polymers, as nanoclay imparts permeability and NFs are susceptible to microorganisms that impart biodegradability.

NF-based packaging materials have many benefits such as specific stiffness and recyclability compared with synthetic packaging materials [43]. The use of renewable resource-based biodegradable polymers in combination with NF and nanofillers such as nanoclay will help in the production of high-performance packaging materials, which may contribute toward demanding food and pharmaceutical packaging applications. In a recent study, it has been concluded that bio-based content and biodegradability are essential elements for single use, short-life disposable packaging, and consumer plastics [45]. With advancement in technology, reinforcing polymers with a hybrid combination of NF/nanoclay fillers might prove to be a suitable technique that can give rise to new materials that are suitable for food packaging, contributing to sustainability and reduction in environmental hazards associated with disposal of synthetic polymer-based packaging materials.

Various companies have brought disposable, eco-friendly cutlery products into the market. Biosphere® packaging products (Figure 12.9) are commercially available in the market and are ultra earth friendly starch-based packaging systems [46]. They are made from renewable resources such as tapioca and potato starch and a small amount of grass fibers. The plant starch and grass fibers are baked into various rigid shapes. They are compostable, biodegradable, and sustainable by using mostly yearly harvested materials. These products may be disposed of in the paper recycling stream. They also degrade in marine environments with no toxicity. Earthcycle makes compostable packaging products from palm fiber, a renewable resource, which gets discarded when the fruit from the palm husk is harvested throughout the year for its oil [47]. Once finished, these packaging materials decompose in any backyard or balcony compost within 90 days, and they make a healthy contribution to the soil.

London Bio Packaging is producing sustainable, high-quality, and affordable packaging materials [48]. It has two brands of packaging products: Sustain™ is renewable and compostable and made from plants and Revive™ is made from recycled materials. They use Ingeo PLA and Mater-Bi for Sustain. They use plant starch, bagasse fiber, and palm leaves to make sustainable, biodegradable products. Another commercially available packaging product made of NFC is KUPILKA, which is made from 50% recycled pine wood fiber and 50% PP [49]. The materials are used for cutlery, have perfect resonance, and are used in musical instruments (Figure 12.10). This was developed by the European Forest Institution. A unique

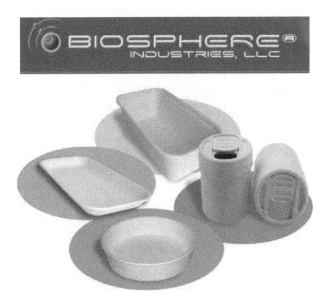

FIGURE 12.9   Eco-friendly products from Biosphere Industries. (From Biosphere Industries Ltd, http://www.biosphereindustries.com/, accessed July 20, 2014. With permission [46].)

KUPILKA

FIGURE 12.10   Eco-friendly products from KUPILKA. (From KUPILKA, http://www .kupilka.fi/en/material, accessed July 20, 2014. With permission [49].)

attribute of KUPILKA is that at the end of its lifespan, the product can be safely disposed of by burning.

Bamboo is now used as a sustainable packaging solution. Dell uses it to cushion many of their lightweight products, including all laptops produced in China [50]. After bamboo is harvested, it is mechanically pulped when 70% of the water is reclaimed and used in the process and the other 30% is lost due to vaporization. The pulp is dried by the sun, reducing electricity use. Dell bamboo packaging has been certified as compostable as per ASTM International standards. This certification ensures that the bamboo will compost satisfactorily and biodegrade at a rate comparable to known compostable materials. It also ensures that the compost resulting from the degradation process of packaging is of good quality and can sustain plant growth.

12.3.5 MEDICAL USES

Natural fibers, when extracted from their sources (bast, leaf, etc.) by various chemical and enzymatic routes, get contaminated and they no longer remain suitable for medical applications that demand a high level of purity. Hence, no NFCs are commercially available in the market for medical applications. However, advanced fiber extraction techniques in the future might open some new avenues in this area. There has been considerable research in the field of medical applications of pure cellulose-reinforced composites. Microbial cellulose (a pure form of cellulose)-based composites have recently been used for the development of tissue-engineered constructs due to their unique nanostructure that closely resembles the structure of native extracellular matrices [51–53]. Different approaches are found in the literature regarding the use of composites with microbial cellulose for medical applications (Table 12.2) [43,54–57]. Hydroxyapatite-based biocomposites have not been discussed here, as

TABLE 12.2
Application of Natural Fiber Composites for Biomedical Applications

Biocomposite	Reinforcement Type	Applications	Reference
Poly(3-hydroxubutyrate-co-4-hydroxubutyrate)/cellulose	Microbial cellulose fibers	Tissue engineering	[54]
Nanocrystalline cellulose (NCC)/polyvinyl alcohol (PVA)	Commercially available rod-shaped nanocrystalline cellulose (NCC)	Tissue regeneration	[55]
Cellulose/PVA	Microbial	Tissue regeneration	[43]
All-cellulose composite	Commercial cellulose nanowhiskers	Small grafts	[56]
All-cellulose composite	Norway spruce cellulose fibers	Ligament or tendon substitute	[57]

Source: Tserki, V., Matzinos, P., Zafeiropoulos, N.E., and Panayiotou, C., Development of biodegradable composites with treated and compatibilized lignocellulosic fibers, *J. Applied Polym. Sci.*, 100, 4703–10, 2006. With permission [44].

they are beyond the scope of this chapter. Although several research activities are ongoing, commercially available NF or cellulose fiber-reinforced composite products for medical applications are yet to become commercially available.

12.4   CHALLENGES OF RESEARCH AND DEVELOPMENT

There are considerable challenges in applying NFC for extensive industrial applications, and success in the future depends on both advancing technology and finding a design approach that best suits the particular performance characteristics of the biocomposite product. A range of nonstructural and semistructural applications are already evolving and replacing wood, plastics, and similar materials with moderate performance improvement and competitive cost.

The major technical barriers to the widespread commercial use of NFC include the current physical limitations of the raw materials. The nonuniformity in fiber properties arising from one harvest to another at different times of the year and at different geographic locations bring nonuniformity in the quality and performance of the finished composite products, which adversely affect their performance, and the products lose their reliability in the market. The lack of suitable reinforcement textiles is also currently hampering the larger-scale use of NF as a reinforcement in thermosetting fiber-reinforced composites. Currently, NF-based reinforcements are mostly limited to nonwoven mats, which are not very suitable for high-performance composite applications. However, in recent years, high-quality woven NF reinforcements have arrived in the market [13–14], which may find application in high-end composites. Along with NFs such as flax, hemp, jute, and so on, a second generation of cellulose fibers has recently been introduced in the market called Biomid [58]. Unlike the discontinuous NF stripped from plant stalks, which vary in length from inches to a few feet, the BioMid fibers are spun and, therefore, can be produced continuously. They need not be twisted into bundles, but instead they can be used in parallel, enabling much higher properties and producing thinner fabrics than twisted yarns. This helps reduce resin puddling at the warp–weft interstices, resulting in a more even resin-to-fiber distribution and thus avoiding resin pockets [59]. However, there is a growing need for more research and development investment to help establish a stable supply infrastructure, including processing of technical textiles.

Another challenge for NFC is the restriction in processing temperature. As NFs degrade above 200°C, only thermoplastic polymers that are suitable for processing below 200°C can be combined with them. That is the reason that thermoplastic biocomposites are restricted to PE, PP, PLA, and a few more biocopolyesters, which have melting points below 200°C.

In addition, only short-length NFs are melt-mixed or extruded with thermoplastic polymers, which are then injection molded into products of moderate to low strength. Long fiber-reinforced thermoplastic composites can offer higher mechanical strength but are difficult to achieve due to poor resin impregnation caused by high viscosity of the thermoplastic polymers. *In situ* polymerization can be an effective route to manufacture long fiber or fabric-reinforced thermoplastic composites. Liquid thermoplastic monomers can effectively impregnate the fabrics, as they have very low viscosity and a thermoplastic, recyclable NFC can be manufactured after *in situ* polymerization of the monomer [60–61]. This *in situ* route is also likely to reduce moisture absorption, which is a serious concern for NFC [62–63].

Another big obstacle is the compatibility between the hydrophobic polymers and hydrophilic NF. It is absolutely necessary to surface treat NF with compatibilizing agents that will bond well with the hydrophobic polymers. For hydrophilic thermo-plastic polymers or for thermosetting resins, coupling agents are required to ensure strong fiber–matrix bonding at the interface to achieve high mechanical strength. It is well known that different types of silane coupling agents are used for all commercial glass fiber reinforcements. For NF, different chemical treatments have been attempted by researchers in the laboratory, but none have finally arrived in the market as a commercial product. There is an urgent need to develop commercially viable techniques and materials, costwise as well as processwise, to modify the surface properties of NF to enhance compatibility with the polymer matrix, which can lead to the development of high-performance NFC.

If NFCs are to be sold on their environmental credentials, it is important that they are substantiated by life cycle analysis (LCA). The results of LCA can be revealing and it is by no means certain that an NF alternative product will be greener. The greatest impact in environmental terms often arises from the polymer matrix and thus significant interest is being directed toward the development of bio-based ther-mosetting and thermoplastic resins. There is also a need for significant investment in production capability and new processing technologies to overcome the technical hurdles and make commercial biocomposites a reality. Suitable test standards need to be developed that will support the growth of such materials and facilitate their entry into the commercial market.

12.5 FUTURE TRENDS

The recent interest in biocomposites has been driven by government policies seek-ing lighter weight materials that are not reliant on petroleum resources. Currently, the main markets for biocomposites are in the construction and automotive sectors. With further developments and improvements in performance, newer opportuni-ties and applications will likely arise. Along with improvements in the mechani-cal performance of existing biocomposites through the introduction of new fiber types, advanced processing technologies may result in an expansion of their use into more diverse, and technically demanding, high-performance application areas. Future growth is pointing toward eco-friendly, sustainable materials from renew-able resources for various applications. Natural insulation materials to improve energy efficiency of buildings; plant-derived products acting as carbon sinks to lock up carbon dioxide; recyclable or compostable materials to reduce the landfill cri-sis; and lightweight green automotive components increasing fuel efficiency of cars and reducing carbon emissions are some of the materials of immense interest in the future. In recent times, to meet with such emerging market demands, various new NF-based reinforcements, bioresins, and bio-based prepregs have been introduced in the market.

Biotechnology may be helpful in the future in genetically modifying the fiber types or crop types that can be used as precursors for preparing bioresins.

Currently, the bioresins available in the market are much more expensive than their synthetic counterparts. To increase their acceptance in industries, their price should be comparable to that of the synthetic ones. Research is being conducted at different research laboratories to develop new pathways to synthesize inexpensive biodegradable resins from nonfood crop products with better mechanical properties. These inexpensive bioresins and improved-quality NF with proper coupling agents hold great promise for replacing many of the synthetic advanced composites currently in use. There are also opportunities for hybrid composites using natural mineral fibers such as basalt fibers in combination with plant fibers to increase overall performance, although at the cost of weight. There are also good prospects for using recycled carbon fibers, reclaimed fibers from products such as MDF (medium density fiberboard), or other waste streams from the pulp and paper industry to manufacture a range of sustainable, environmentally effective products. However, the success of these materials will depend on the acceptance of such products in our daily lives. Promoting the use of these materials through more widespread training and education is required if they are to be commercially successful. An increase in consumer awareness on the impact that materials have on the environment will play a key role in the adoption of these novel materials. A greater understanding of the NFC materials by researchers will also contribute to a greater interest and uptake in these NF-based composite systems by the industry that will continue to result in more products in the future [64].

12.6 CONCLUSIONS

Renewable bioresources offer a limitless supply of potentially sustainable raw materials for the production of biocomposites. Although in its infancy, there is a growing market for biocomposite products and with further technology development, a host of new applications can be anticipated. Along with NFs such as flax, hemp, wood fibers, and so on, a second generation of cellulose fibers have arrived in the market; these are spun in continuous filaments and can offer much improved mechanical performance. In parallel, in recent years, significant developments are ongoing in the field of biopolymers. These combined initiatives ensure that biocomposites are likely to see a period of sustained growth; however, attention and investment are needed in research and development if a sustainable biocomposites industry is to be established. Inexpensive, environmentally friendly, and easy production of NFC are attractive benefits for design development and applications. An effective collaboration among scientists, engineers, and designers is important to achieve quality materials and produce a good design, which has benefits not only for the companies but also for consumers and society.

ACKNOWLEDGMENTS

I would like to acknowledge the Irish Center for Composites Research (IComp) funded by Enterprise Ireland and IDA, Ireland; the Department of Mechanical, Aeronautical, and Biomedical Engineering (MABE Dept.); and the Materials and Surface Science Institute (MSSi), University of Limerick, for their support in writing this chapter.

REFERENCES

1. Biobased News. 2013. Biocomposites: 350,000 t production of wood and natural fiber composites in the European Union in 2012. http://bio-based.eu/news/biocomposites/ (accessed January 10, 2014)
2. La Mantia, F.P., and Morreale, M. 2011. Green composites: A brief review. *Composites: Part A* 42:579–88.
3. Technologies and Products of Natural Fiber Composites, CIP-EIP-Eco-Innovation. 2008. Pilot and market replication projects - ID: ECO/10/277331, CELLUWOOD. http://www.celluwood.com/LinkClick.aspx?fileticket=F6U7DtrDqow%3D&tabid=465&mid=2217 (accessed January 8, 2014).
4. Suddell, B.C. 2009. Industrial fibers: Recent and current developments. Proceedings of the Symposium on Natural Fibers, 2009. ftp://ftp.fao.org/docrep/fao/011/i0709e/i0709e10.pdf (accessed January 10, 2014).
5. Hemp materials. http://hempmaterials.com/hempcrete/ (accessed April 1, 2014).
6. Hempcrete: The best concrete is made from hemp. http://humansarefree.com/2013/10/hempcrete-best-concrete-is-made-from.html (accessed April 1, 2014).
7. The Limecrete Company Ltd.: The sustainable construction. http://limecrete.co.uk/ (accessed April 1, 2014).
8. Biocomposites for the construction materials and structures. http://www.academia.edu/1266940/BIOCOMPOSITES_FOR_THE_CONSTRUCTION_MATERIALS_AND_STRUCTURES (accessed April 7, 2014).
9. Earth-wood: Inspired by nature, engineered by man. http://www.earthnwood.com/ (accessed August 4, 2014).
10. Composite decking. http://www.betterdeck.ie/compositedecking.php (accessed August 10, 2014).
11. Specialist timber supplier nationwide. http://www.timberireland.ie/composite.html (accessed August 1, 2014).
12. Composites.http://www.upm.com/EN/PRODUCTS/composites/upm-profi/Pages/default.aspx (accessed August 10, 2014).
13. Natural fiber composites. http://www.bcomp.ch/ (accessed August 13, 2014).
14. Composites evolution. http://www.compositesevolution.com/ (accessed August 12, 2014).
15. TenCate: Amber composites. http://www.ambercomposites.com/biotex-flax-fiber (accessed August 9, 2014).
16. Amber Composites offers flax fiber prepreg. http://www.plasticstoday.com/articles/amber-composites-offers-new-natural-flax-fiber-prepreg (accessed August 9, 2014).
17. GreenCore: Naturally strong. http://greencorenfc.com/products.htm (accessed July 8, 2014).
18. NPSP: Solution in composites. http://www.npsp.nl/ (accessed August 10, 2014).
19. Lineo: Flax fiber impregnation. http://www.lineo.eu/ (accessed May 5, 2014).
20. European linen and hemp. http://www.mastersoflinen.com/eng/technique/21-vos-solutions-lin-et-chanvre (accessed May 5, 2014).
21. Fiber Research Development. http://www.f-r-d.fr (accessed May 5, 2014).
22. Flax technic Groups Dehondt http://www.flaxtechnic.fr/ (accessed May 5, 2014).
23. EcoTechnilin. http://www.eco-technilin.com (accessed May 5, 2014).
24. Norafin Industries (Germany) GmbH. www.norafin.com (accessed May 5, 2014).
25. Holbery, J., and Houston, D. 2006. Natural-fiber-reinforced polymer composites in automotive applications: Overview low-cost composites in vehicle manufacture. *The Journal of the Minerals, Metals & Materials Society* 58:80–6.
26. Directive 2000/53/EC of the European Parliament and of the Council of 18 September 2000 on end-of-life vehicles. *Official Journal of the European Communities* (October 21, 2000).

27. Biocomposites in automotive manufacturing: Replacing one car part at a time. http://www.bioproductsatguelph.ca/newsevt/colloquium/assets/flashpaper/r_campbell.swf (accessed February 11, 2014).

28. Hemp plastic. http://www.hempplastic.com (accessed April 1, 2014).

29. Benson Ford Research Centre. http://www.thehenryford.org/research/soybeancar.aspx (accessed April 1, 2014).

30. Brouwer, W.D. 2001. Natural fiber composites: Saving weight and cost with renewable materials. http://www.iccm-central.org/Proceedings/ICCM13proceedings/SITE/PAPERS/Paper-1414.pdf (accessed April 3, 2014).

31. Use of renewable resources: Ecoplastics. http://www.toyota-global.com/sustainability/environmental_responsibility/automobile_recycling/design_for_recycling/use_of_renewable_resources.html (accessed May 8, 2014).

32. Cleantech Canada: Magna joins growing list of auto firms pursuing biocomposites. http://www.canadianmanufacturing.com/technology/magna-joins-growing-list-of-auto-firms-pursuing-biocomposites-132702/ (accessed May 12, 2014).

33. Engineering materials: Volkswagen Golf to use bio-composites from next year. http://www.materialsforengineering.co.uk/engineering-materials-news/volkswagen-golf-to-use-bio-composites-from-next-year/61372/#sthash%2E3zxB0ia4%2Edpuf (accessed May 15, 2014).

34. Sue, E.S. 2005. Special report: Cars made of plants. http://www.edmunds.com/fuel-economy/special-report-cars-made-of-plants.html (accessed April 12, 2005).

35. Motive Industries. http://www.motiveindustries.com/ (accessed June 6, 2014).

36. Walia, A. 2013. The world's most eco-friendly car: It's made entirely from HEMP. http://www.collective-evolution.com/2013/11/01/the-worlds-most-eco-friendly-car-its-made-entirely-from-hemp/ (accessed July 10, 2014).

37. Biofore company. http://www.upm.com/upmcc-en/Pages/default.aspx (accessed August 5, 2014).

38. Le Duigou, A., Deux, J.M., Davies, P., and Baley, C. 2011. Protection of flax/PLLA biocomposites from seawater ageing by external layers of PLLA. *International Journal of Polymer Science* 2012:1–8.

39. NetComposites. http://www.netcomposites.com/news/eco-friendly-mini-transat-650-sailing-boat-prototype/6298 (accessed April 28, 2014).

40. E Tech Surfboards. http://www.surfboard-tracker.com/shapers-index/e-tech-surfboards/582 (accessed April 29, 2014).

41. LeGault, M. Composites World 2013. http://www.compositesworld.com/articles/biocomposites-update-beyond-eco-branding (accessed April 29, 2014).

42. Sustainable Composites Ltd. http://www.suscomp.com/ (accessed July 30, 2014).

43. Majeed, K., Jawaid, M., Hassan, A., Abu Bakar, A., Abdul Khalil, H.P.S., Salema, A.A., and Inuwa, I. 2013. Potential materials for food packaging from nanoclay/natural fibers filled hybrid composites. *Materials and Design* 46:391–10.

44. Tserki, V., Matzinos, P., Zafeiropoulos, N.E., and Panayiotou, C. 2006. Development of biodegradable composites with treated and compatibilized lignocellulosic fibers. *Journal of Applied Polymer Science* 100:4703–10.

45. Narayan, R. 2006. Biobased and biodegradable polymer materials: Rationale, drivers, and technology exemplars. *Degradable Polymers and Materials: American Chemical Society.* 18:282–06.

46. Biosphere Industries Ltd. http://www.biosphereindustries.com/ (accessed July 20, 2014).

47. Earthcycle. http://www.earthcycle.com/products/index.html (accessed July 20, 2014).

48. London Bio Packaging. http://www.londonbiopackaging.com/about-our-products/our-materials/ (accessed July 22, 2014).

49. KUPILKA. http://www.kupilka.fi/en/material (accessed July 20, 2014).

50. Dell: Green packaging & shipping. http://www.dell.com/learn/us/en/uscorp1/corp-comm/bamboo-packaging (accessed July 25, 2014).

51. Baptista, A., Ferreira, I., and Borges, J.P. 2013. Cellulose-based composite systems for biomedical applications. http://www.smithersrapra.com/SmithersRapra/media/Sample-Chapters/Biomass-based-Biocomposites.pdf (accessed July 3, 2015).

52. Wan, Y.Z., Huang, Y., Yuan, C.D., Raman, S., Zhu, Y., Jiang, H.J., He, F., and Gao, C. 2007. Biomimetic synthesis of hydroxyapatite/bacterial cellulose nanocomposites for biomedical applications. *Materials Science and Engineering: C* 27:855–64.

53. Millom, L.E., and Wan, W.K. 2006. The polyvinyl alcohol-bacterial cellulose system as a new nanocomposite for biomedical applications. *Journal of Biomedical Materials Research Part B: Applied Biomaterials* 79:245–53.

54. Zhijiang, C., Chengwei, H., and Guang, Y. 2012. Poly(3-hydroxubutyrate-co-4-hydroxubutyrate)/bacterial cellulose composite porous scaffold: Preparation, characterization and biocompatibility evaluation. *Carbohydrate Polymers* 87:1073–80.

55. Li, W., Yue J., and Liu, S. 2012. Preparation of nanocrystalline cellulose via ultrasound and its reinforcement capability for poly(vinyl alcohol) composites. *Ultrasonics Sonochemistry* 19:479–85.

56. Pooyan, P., Tannenbaum, R., and Garmestani, H. 2012. Mechanical behavior of a cellulose-reinforced scaffold in vascular tissue engineering. *Journal of the Mechanical Behavior of Biomedical Materials* 7:50–9.

57. Mathew, A.P., Oksman, K., Pierron, D., and Harmand, M.F. 2012. Fibrous cellulose nanocomposite scaffolds prepared by partial dissolution for potential use as ligament or tendon substitutes. *Carbohydrate Polymers* 87:2291–98.

58. BioMid fibers. http://www.biomidfiber.com/ (accessed February 8, 2014).

59. New cellulose-based fiber. Composites World 2012. http://www.compositesworld.com/products/new-cellulose-based-fiber (accessed February 8, 2014).

60. ALTUGLAS. http://www.altuglas.com/en/ (accessed January 15, 2014).

61. Ensinger. http://www.ensinger.at (accessed January 15, 2014).

62. Banerjee, M., Sain, S., Mukhopadhyay, A., Suparna Sengupta, S., Kar, T., and Ray, D. 2013. Surface treatment of cellulose fibers with methylmethacrylate for enhanced properties of in situ polymerized PMMA/cellulose composites. *Journal of Applied Polymer Science* DOI: 10.1002/app.39808.

63. Sain, S., Bose, M., Ray, D., Mukhopadhyay, A., Sengupta, S., Kar, T., Ennis, C.J., Rahman, P.K.S.M., and Misra, M. 2013. A comparative study of polymethylmethacrylate/cellulose nanocomposites prepared by in situ polymerization and ex situ dispersion techniques. *Journal of Reinforced Plastics and Composites* 32:147–59.

64. Fowler, P.A., Mark Hughes, J., Elias, R.M. 2006. Biocomposites: Technology, environmental credentials and market forces. *Journal of the Science of Food and Agriculture* 86:1781–89.

Index